LabStudio

LabStudio: Design Research between Architecture and Biology introduces the concept of the research design laboratory in which funded research and trans-disciplinary participants achieve radical advances in science, design, and applied architectural practice. The book demonstrates to natural scientists and architects alike new approaches to more traditional design studio and hypothesis-led research that are complementary, iterative, experimental, and reciprocal. These originate from studying 3-D tissue architecture in biology and generative design in architecture, creating philosophies and practices that are high-risk, nonlinear, and both hypothesis- and design-driven for often surprising results.

Authors Jenny E. Sabin, an architectural designer, and Peter Lloyd Jones, a spatial biologist, present case studies, prototypes, and exercises from their practice, LabStudio, illustrating in hundreds of color images a new model for seemingly unrelated, open-ended, data-, systems, and technology-driven methods that you can adopt for incredible results.

Jenny E. Sabin is an architectural designer whose work is at the forefront of a new direction for twenty-first-century architectural practice—one that investigates the intersections of architecture and science, and applies insights and theories from biology, emerging technologies, materials science, physics, fiber science, electrical and systems engineering, and mathematics to the design of adaptive material structures and spatial interventions.

Peter Lloyd Jones is a scientist, whose work in 3-D architectural biology has uncovered fundamental mechanisms in embryonic development and human disease. Jones's post-disciplinary research actively seeks and finds new solutions to complex problems in medicine and design by forging extreme collaborations between seemingly unrelated fields, including stem cell biology, biocomputation, fashion, smart textiles, aerospace, and architecture.

LabStudio

Design Research between Architecture and Biology

Jenny E. Sabin and
Peter Lloyd Jones

NEW YORK AND LONDON

First published 2018
by Routledge
711 Third Avenue, New York, NY 10017

and by Routledge
2 Park Square, Milton Park, Abingdon, Oxon, OX14 4RN

Routledge is an imprint of the Taylor & Francis Group, an informa business

© 2018 Taylor & Francis

The right of Jenny E. Sabin and Peter Lloyd Jones to be identified as authors of this work has been asserted by them in accordance with sections 77 and 78 of the Copyright, Designs and Patents Act 1988.

All rights reserved. No part of this book may be reprinted or reproduced or utilized in any form or by any electronic, mechanical, or other means, now known or hereafter invented, including photocopying and recording, or in any information storage or retrieval system, without permission in writing from the publishers. Printed in Canada.

Trademark notice: Product or corporate names may be trademarks or registered trademarks, and are used only for identification and explanation without intent to infringe.

Library of Congress Cataloging-in-Publication Data
Names: Sabin, Jenny E., author. | Jones, Peter Lloyd (Spatial biologist), author.
Title: LabStudio : design research between architecture and biology / Jenny E. Sabin and Peter Lloyd Jones.
Description: New York : Routledge, 2017. | Includes bibliographical references and index.
Identifiers: LCCN 2016047495| ISBN 9781138783966 (hb : alk. paper) | ISBN 9781138783973 (pb : alk. paper) | ISBN 9781315768410 (eb)
Subjects: LCSH: Architecture and biology. | LabStudio.
Classification: LCC NA2543.B56 S23 2017 | DDC 720/.47--dc23
LC record available at https://lccn.loc.gov/2016047495

ISBN: 978-1-138-78396-6 (hbk)
ISBN: 978-1-138-78397-3 (pbk)
ISBN: 978-1-315-76841-0 (ebk)

Typeset in Minion by
Servis Filmsetting Ltd, Stockport, Cheshire

Contents

Preface viii

Foreword: Reinventing Nature xii
ANTOINE PICON

Acknowledgments xv

PART I
Design Research in Practice: Methodology and Approach with Historical Precedents and Case Studies 1

1 Bioconstructivisms 3
DETLEF MERTINS

Trans-Disciplinary Research Practice: Contemporary Case Studies 17
JENNY E. SABIN

2 Design Research in Practice: A New Model 31
JENNY E. SABIN AND PETER LLOYD JONES

PART II
Design Computation Tools for Architecture and Science: New Tools and Forms 45

Introduction to Design Computation Tools for Architecture and Science: New Tools and Forms 47
JENNY E. SABIN AND PETER LLOYD JONES

3 Networking: Elasticity and Branching Morphogenesis 55
JENNY E. SABIN AND PETER LLOYD JONES

Comments on the Role of the Matrix 79
SANFORD KWINTER AND JENNY E. SABIN

New Architectural Concerns 106
ROLAND SNOOKS

4 Motility: Adaptive Architecture and Personalized Medicine 107
JENNY E. SABIN AND PETER LLOYD JONES

Topologically Free Cells 136
ROLAND SNOOKS

Avoiding Biomimicry 136
SHAWN SWEENEY

Visualizing in Another Dimension 148
SHU YANG

Positioning Mechanism 155
PETER F. DAVIES

Newness 155
RONNIE PARSONS

5 Surface Design: The Mammary Gland as a Model of Architectural Connectivity 157
JENNY E. SABIN AND PETER LLOYD JONES

On Geometry and Cellular Mechanics 186
SANFORD KWINTER, PETER F. DAVIES, AND PETER LLOYD JONES

Biological Data and Intuition 199
SANFORD KWINTER

Case Study: Understanding Behavioral Rule Sets through Cell Motility 201
ERICA S. SAVIG, MATHIEU C. TAMBY, JENNY E. SABIN, AND PETER LLOYD JONES

Case Study: Motility and the Observation of Change 215
ANDREW LUCIA WITH JENNY E. SABIN AND PETER LLOYD JONES

Case Study: Microfabrication: Spatializing Cell Signaling and Sensing Mechanisms 231
KEITH NEEVES

6 BioInspired Materials and Design 239
JENNY E. SABIN

PART III
Architectural Prototyping: Human-Scale Material Systems and Big Datascapes 257

7 The New Science of Making 259
MARIO CARPO

8 Matter Design Computation: Biosynthesis and New Paradigms of Making 265
JENNY E. SABIN

Project: Branching Morphogenesis, 2008 273
JENNY E. SABIN, ANDREW LUCIA, AND PETER LLOYD JONES

Project: Ground Substance, 2009 290
JENNY E. SABIN AND PETER LLOYD JONES

Workshop: Nonlinear Systems Biology and Design, 2010 299
JENNY E. SABIN AND PETER LLOYD JONES

PART IV
Personalized Architecture and Medicine 311

9 eSkin: BioInspired Adaptive Materials 313
JENNY E. SABIN AND ANDREW LUCIA WITH JAN VAN DER SPIEGEL,
NADER ENGHETA, KAORI IHIDA-STANSBURY, PETER LLOYD JONES,
AND SHU YANG

10 myThread Pavilion 335
JENNY E. SABIN
Interview: RE(IN)FORM(ULAT)ING Health Care via Medicine + Creativity 352
MARK L. TYKOCINSKI, PETER LLOYD JONES, AND JENNY E. SABIN

Conclusion 361
JENNY E. SABIN AND PETER LLOYD JONES

Notes on Contributors 365
Image Credits 371
Index 385

Preface

The ability to forgo disciplinary boundaries allows for unique views of similar issues, even at radically different physical and temporal scales. Whether clinicians, scientists, architects, artists, engineers, musicians, or mathematicians, we are all bound to deal with matter and its effects. Technology has afforded these fields an extraordinary ability to generate information, yet this has resulted in an ever-increasing inability to organize, visualize, and model diverse systems. As we become faced with exabyte datasets and beyond, it will become increasingly challenging to view and comprehend 4-D bio/medical data using existing means. While the end goals may differ in science and architecture, there is a driving necessity in both disciplines to spatialize, organize, contextualize, model, and fabricate complex, emergent, and self-organized nonlinear systems. We ask, "How can we intuit, see, and understand complex wholes that are often indiscernible from their individual parts?" The collaborations between architects, scientists, clinicians, and engineers launched by the Sabin and Jones LabStudio (a hybrid research unit exploring intersections at the forefront of design research and founded by Sabin and Jones in 2006 at the University of Pennsylvania) offer new venues for productive exchange in biology and design, while revealing powerful models for visualizing and materializing the complex and seemingly intangible. Information gathers meaning when filtered through multiple modes of expression, and new models for visualization and interpretation will be essential to meet this need. Intuitive pattern recognition and alternative representations of complex systems that can be tasted, smelled, seen, heard, and even held (e.g. via rapid prototyping), are and will be essential. Using logic, rigorous design, and intuition as guiding devices, relationships can be explored between the organizational properties of the biological design problem being studied and the resulting captured datasets. In the production of relationships and correspondences, "tools," which may originate from molecular nanoprobes applied to living systems to inspire writing of customized computer scripts, are developed to enable reversible and scale-free movement between multiple modes of working in vivo, in vitro and in silico, and back again. As will be seen, a rigorous understanding and analysis of these types of models will allow scientists and architects to retool and reevaluate how to negotiate visualization and quantification within complex 3-D biological phenomena. Translation will require all our senses and intuitions, and this can only be achieved using unconventional approaches.

In the last decade, digital practices and the subsequent ability to explore iterative, systems-based and complex formal strategies have transformed architectural research and design. Concurrently, a post-genomic revolution in digital, systems-based thinking has occurred in the natural and medical sciences. Given these overlapping philosophies and practices, the potential arises to reinvigorate (reinvent/discover) dialog between these separate fields with the prospect that a mutually beneficial hybrid field might emerge that is centered between biology and architecture. Certainly, the question "What is design research?" is stirring pedagogical and infrastructural debate within architecture departments and schools. The concept of a design research laboratory replete with funded research and trans-disciplinary participants offers a challenging, yet fruitful addition to the design studio, in terms of productive discovery, research, and learning. Similarly, in the sciences, and despite mounting evidence to the contrary, data-driven or "inductive" or intuitive advances in scientific knowledge, especially via collaborations with specialists in the arts and humanities, are still viewed as marginal or irrelevant. Our experience in LabStudio, however, clearly demonstrates that seemingly unrelated, open-ended, data-, systems- and technology-driven programs are not mere alternatives to the more traditional design studio or hypothesis-led research, but can represent complementary, iterative, and reciprocal approaches that benefit all participants. LabStudio and its current offspring: Sabin Design Lab at Cornell Architecture and MedStudio@JEFF at The Sidney Kimmel Medical College of Thomas Jefferson University in Philadelphia, are now viewed as a new paradigm for thinking in biology and architectural design, pedagogy, and collaboration, and have since inspired other schools of architecture and colleges of medicine.

In support of the aforementioned concepts and models, *LabStudio: Design Research between Architecture and Biology* provides specific examples of parallel and cross-over thinking between architectural research, cell and systems biology, materials science, and engineering that have been garnered over 11 years of research and education in LabStudio and its current formations. Another aim of this book is to argue that current generative and systems-based approaches in design and science will also remain incremental, subject to shifting trends, and within the rhetorical realm unless high-risk, rigorous dialogs are (re-)established between two cultures that are currently separate. The aim of this book is not only to report upon a radical and often extreme ongoing experiment between architectural designer Jenny E. Sabin and biologist Peter Lloyd Jones, but to demonstrate to scientists, clinicians, and architects alike how high-risk, nonlinear, design-driven philosophies and practices emanating from 3-D spatial biology and generative design in architecture can result in radical advances in both scientific and design research and applied architectural practice. Overall, this book presents methods, tools, processes, and prototypes that are the result of a new and hybrid form of thinking in design at the nexus of architecture, science, medicine, and technology.

Experimental applications and approaches emerge in the areas of biological data visualization and simulation, ecological building design, new materials, adaptive architecture, rapid prototyping, experimental structures, and personalized diagnostic tools in medicine.

The book is organized into four Parts with ten chapters that include case studies, workshops, projects, and interviews. The primary topics include:

▶ *Part I Design Research in Practice: Methodology and Approach with Historical Precedents and Case Studies.* Part I will discuss the dialog preceding the formalization of LabStudio, including our working methodology for conducting design research at the lab bench side and in the design studio. Relevant contemporary case studies in design research will also be presented. We will address how areas of interdisciplinary study were selected, how we familiarize ourselves with research from other disciplines and how we translate and transform such research into architectural and biomedical terms.

▶ *Part II Design Computation Tools for Architecture and Science: New Tools and Forms.* This section will explore novel methodologies in computational design and to discover new, formerly unseen relationships in dynamic biological systems. Part II will demonstrate how the design and production of catalogs of visualization and simulation tools are used to discover new behaviors in geometry and matter. Project work from LabStudio and two co-taught seminars entitled *Nonlinear Systems Biology and Design* (2007–2010, Graduate Department of Architecture, School of Design, University of Pennsylvania) and *BioInspired Materials and Design* (2011, Department of Architecture, Cornell University, New York) will form the basis for discussion and exploration. Through the analysis of biological design problems in specialized 3-D designer microenvironments, students and researchers in both venues were exposed to new modes of thinking about design ecology through an understanding of how dynamic and environmental feedback specifies structure, function, and form. Specific physical (including 3-D printing and various digital fabrication outputs) and visual exercises (generative and algorithmic studies) will be explained to familiarize the reader with the topics in architectural and scientific terms.

▶ *Part III Architectural Prototyping: Human-Scale Material Systems and Big Datascapes.* This section will explore approaches to applied research-based architectural design projects across multiple length scales, mediums, and software. Part III will highlight adaptive structures and material assemblages designed and produced in LabStudio, Sabin Design Lab, and Jenny Sabin Studio. The material and ecological potentials of visualization and simulation tools are explored through the production of experimental structures and material systems. Work from the 2010 Smart Geometry Workshop (where LabStudio participated as one of ten selected clusters) in Barcelona, Spain, at

IaaC will offer additional extended and advanced examples that examine the nature of nonlinearities, emergent properties, and loosely coupled modules that are cardinal features of complexity.

- ▶ *Part IV Personalized Architecture and Medicine*. The final section will highlight advanced prototypes and approaches to ecological building design, adaptive architecture, and personalized medicine. Through project work, this section will highlight research-based approaches that may be missing from a typical design or hypothesis-driven scientific process and how they are implemented in architecture, science, medicine, and engineering.

Throughout the book, contributions by historians, theorists, LabStudio participants, peers, collaborators, and mentors bolster and augment the featured sections and topics.

Foreword

Reinventing Nature

Antoine Picon

Path-breaking contributions to the sciences and the arts are generally rooted in a tradition, while challenging some of its funding tenets. This book by Jenny E. Sabin and Peter Lloyd Jones provides a striking illustration of such duality. On the one hand, their ideas, explorations, and experiments retraced here may appear as a new episode in the long history of the relations between architecture and nature; on the other hand, they bring new and sometimes revolutionary propositions to the table, propositions that tend to reframe radically the way these relations must be envisaged.

The attempt to relate architectural design to natural principles is almost as old as the architectural discipline itself. In his *De Architectura* or *Ten Books on Architecture*, the only architectural treatise from Greco-Roman antiquity to reach us, the Roman architect and engineer Vitruvius gives seminal examples of this attempt, first, when he tries to derive the use of ratios in buildings, for the orders of columns in particular, from the observation of the system of proportions followed by nature in the human body. This parallel was to exert a lasting influence on architectural theorists and practitioners from the Renaissance onwards. In *De Architectura*, another striking example of reference to nature occurs when its author accounts for the origin of the Corinthian capital. Its decor of acanthus leaves is explained by the fact that the ornamentation was derived from the observation made by the Greek sculptor Callimachus of how an acanthus plant had wrapped itself around a vase left on the grave of a young girl from the city of Corinth who had recently died.

The two previous examples are definitely not the only ones in Vitruvius's text, but they allow us to identify an important duality between references made to the principles at work in nature and the mere imitation of natural objects. For proportions were not seen by Vitruvius as a static set of recipes; he interpreted them rather as rules governing natural processes. By contrast, the acanthus leaves of the Corinthian capital appeared as copies in stone of organic forms observed in nature. More broadly, nature can be envisaged either as an active principle of change or as a collection of productions. Because it was overall considered as a non-imitative art, contrary to painting or sculpture that derived part of their inspiration from surrounding natural objects, architects were usually tempted by the first of the two alternatives. Architectural design seldom imitated the forms of natural objects—the acanthus leaves of the Corinthian capital remained somewhat exceptional in this respect—its ambition was to be in tune with the fundamental rhythms that animated the natural world.

Despite this ambition, the risk was always to fall back into the imitation of natural forms rather than mobilize natural dynamics in order for design to appear almost literally as an extension of the spontaneous creativity of nature. Episodes like Art Nouveau are typical of the ambiguities that arose from such a situation. Were the organic forms adopted by designers like Victor Horta or Hector Guimard truly dictated by a thorough understanding of the natural world? Revealingly, many Art Nouveau designers took their inspiration from the strange forms revealed by German biologist Ernst Haeckel in his lavishly illustrated book *Kunstformen der Natur* [Art Forms of Nature], without further inquiry into the processes that shaped them.

One of the most salient features of the research led by Sabin and Jones lies in their dramatic departure from this traditional tension. Indeed, the approach revealed in this book ignores the distinction between processes of formation and form. Computation and the possibility of investigating natural as well as artificial structures through visualization and simulation reveal a nature populated with complex, dynamic, and ever-evolving systems. Information relates both to process and to the patterns and shapes that they produce. Advocated at the dawn of the twentieth century by the Scottish biologist, D'Arcy Wentworth Thompson, in his seminal treatise *On Growth and Form*, the blurring of the distinction between formation and form has a series of fundamental consequences.

The first is to contribute to a profound redefinition of the relation between architectural design and science. For centuries, these relations had been asymmetrical. Whereas scientific theories and results impacted architecture often directly, the influence of the latter on scientific thought remained limited. For scientists, architecture represented merely a source of convenient metaphors. The cooperation between Sabin and Jones tells a completely different story in which design becomes a genuine method of scientific investigation. No hierarchy is involved in a partnership that does not distinguish between exploring nature and mobilizing some of the processes at work within it in order to promote more efficient solutions to human problems, such as the quest for more energy-efficient buildings.

A second consequence has to do with the changing relation to nature implied by this innovative partnership between design and science. Indeed, it erodes the traditional notion of a nature "already there," waiting passively to be discovered. It implies instead a more dynamic approach, according to which nature is as much invented as discovered. But it is perhaps more accurate to evoke a reinvention than an invention since science studies and authors like Bruno Latour have abundantly glossed upon the fact that nature has always been, to a certain extent, culturally constructed. If we are to adopt such a perspective, the main difference between past attitudes toward nature and the path investigated by Sabin and Jones lies in the ever-increasing consciousness of what it means in practice to be simultaneously investigating and designing nature.

In this process, received boundaries such as those which separate architecture from other design disciplines become more porous. Prior to the rise of the digital age, twentieth-century pioneers like Buckminster Fuller had already envisaged the possibility of a "design science revolution" that would make many disciplinary frontiers obsolete. But they were unable to validate their intuition beyond the conception of a few key objects such as Fuller's celebrated geodesic domes. Sabin and Jones' book provides a more substantial set of examples going in this direction. Is their eSkin project with Dr. Shu Yang architecture, even if it tries to understand the effect of nano-to-micro material properties at an architectural scale? It does not matter in reality. Permeated by materials science, physics, chemistry, mechanical engineering, and above all biology, design becomes a vector of change indifferent to existing professional domains.

Ultimately, feeding extensively on contemporary biological research, their approach is inseparable from a profound evolution of our vision of life. With the discovery of DNA in the 1950s, life became synonymous with code. Some of its key processes were envisaged as a univocal series of encryption and decryption steps comparable to what was taking place at the same moment in computing science. Contemporary research has become much more receptive to the complexity of the processes involved in the encryption and decryption of DNA information. Above all, these processes are now seen as inherently spatial. This spatial character ranks among the fundamental justifications of the productivity of the use of design as an investigative tool. Being able to decode long strings of DNA is no longer enough; topological configurations, structural frameworks guiding the formation of RNA and protein need to be understood. Conversely, much is to be gained from such an understanding in order to improve the way artifacts are conceived. Within the new nature that Sabin and Jones are exploring/reinventing, information can no longer be envisaged independently from space.

Acknowledgments

We would like to thank the following mentors, colleagues, collaborators, friends, and peers who contributed to the formation of the Sabin+Jones LabStudio and its current offspring and to those who have provided tremendous support and critique along the way. This work and this book would not have been possible without the voices, contributions, and support from the following people:

Cecil Balmond, Jan Baranski, Philip Beesley, Mina J. Bissell, Nancy Boudreau, Marie-Ange Brayer, Andrea Burry, Bernard Cache, Mario Carpo, David Chalmers, A.J. Chan, Natalie Charles, Christina Cogdell, Mark Cruvellier, Dido Davies, Peter F. Davies, Genevieve Dion, Keller Easterling, Nader Engheta, Annette Fierro, J.J. Fox, Eliza Fredette, Helene Furjan, Mark Gallini, Maria Paz Gutierrez, Andrea Hemmann, Lily Jencks, Brian Jones, Frederick S. Jones, Gwyneth Jones, Neil H. Jones, Randell Kamien, Vanesa Karamanian, Kent Kleinman, James Knight, Mike Koerner, Ferda Kolatan, Sanford Kwinter, James Knight, Mike Koerner, Aaron Levy, Andrew Lucia, Dan Luo, Manuela Martins-Green, Alexander Masters, Peter McCleary, Patrick McMullan, Matilda McQuaid, Detlef Mertins, Achim Menges, Frédéric Migayrou, Martin Miller, Mark Mondrinos, Keith Neeves, Carole Nehez, Raymond Neutra, Lira Nikolovska, Antoine Picon, Anne Plant, Marlene Rabinovitch, David Ruy, Katherine Sender, Erica Swesey Savig, Andrea Simitch, Roland Snooks, Lars Spuybroek, Theodore Spyropoulos, Margaret Stanley, Kaori-Ihida Stansbury, Shawn Sweeney, Mathieu Tamby, Agne Taraseviciute, Mark L. Tykocinski, Jan Van der Spiegel, Yoshiko Wada, Trudy A. Watt, Marion Weiss, Zena Werb, and Shu Yang.

Thank you to our families, friends, and students and to our colleagues at Routledge Taylor and Francis for their support and enthusiasm for this book.
    Special thanks to Jordan Berta for his tireless efforts in the coordination of content.

Supporting Institutions
    Cornell University, Ithaca
    University of Pennsylvania, Philadelphia
    Thomas Jefferson University, Philadelphia
    Funding Support: Graham Foundation, Production and Presentation Grant

Content in this book is partially supported by The American Heart Association, National Institutes of Health, CMREF, and the National Science Foundation under NSF EFRI. Any opinions, findings, conclusions, or recommendations expressed in this material are those of the authors and do not necessarily reflect the views of the National Science Foundation.

PART I
Design Research in Practice

Methodology and Approach with Historical Precedents and Case Studies

Note on the Text

The following text, entitled, "Bioconstructivisms," by Detlef Mertins, originally appeared in *NOX: Machining Architecture* by Lars Spuybroek *et al*. (2004). As Professor and Chair of the Architecture Department at the University of Pennsylvania, starting in 2003, Detlef Mertins was influential in the formation of LabStudio, both through his support and guidance and also through his work and writings. Mertins was instrumental in creating a space for the high-risk experimental collaboration between Sabin and Jones to flourish as formalized through symposia, workshops, lectures, exhibitions, and most importantly through our graduate course entitled, "Nonlinear Systems Biology and Design" (2007–2011), jointly housed in the Department of Architecture, School of Design, and the Institute for Medicine and Engineering, UPenn. The following text unfolds key topics through the work of two important figures: Lars Spuybroek and Frei Otto. These concepts were inspirational to our work and thinking, including: epigenesis, transcendence verses immanence, autopoiesis or self-generation, bioconstructivist theory, analogic modeling, materially-directed generative fabrication, and heterogeneous and divergent methods and techniques. Finally, this text outlines a shift in thought critical to our work and collaboration that embraces experimentation without predetermination across disciplinary boundaries. In the case of the Sabin and Jones collaboration, emphasis is placed upon the dynamics of biological data, of behavior and process where contextual events act upon code and matter in the formation of difference and heterogeneous entities.

1
Bioconstructivisms[1]

Detlef Mertins

On meeting the German structural engineer Frei Otto in 1997, Lars Spuybroek was struck by the extent to which Otto's approach to the design of light structures was resonant with his own interest in the generation of complex and dynamic curvatures. Having designed the Freshwater Pavilion (1994–1997) through geometric and topological procedures, which were then materialised through a steel structure and flexible metal sheeting, Spuybroek found in Otto a reservoir of experiments in developing curved surfaces of even greater complexity through a process that was already material – that was, in fact, simultaneously material, structural and geometric. Moreover, Otto's concern with flexible surfaces not only blurred the classic distinctions between surface and support, vault and beam (suggesting a non-elemental conception of structural functions) but made construction and structure a function of movement, or more precisely a function of the rigidification of soft, dynamic entities into calcified structures such as bones and shells. Philosophically inclined toward a dynamic conception of the universe – a Bergsonian and Deleuzian ontology of movement, time and duration – Spuybroek embarked on an intensive study of Otto's work and took up his analogical design method. A materialist of the first order, Spuybroek now developed his own experiments following those of Otto with soap bubbles, chain nets and other materials as a way to discover how complex structural behaviours find forms on their own accord, which can then be reiterated on a larger scale using tensile, cable or shell constructions.

This curious encounter between Spuybroek and Otto sends us back not only to the 1960s but deeper in time. The recent re-engagement of architecture with generative models from nature, science and technology is itself part of a longer history of architects, engineers and theorists pursuing *autopoiesis* or self-generation. While its procedures and forms have varied, self-generation has been a consistent goal in architecture for over a century, set against the perpetuation of predetermined forms and norms. The well-known polemic of the early twentieth-century avant-garde against received styles or compositional systems in art and architecture – and against style *per se* – may, in fact, be understood as part of a longer and larger shift in thought from notions of predetermination to self-generation, from transcendence to immanence. The search for new methods of design has been integral to this shift, whether it be figured in terms of a period-setting revolution or the immanent production of multiplicity. Although a history of generative architecture has yet to be written, various partial histories in art, philosophy and science may serve to open this field of research.

In his landmark cross-disciplinary study, *Self-Generation: Biology, Philosophy and Literature around 1800* (1997), Helmut Müller-Sievers describes how the Aristotelian doctrine of the epigenesis of organisms – having been challenged in the seventeenth century by the rise of modern sciences – resurfaced in the eighteenth century, as the mechanistic theories of Galileo, Descartes and Newton floundered in their explanations of the appearance of new organisms. Whereas figures such as Charles Bonnet and Albrecht von Haller held that the germs of all living beings had been pre-formed since the Creation – denying nature any productive energy – a new theory of self-generation gradually took shape. An active inner principle was first proffered by Count Buffon and then elaborated by Caspar Friedrich Wolff, explaining the production of new organisms through the capacity of unorganised, fluid material to consolidate itself. Johann Friedrich Blumenbach transformed Wolff's 'essential force' into a 'formative drive' that served as the motive for the successive self-organisation of life forms, understanding this as a transition from unorganised matter to organised corporations.[2]

The biological theory of epigenesis came to underpin the theory of autonomy in the human sphere – in art, aesthetics, philosophy, politics and social institutions such as marriage. As Müller-Sievers has noted, Blumenbach's epigenesis provided a direct model for Kant's deduction of the categories, on which his shift from metaphysics to epistemology relied: 'Only if they are self-produced can the categories guarantee transcendental apriority and, by implication, cognitive necessity and universality.'[3]

In a similar vein, but looking to mathematics and its influence rather than biology or aesthetics, the philosopher David Lachterman characterised the whole of modernity as 'constructivist' and traced its origins further back to the shift in the seventeenth century from ancient to modern mathematics. Where the mathematics of Euclid focused on axiomatic methods of geometric demonstration and the proof of theorems (existence of beings), modern mathematics emphasised geometrical construction and problem-solving.[4] As Lachterman put it, a fairly direct line runs from the 'construction of a problem' in Descartes through the 'construction of an equation' in Leibniz to the 'construction of a concept' in Kant.

Rather than reiterating ontologies of sameness, modern mathematics sought to produce difference through new constructions. In this regard it is telling that, as Lachterman points out, Euclidean geometry arose against a platonic backdrop which understood each of the mathematicals as having unlimited manyness. According to the doctrine of intermediates, 'the mathematicals differ from the forms inasmuch as there are many 'similar' [*homoia*] squares, say, while there is only one unique form'. Lachterman continues:

> The manyness intrinsic to each 'kind' of figure as well as the manyness displayed by the infinitely various images of each kind must somehow be a multiplicity indifferent to itself, a manyness of differences that make no fundamental difference, while nonetheless never collapsing into indiscriminate sameness or identity with one another.[5]

A Euclidean construction, then, does not produce heterogeneity, but rather negotiates an intricate mutuality between manyness and kinship, variation and stability. It is always an image of this one, uniquely determinate, specimen of the kind. 'There is no one perfect square, but every square has to be perfect of its kind, not sui generis.'[6] The quest for autopoiesis has been expressed, then, in a variety of oppositional tropes – creation versus imitation, symbol versus rhetoric, organism versus mechanism, epigenesis versus pre-formation, autonomy versus metaphysics and construction sui generis versus reiteration of Forms. In the nineteenth century, such binary oppositions came to underpin the quest for freedom among the cultural avant-garde. In his *Five Faces of Modernity* (1987), Matei Calinescu recounted that the term 'avant-garde' was first introduced in military discourse during the Middle Ages to refer to an advance guard. It was given its first figurative meaning in the Renaissance, but only became a metaphor for a self-consciously advanced position in politics, literature and art during the nineteenth century. In the 1860s, Charles Baudelaire was the first to point to the unresolved tension within the avant-garde between radical artistic freedom and programmatic political campaigns modelled on war and striving to install a new order – between critique, negation and destruction, on the one hand, and dogma, regulation and system, on the other. An alternative interpretation of what Calinescu calls the aporia of the avant-garde – one that sharpens the implications of this problematic, both philosophically and politically – is suggested by Michael Hardt and Tony Negri's account of the origins of modernity in their book, *Empire* (2000). Their history is even more sweeping than those reviewed above, summarising how, in Europe between 1200 and 1400, divine and transcendental authority over worldly affairs came to be challenged by affirmations of the powers of this world, which they call 'the [revolutionary] discovery of the plane of immanence'. Citing further evidence in the writings of Nicholas of Cusa among others, Hardt and Negri conclude that the primary event of modernity was constituted by shifting knowledge from the transcendental plane to the immanent, thereby turning knowledge into a doing, a practice of transforming nature. Galileo Galilei went so far as to suggest that it was possible for humanity to equal divine knowledge (and hence divine doing), referring specifically to the mathematical sciences of geometry and arithmetic. As Lachterman suggested using somewhat different terms, on the plane of immanence, mathematics begins to operate differently than it does within philosophies of transcendence where it secures the higher order of being. On the plane of immanence, mathematics is done constructively, solving problems and generating new entities. For Hardt and Negri, 'The powers of creation that had previously been consigned exclusively to the heavens are now brought down to earth.'

By the time of Spinoza, Hardt and Negri note, the horizon of immanence and the horizon of democratic political order had come together, bringing the politics of immanence to the fore as both the multitude, in theoretical terms, and a new democratic conception of liberation and of law through the assembly of citizens.[7] The historical process of subjectivisation launched an immanent constitutive power and with it a politics of difference and multiplicity. This in turn sparked counter-revolutions, marking the subsequent history as 'an uninterrupted conflict between the immanent, constructive, creative forces and the transcendent power aimed at restoring order.'[8] For Hardt and Negri, this crisis is constitutive of modernity itself. Just as immanence is never achieved, so the counter-revolution is also never assured.

The conflict between immanence and transcendence may also be discerned in architecture, along with efforts to resolve it through the mediation of an architectonic system for free expression or self-generation. Critical of using historical styles, which were understood as residual transcendent authorities no longer commensurate with the present, progressive architects of the early twentieth century sought to develop a modern style that, in itself, would also avoid the problem of predetermination, which had taken on new urgency under the conditions of industrialisation and mass production. Such a style was conceived more in terms of procedures than formal idioms. For instance, in a piece of history that has received inadequate attention, a number of Dutch architects around 1900 turned to proportional and geometric constructions as generative tools. Recognising that classical, but also medieval and even Egyptian, architecture employed proportional systems and geometric schema, they hoped to discover a *mathesis universalis*, both timely and timeless, for a process of design whose results were not already determined at the outset. The validity and value of such forms were guaranteed, it was thought, by virtue of the laws of geometry, whose own authority was in turn guaranteed by their giveness in nature. Foremost among a group that included J.H. de Groot, K.P.C. de Bazel, P.J.H. Cuypers and J.L.M. Lauweriks, was H.P. Berlage, whose celebrated Stock Exchange in Amsterdam (1901) was based on the Egyptian triangle.

In lectures and publications of around 1907 – synopses of which were translated and published in America in 1912 – Berlage articulated his theory of architecture based on the principles and laws of construction. Taking issue with the growing pluralism of taste-styles, he sought an objective basis for design – including the peculiarities of construction and the arrangement of forms, lines and colours – in the laws of nature. He described these as 'the laws under which the Universe is formed, and is constantly being reformed; it is the laws which fill us with admiration for the harmony with which everything is organised, the harmony which penetrates the infinite even to its invisible atoms.'[9] He went on to argue that adherence to nature's laws and procedures need not lead to mindless repetition and sameness, since nature produces a boundless variety of organisms and creatures through the repetition of basic forms and elements. Similarly, he considered music a paradigm, since here too creativity appeared unhampered in the adherence to laws. Citing Gottfried Semper, Berlage extended this analogy to suggest that even evolution is based on 'a few normal forms and types, derived from the most ancient traditions'. They appear in an endless variety that is not arbitrary but determined by the combination of circumstances and proportions, by which he meant relations or, more precisely, organisation. For Berlage, this led directly – for both practical and aesthetic reasons – to mathematics in art as in nature. He wrote:

> I need only remind you in this connection of the stereometric-ellipsoid forms of the astral bodies, and of the purely geometrical shape of their courses; of the shapes of plants, flowers and different animals, with the setting of their component parts in purely geometrical figures; of the crystals with their purely stereometric forms, even so that some of their modification remind one especially of the forms of the Gothic style; and lastly, of the admirable systematism of the lower animal and vegetable orders, in latter times brought to our knowledge by the microscope, and which I have myself used as motif for the designs of a series of ornaments.[10]

It is worth noting that, as Berlage was putting forward a constructivist cosmology of architecture, Peter Behrens in Germany drew on some of the same proportional systems but with a more conservative agenda, reiterating the transcendent claims of classicism through a neo-Kantian schematism. For Behrens, geometry constituted an *a priori* architectonic system that was to be applied across buildings, landscapes and furniture to raise the material world to the higher plane of *Kultur*, while for Berlage a living geometry, in itself heterogeneous rather than homogeneous, was the basis for producing novel astylar forms that belonged to this world.[11] Behrens' pursuit of the 'great form' – symbol of the transcendence of pure mind and spirit – privileged architectonics over construction and maintained a clear hierarchy between the material and the ideal. For Berlage, by contrast, architecture was at once geometric, material, technological and biological. He understood beauty to be immanent to the self-actualisation of material entities, contingent only on the rational (*sachlich*) use of means and the laws of geometry.

In citing the 'admirable systematicalness of the lower animal and vegetable orders', Berlage alluded to the microscopic single-cell sea creatures studied by the German zoologist Ernst Haeckel in the 1880s and popularised in his book of 1904, *Kunstformen der Natur* [The Art Forms of Nature], as well as other writings, including his *Report of the Scientific Results of the Voyage of HMS Challenger* (London, 1887). Haeckel estimated that there were 4,314 species of radiolarians included in 739 genera found all over the world, without any evident limitations of geographical habitat.[12] He also noted that the families and even genera appear to have been constant since the Cambrian Age. This unicellular species of organisms

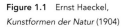
Figure 1.1 Ernst Haeckel, *Kunstformen der Natur* (1904)

became an exemplar for those interested in learning from how self-generation in nature could produce seemingly endless variety – if not multiplicity *per se* – in complex as well as simple forms of life. Haeckel hoped that knowledge of ur-animals (protozoa such as radiolarians, thalamophora and infusorians) and ur-plants (protophytes such as diatoms, rosmarians and veridienians) 'would open up a rich source of motifs for painters and architects' and that 'the real art forms of Nature not only stimulate the development of the decorative arts in practical terms but also raise the understanding of the plastic arts to a higher theoretical level'.[13]

In his own landmark book, *On Growth and Form* (1917), the Scottish biologist D'Arcy Wentworth Thompson developed science's understanding of form in terms of the dynamics of living organisms, their transformation through growth and movement.[14] In considering the formation of skeletons, he recounted Haeckel's theory of 'bio-crystallisation' among very simple organisms, including radiolarians and sponges. While the sponge-spicule offered a simple case of growth along a linear axis, the skeleton always beginning as a loose mass of isolated spicules – the radiolarian provided a more complex case among single-cell organisms, exhibiting extraordinary intricacy, delicacy and complexity as well as beauty and variety, all by virtue of the 'intrinsic form of its elementary constituents or the geometric symmetry with which these are interconnected and arranged'.[15] For Thompson, such 'biocrystals' represented something 'midway between an inorganic crystal and an organic secretion'.[16] He distinguished their multitudinous variety from that of snowflakes, which were produced through symmetrical repetitions of one simple crystalline form, 'a beautiful illustration of Plato's *One among the Many*'.[17] The generation of the radiolarian skeleton, on the other hand, is more complex and open-ended, for it 'rings its endless changes on combinations of certain facets, corners and edges within a filmy and bubbly mass'. With this more heterogeneous technology, the radiolarian can generate continuous skeletons of netted mesh or perforated lacework that are more variegated, modulated and intricate – even more irregular than any snowflake.[18]

For enthusiasts of bio-crystallisation, one of the key features of the radiolarians was the apparently perfect regularity of their form, or more precisely of their skeleton and the outer surface layer of froth-like vesicles, 'uniform in size or nearly so', which tended to produce a honeycomb or regular meshwork of hexagons. The larger implications of this regularity were made explicit in scientific cosmologies of the early twentieth century, such as Emmerich Zederbauer's *Die Harmonie im Weltall in der Natur und Kunst* (1917) and Ernst Mössel's *Vom Geheimnis der Form und der Urform des Seines* (1938). Supported by the evidence of ever more powerful microscopes and telescopes, these authors sought to confirm that the entire universe was ordered according to the same crystalline structural laws – establishing continuity from the structure of molecules and microscopic radiolarians to macroscopic celestial configurations, between the organic and the inorganic, nature and technology.

Perhaps the most sweeping statement of platonic oneness at mid-century – embracing industrialised structures as well as natural ones – was provided by R. Buckminster Fuller when he wrote that the 'subvisible microscopic animal structures called *radiolaria* are developed by the same mathematical and structural laws as those governing the man-designed geodesic and other non-man-designed spheriodal structures in nature'.[19] This similarity of underlying laws gave the radiolarians, like the geodesic domes that Fuller designed, the character of an exemplar for fundamental structures, which, he explained, were not in fact things but rather 'patterns of inherently regenerative constellar association of energy events'.[20] As if to substantiate Fuller's point, Paul Weidlinger illustrated his own account of isomorphism in organic and inorganic materials, as well as microscopic and macroscopic events, by comparing Haeckel's drawing of a radiolarian with a magnified photograph of soap bubbles, the stellate cells of a reed and one of Fuller's

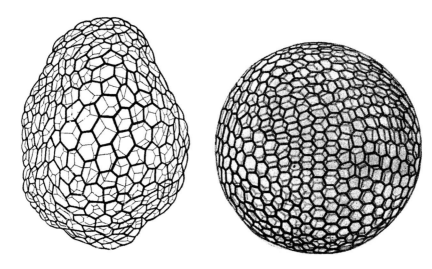

Figure 1.2 Two radiolarians, the *Reticulum plasmatique*, after Carnoy, and the *Aulonia hexagona*, as depicted by Ernst Haeckel

geodesic domes, replete with tiny spikes that reinforce its resemblance to the radiolarian.[21]

Yet Thompson's lengthy effort to account for the diversity of the tiny creatures ultimately ran aground on the impossible mathematics of Haeckel's theory of bio-crystallisation. Not only did Thompson find it necessary to acknowledge and examine less perfectly configured specimens, such as the *reticulum plasmatique* depicted by Carnoy, but in comparing them with Haeckel's Aulonia – 'looking like the finest imaginable Chinese ivory ball' – he invoked Euler to explain that 'No system of hexagons can enclose space; whether the hexagons be equal or unequal, regular or irregular, it is still under all circumstances mathematically impossible ... the array of hexagons may be extended as far as you please, and over a surface either plane or curved, but it never closes in.'[22] Thompson pointed out that Haeckel himself must have been aware of the problem for, in his brief description of the *Aulonia hexagona*, he noted that a few square or pentagonal facets appeared among the hexagons. From this Thompson concluded that, while Haeckel tried hard to discover and reveal the symmetry of crystallisation in radiolarians and other organisms, his effort 'resolves itself into remote analogies from which no conclusions can be drawn'. In the case of radiolarians, 'Nature keeps some of her secrets longer than others.'[23]

During the 1960s, armed with evidence from advanced microscopes that the surface meshworks of radiolarians were in fact irregular, Phillip Ritterbush underscored the problem of regularity and biaxial symmetry when he suggested that Haeckel had altered his drawings of the radiolarians 'in order for them to conform more precisely to his belief in the geometric character of organisms'.[24] Ritterbush pointed out that Haeckel's appreciation of the regularities and symmetries of the skeletons of living organisms – and, by extension, Fuller's conception of geodesic domes as manifesting patterns of 'constellar' associations – relied on a permutation of the analogy with the crystal, which had been employed in biology since the seventeenth century. Nehemiah Grew (1628–1712), for instance, was an early plant anatomist who regarded regularities in natural forms as evidence that the processes of growth consisted of the repetition of simple steps, in which forms might be successfully analysed.

Fuller assumed that the modular regularity of the radiolarians demonstrated the existence of a universal transcendental order, and so reiterated it in the combinatorial logic of irreducible struts and universal joints that comprised his geodesic domes. In contrast, the botanist and popular science writer, Raoul H. Francé, had already in the 1920s interpreted the radiolarians within a cosmology of composite assemblages that understood all of creation to be constructed not of one ur-element but of seven. In his *Die Pflanze als Erfinder* [The Plant as Inventor] (1920) Francé argued that the crystal, sphere, plane, rod, ribbon, screw and cone were the seven fundamental technical forms employed 'in various combinations by all world-processes, including architecture, machine elements, crystallography, chemistry, geography, astronomy and art – every technique in the world'. Comparing what he called the 'biotechnics' of maple keys and tiny flagellates moving through rotation with ships' propellers underscored the isomorphism between human and natural works, inspiring the Russian artist-architect El Lissitzky to denounce the fixation with machines in the early 1920s in favour of constructing 'limbs of nature'.[25]

Francé was read enthusiastically in the mid-1920s by artists and architects whom we associate with 'international constructivism' – not only El Lissitzky, but also Raoul Hausmann, László Moholy-Nagy, Hannes Meyer, Siegfried Ebeling and Ludwig Mies van der Rohe. So extensive was this reception of biotechnics or 'cosmobiotechnics', as Hausmann put it, that we may well refer to this orientation within constructivism as 'bioconstructivist'. Looking back, we may also recognise Berlage as providing an earlier iteration of bioconstructivist theory.

Lissitzky paraphrased Francé in his 'Nasci' issue of *Merz* in 1924, which he co-edited with Kurt Schwitters. It was there that Lissitzky gave a constructivist – and now scientific – twist to the idea of becoming that had saturated the Weimar artistic culture, associated with both Expressionism and Dada. The word *nasci* is Latin for 'begin life' and approximates *Gestaltung*, which was used in technical discourse as well as aesthetics and biology and referred simultaneously to form and the process of formation. It implied a self-generating process of form-creation through which inner purposes or designs became visible in outer shapes. Having reiterated Francé's theory of biotechnics in their introduction to the journal, Lissitzky and Schwitters then provided a portfolio of modern artworks that can only be interpreted as demonstrations of the theory. What is remarkable in this collection is the diversity produced with the seven technical forms. Beginning with Kasimir Malevich's *Black Square*, the folio then features one of Lissitzky's own Prouns; additional paintings by Piet Mondrian and Fernand Léger; collages by Schwitters, Hans Arp and Georges Braque; sculpture by Alexander Archipenko;

photograms by Man Ray; architecture by Vladimir Tatlin, J.J.P. Oud and Ludwig Mies van der Rohe; and several phenomena from nature. The sequence concludes with an unidentified microscopic image punctuated by a question mark, suggesting something of the formlessness from which all form emerges or, perhaps, to which biotechnics might lead.

By the 1960s, scientists sought to come to terms with the limitations of the crystal metaphor for living phenomena. While Kathleen Lonsdale, for instance, attempted to shore up the transcendental authority of the crystalline by defining it more broadly as arrangements of atoms in repeating patterns,[26] the animal geneticist Conrad Waddington turned to other concepts to account for irregularities. Waddington used radiolarians to discuss not the similarities between organic forms and technological objects but the difference between them, characterising man-made objects as reductive, simplistic and monofunctional in comparison with the complex, varied and multipurpose nature of living organisms. For him, organic form 'is produced by the interaction of numerous forces which are balanced against one another in a near-equilibrium that has the character not of a precisely definable pattern but rather of a slightly fluid one, a rhythm'.[27] Invoking Alfred North Whitehead's conception of rhythm to address the irregularities with which Thompson had already struggled, Waddington wrote:

> It is instructive to compare the character of the variations from the ideal form in an organic and in human creation. The shell of the minute unicellular organism Aulonia hexagona is one of those animal structures whose functions are simple enough for it to approximate to a simple mathematical figure, that of a sphere covered by almost regular hexagons. It will be seen that the hexagons are in practice not quite regular; they do not make up a rigidly definable pattern, but rather a rhythm, in the sense of Whitehead, who wrote: 'A rhythm involves a pattern, and to that extent is always self-identical. But no rhythm can be a mere pattern; for the rhythmic quality depends equally upon the differences involved in each exhibition of the pattern. The essence of rhythm is the fusion of sameness and novelty; so that the whole never loses the essential unity of the pattern, while the parts exhibit the contrast arising from the novelty of the detail. A mere recurrence kills rhythm as surely as does a mere confusion of detail.'[28]

Like Waddington, the French-American structural engineer Robert Le Ricolais – a pioneer of the space frame – insisted on distinguishing natural and man-made objects and on the limits of instrumental knowledge. While 'amazed' by the coherence and purity of design that the radiolarians represented, he also characterised it as 'frightening'. 'What man makes', he wrote, 'is usually single-purposed, whereas nature is capable of fulfilling many requirements, not always clear to our mind.'[29] Where engineers had been speaking about space frames for only 25 or 30 years, the radiolarians were, he explained, three hundred million years old. 'Well, it's not by chance, and I'm glad that I saw the Radiolaria before I saw Mr Fuller's dome.' Acknowledging that analogies with natural phenomena could help resolve some problems, he held that 'it's not so important to arrive at a particular solution as it is to get some general view of the whole damn thing, which leaves you guessing'.[30] Ricolais' use of material experiments was consistent with such scepticism, privileging specificity and concreteness over universal *mathesis*. Fascinated by the 'fantastic vastitude' of the radiolarians, neither Ricolais nor Frei Otto treated them as synecdoches for the entire universe.[31] They were merely one among many phenomena from which an engineer could learn.

During this period Frei Otto also took up the notion of self-generation and the analogy between biology and building, but eschewed the imitation of nature in favour of working directly in materials to produce models that were at once natural and artificial. At the same time, he also eschewed their translation into a universalising *mathesis*. Rather than focusing on form or formula, he took the idea of analogy in an entirely different direction, preferring to stage experiments in which materials find their own form. Where the theory of *Gestaltung* in the 1920s posited the unfolding of an essential germ from within, understanding external form as an expression of inner purpose, in the 1960s, autogenesis was redefined through cybernetics and systems theory as a function of dynamic, open systems of organisation and patterning. In this context, Otto's experiments in the form-finding potential of material process sidestep purist essentialism to open up a world in which unique and complex structures result immanently from material exigencies without being subject to any transcendent authority, either internal or external. Otto's analogical models involve iterations on different scales and in different materials, but without positing an overarching totality, reductive universality or optimised homogeneity. Open to the air, rambling and polycentric, Otto's tensile structures operate demonstrably outside the terms of physiognomic and formal expression, leaving behind the problematics of inner–outer identity, closure and unity that had been integral to the modernist conception of the autonomous organism and of *autopoiesis* in human works.

It is telling that an entire issue of the *IL* journal of Otto's Institute for Lightweight Structures was devoted to radiolarians, whose composite of pneumatic and net structures intrigued Otto and his research group just as they did Ricolais. Unlike other admirers, however, Otto's group did not take these creatures as models for engineering, but rather sought to explain their self-generation with analogical models. Situated between natural phenomena and engineering, the isomorphic character of Otto's analogical models gives them not only instrumental value for new constructions but also explanatory power for natural phenomena.[32]

Spuybroek too is fascinated by how complex surfaces in nature result from the rigidification of flexible structures, a process so intricate as to elude precise theoretical or mathematical analysis. Like Otto, he uses a varied repertoire of analogical material models that are deceptively simple but remarkably effective for generating complex structures and tectonic surfaces. In his hands, radiolarians are no longer emblems of universal order, their imperfections corrected into the perfect regularity of crystalline spheres. 'What is so interesting about radiolarians', he writes, 'is that they are never spheres, though they tend towards the spherical. They are all composite spheres – tetrahedral, tubular, fan-shaped, etc.' Focusing on examples different from the perfect spheres singled out by Fuller, Spuybroek sees radiolarians not as homogeneous forms but as material technologies that produce hybrid tectonic surfaces – part pneumatic, part net structures – which are flexible in contour and shape. The rhythmic variability of these surfaces is achieved by changes in the size of openings and the thickness of the net fibres between them. With this shift from form to surface, Spuybroek leaves behind the modernist quest for the supposed self-same identity of the organism in favour of a surface that can be modulated to assume different shapes and sizes as well as various architectural roles – from facades to roofs and from towers to vaults, halls and edges. While Spuybroek's bundle of interwoven towers for the World Trade Center in New York demonstrates the flexibility of radiolarian technology, the more recent project for the European Central Bank realises its potential to operate simultaneously in a multitude of ways. More importantly still, Spuybroek's radiolarian tectonic surface is but one of an increasing repertoire of analogical models with which he works. Like Berlage and Francé, his organon of techniques is heterogeneous and divergent rather than homogeneous and convergent. Unlike them, however, he is no longer concerned with the elemental in any way, nor with unifying underlying laws, be they mathematical or biological or both. Although he employs the radiolarian technology to achieve what he calls 'a strong expression of wholeness and pluriformity at the same time', his ECB is radically asymmetrical and irregular, polycentric and contingent. And while its pattern-structure implies repetition and extension, the buildings produced with it remain singular entities.

In taking over Otto's method, Spuybroek uses it as an abstract machine, understanding this term – and the broader pragmatics of which it is a part – through Gilles Deleuze and Félix Guattari.[33] In discussing regimes of signs in *A Thousand Plateaus* (1987), Deleuze and Guattari isolate four components of pragmatics: the generative, the transformational, the abstract machine, and the machinic. The generative, they say, 'shows how the various abstract regimes form concrete mixed semiotics, with what variants, how they combine, and which one is predominant'.[34] The transformational component, on the other hand, 'shows how these regimes of signs are translated into each other, especially when there is a creation of a new regime'.[35] However, they foreground the abstract machine, with its diagrammatic mode of operation, since it deterritorialises already established semiotic formations or assemblages, is 'independent of the forms and substances, expressions and contents it will distribute',[36] and plays a 'piloting role' in the construction of new realities. The machinic component, they conclude, shows 'how abstract machines are effectuated in concrete assemblages'.[37]

While their understanding of the generative is recombinatory and thus avoids implications of beginning from nothing, rethinking the generative impulse of the historical avant-garde in terms of the abstract machine helps to discharge any residual transcendentalism that continues to attend narratives of self-generation, which appears so anachronistic when reiterated by architects today. It offers a stronger and sharper version of *Gestaltung*, detaching process now entirely from form and dynamic organisation from *Gestalt*. Alternatively, we could say, with Zeynep Mennan, that it could lead to a *Gestalt* switch, a new theory of *Gestalt* that would be adequate to complex, rhythmic and modulated forms of heterogeneity.[38] Rather than settling chaos into an order that presumes to transcend it, Spuybroek generates an architecture that is self-estranging and self-different, in which identity is hybrid, multiple and open-ended. If cosmological wholeness is an issue at all, it may now be assumed as given, no longer something lost and needing to be regained, as the Romantics thought. Art need no longer dedicate itself to the production of wholeness, since it is inherently part of the cosmos, whatever limited understanding of it we humans may achieve. As Keller Easterling has argued in another context, we need no longer worry about the One, but only the many.[39] There is no need for closure, unity or a system that assimilates everything into One. Extending the bioconstructivism of Berlage, Francé, Lissitzky and Otto, Spuybroek now engages only in endless experiments with materials, their processes and structural potentials. What he repeats are not entities or forms but techniques, developing a new *modus operandi* for acting constructively in the world. Rather than seeking to overcome the world or to assimilate difference to the sameness of underlying laws, he works to produce new iterations of reality, drawing on the potentials of matter for the ongoing production and enjoyment of heterogeneous events.

Acknowledgments
"Bioconstructivisms" by Detlef Mertins was originally published in *NOX: Machining Architecture* by Lars Spuybroek. Reproduced by kind permission of Thames and Hudson Ltd., London, 2004, pp. 360–369.

Notes
1 Published with permission from Thames and Hudson, Architectural Association Publications and Lars Spuybroek. Originally published in *Bioconstructivisms* by Detlef Mertins from *NOX: Machining Architecture* by Lars Spuybroek. Copyright 2004 Lars Spuybroek. Later published in *Modernity Unbound* by Detlef Mertins as part of Architecture Words 7. Reproduced by kind permission of Thames and Hudson Ltd., London and Architectural Association Publications, London.
2 See Helmut Müller-Sievers, *Self-Generation: Biology, Philosophy and Literature around 1800* (Stanford, CA: Stanford University Press, 1997).
3 Ibid., p. 46.
4 David Rapport Lachterman. *The Ethics of Geometry: A Genealogy of Modernity* (New York: Routledge, 1989), p. vii.
5 Ibid., pp. 117–118.

6 Ibid., p. 118.
7 Michael Hardt and Antonio Negri, *Empire* (Cambridge, MA: Harvard University Press, 2000), p. 73.
8 Ibid., p. 77.
9 H.P. Berlage, "Foundations and Development of Architecture (Part 1)", *The Western Architect*, 18(9) (September 1912): 96–99; quotation is on page 96. Part 2 appears in 18(10) (October 1912): 104–108. These articles were based on his *Grundlagen der Architektur* (Berlin: Julius Bard, 1908), a series of five illustrated lectures delivered at the Kunstgewerbe Museum in Zurich in 1907.
10 Berlage, "Foundations," part 1, p. 97.
11 Berlage, *Grundlagen der Architektur* (Berlin: Julius Bard, 1908), pp. 7, 16, 38–39.
12 Ernst Haeckel, "Report of the Deep-Sea Keratosa," [collected by *HMS Challenger* during the years 1873–76] (London: Eyre and Spottiswoode, 1889), p. clxxxviii.
13 Haeckel, *Kunstformen der Natur* (Leipzig and Vienna: Verlag des Bibliographischen Instituts, 1904), unpaginated introduction, translation by author.
14 D'Arcy Wentworth Thompson, *On Growth and Form* (New York: Dover, 1992). originally published in 1917 and revised in 1942.
15 Ibid., p. 645.
16 Ibid., p. 691.
17 Ibid., p. 695.
18 Ibid., p. 698.
19 R. Buckminster Fuller, 'Conceptuality of Fundamental Structures', in György Képes (ed.), *Structure in Art and in Science* (New York: George Braziller, 1965), pp. 66–88. The quotation is on p. 80.
20 Ibid., p. 66.
21 Paul Weidlinger, 'Form in Engineering', in György Képes (ed.), *The New Landscape in Art and Science* (Chicago: Paul Theobald, 1956), pp. 360–365.
22 Thompson, *On Growth and Form*, p. 708.
23 Ibid., p. 732.
24 Phillip Ritterbush, *The Art of Organic Forms* (Washington, DC: Smithsonian Institution Press, 1968), p. 8. See also Donna Haraway, *Crystals, Fabrics, and Fields: Metaphors of Organicism in Twentieth-Century Developmental Biology* (New Haven and London: Yale University Press, 1976), p. 11.
25 El Lissitzky and Kurt Schwitters (eds.), "Nasci," *Merz* 8/9 (April/July 1924). See translation of text in Sophie Lissitzky-Küppers, *El Lissitzky: Life, Letters, Works* (London: Thames and Hudson, 1992), p. 351.
26 Kathleen Lonsdale, "Art in Crystallography," in Képes, *The New Landscape in Art and Science*, p. 358.
27 C.H. Waddington, "The Character of Biological Form," in Lancelot Law Whyte (ed.), *Aspects of Form* (London: Lund Humphries, 1951, 1968), pp. 43–52.
28 Ibid., p. 26. Waddington's citation is from Alfred North Whitehead, *An Enquiry Concerning the Principles of Natural Knowledge* (Cambridge: Cambridge University Press, 1925), p. 198.
29 Robert Le Ricolais, "Things themselves are lying, and so are their images," *VIA 2: Structures Implicit and Explicit* (University of Pennsylvania, 1973), p. 91.
30 Ibid.
31 Ricolais used the term 'vastitude' in describing the variety of radiolarians.
32 The 1990 issue of *The Journal of the Institute for Lightweight Structures*, IL 33, was dedicated to explaining the self-generation process in the skeletons of some radiolarians.
33 Gilles Deleuze and Félix Guattari, *A Thousand Plateaus: Capitalism and Schizophrenia*, translation and foreword by Brian Massumi (Minneapolis: University of Minnesota Press, 1987).
34 Ibid., p. 139.
35 Ibid.
36 Ibid., p. 141.
37 Ibid., p. 146.
38 Zeynep Mennan, "Des formes non standard: un 'Gestalt Switch'," in *Architectures non standard* (Paris: Centre Pompidou, 2003), pp. 34–41.
39 Keller Easterling presented this argument in a lecture at the University of Pennsylvania on 19 November, 2003. See her book, *Enduring Innocence. Global Architecture and its Political Masquerades* (Cambridge, MA: MIT Press, 2005).

Trans-disciplinary Research Practice: Contemporary Case Studies[1]

Jenny E. Sabin

Why is trans-disciplinary research practice important? Why would an architect and a cell and molecular biologist engage in collaboration now? At its inception, the structure and trajectories of LabStudio were purposefully open-ended. We had a shared belief that through the coupling of architecture students, designers, and scientists in the lab and studio settings that new questions would arise and at best, new applications and hypotheses would be generated. This shared commitment towards the development of a truly shared conceptual space across disciplinary boundaries sparked new structured research projects spanning embryonic development and human disease in biomedicine, to novel data visualization tools and adaptive materials, to issues of sustainability and ecology in architecture. By directly resisting the formation of goal-driven trajectories where problems and projects are solved and produced, we instead generated a creative thinking space that spawned important and radical transformative models to pressing issues in both of our fields. One area of investigation concerns sustainability and building energy in the context of adaptive architecture and responsive materials.

It is well known that buildings in the United States alone account for nearly 40 percent of the total national energy consumption. Currently, most contemporary sustainable approaches to the problem offer technological solutions through sanctioned rating systems, such as LEED, a rating system launched by the U.S. Green Building Council for new construction and existing building renovations. LEED takes into account five key measurements when evaluating new construction projects and building renovations: (1) sustainable site development; (2) water savings; (3) energy efficiency; (4) materials selection; and (5) indoor environmental quality. Additional points may be obtained through innovation in design and regional priority. While these measures adequately address issues of resource consumption in buildings, they do not address the systemic ecology of the built environment over the long term. How might we rethink our conceptual approach toward the problem of sustainability in architecture? Are there design research models and methods that may counteract this emphasis upon solutionism in favor of transformative practices that engage a dynamic reciprocity between form and environment, placing emphasis upon behavior over technology? More specifically, are there *affordances* within the environment that we may use as design drivers toward a transformative and ecological architecture? This chapter highlights three case studies; design research models that address these questions through applied projects across disciplinary boundaries. The intent is twofold: (1) to expose the reader to high-risk experimental models for conducting design research that in turn produce functional and radical applications in the context of sustainable and adaptive architecture; and (2) to highlight the relevance of and to position the work of LabStudio and the Sabin Design Lab at Cornell Architecture in this broader context.

There are now certainly many established and internationally recognized research and design units engaged in these topics and questions: Terreform ONE led by Mitchell Joachim and Maria Aiolova; the work of Michelle Addington, who is the Hines Professor of Sustainable Architectural Design at Yale; the Mediated Matter Group at MIT, led by Neri Oxman; the Self-Assembly Lab at MIT, led by Skylar Tibbits; Epiphyte Lab, led by Dana Cupkova; CITA at the Royal Danish Academy of Fine Arts, led by Mette Ramsgard Thomsen; ecoLogicStudio, led by Marco Poletto and Claudia Pasquero; and The Living, led by David Benjamin, are but a few influential research practices that immediately come to mind. The following surveyed practices offer three intersecting trajectories that touch upon a larger collective body of work engaged in adaptive architecture, energy systems, programmable matter, bioinspired design, and material computation. They are Philip Beesley Architect Inc. at Waterloo Architecture, the BIOMS group at UC Berkeley, and the Institute for Computational Design at the University of Stuttgart.

These research practices generate transformative research models that address the topic of sustainability in buildings through conceptual approaches that do not merely offer solutions, but afford new modes of thinking and research across disciplines. This requires a radical departure from traditional research and design models in architecture and science with a move toward hybrid, transdisciplinary concepts and new models for collaboration. Three interdisciplinary research practices are surveyed with emphasis upon innovation and architectural prototypes that actuate affordances within synthetic and natural environments.[2] The word 'affordance' refers to James Gibson's "Ecological Approach to Visual Perception" and more specifically to the development of his argument pertaining to "The Theory of Affordances." Here 'affordance' refers to how context may specify constraints and thus contribute to emergent and transformative relational models for design through notions of feedback and ecology as opposed to symbolic or function-based solutions. In general, an affordance gives rise to the possibility of an action or series of actions, a relationship between environment and organism. This section explores three bodies of work that exhibit architectural affordances that emerge through dynamic exchanges between environment, technology, biology, human engagement, and form. First, we will situate these new conceptual frameworks within the broader context of responsive and adaptive architecture.

Rachel Armstrong generates near living adaptive materials and is a leading innovator in the realm of sustainability. She states:

> While conservation of energy and frugal use of natural reserves may buy us time to develop new paradigms to underpin human development, they are not sustainable in the long term, as they continue to operate according to the laws of resource consumption.[3]

To this end, sustainable building practices should not simply be technical endeavors. They should include the transformation of existing built fabric into sustainable models that inspire both positive socio-cultural change and innovation in design, science, and technology. Professor and architect, Michael Hensel at a recent symposium hosted by the Department of Architecture at Cornell University entitled "Sustaining Sustainability," underscored this notion.[4] The symposium featured lectures by a diverse group of researchers and practitioners spanning multiple disciplines from biology to architecture, who share a common concern for what Hensel has labeled "sustainability fatigue." This symposium was not centered upon exhausted issues involving energy, optimization, and performance, which tend to dominate most conferences on sustainability in architecture today, but was instead focused on re-thinking the entire conceptual foundation for the project, one which fundamentally examines our relationship with nature and nature's relationship with humans. Important to this shift is a move away from purely technical solutions to environmental sustainability toward an understanding that our built and natural environments are equally becoming the contexts for thriving hybrid ecosystems. As Maria Paz Gutierrez, Director of the BIOMS research group at UC Berkeley states, "The reinvention of conceptual frameworks and processes of technologies becomes transformative when it situates itself beyond the introduction of new productions. Trans-disciplinary research in building technology can craft new habits of thought; it reorients innovation."[5] Clearly, the design and production of new energy efficient technologies are crucial to successfully meeting goals such as the Net-Zero Energy Commercial Building Initiative (CBI) put forward by the U.S. Department of Energy (DOE), which aims to achieve zero-energy commercial buildings by 2025, but as Gutierrez points out, these technological imperatives are largely based upon resource consumption. The discipline of architecture needs to move away from reactionary responses to the problem of sustainability and toward new habits of thought that question, actuate, and redefine relationships between environment and form. Trans-disciplinary models afford such a dynamic reciprocity. How is the notion of change and adaptation explored in architecture?

Popular examples of responsive architecture include Galleria Hall West, Institut Du Monde Arabe, Aegis Hypo-Surface, POLA Ginza Building Façade, and SmartWrap™. Most of these examples, however, rely heavily upon the use of mechanically driven units that communicate through a mainframe and are nested

Figure 1.3 Epiphyte Chamber. The Epiphyte Chamber is envisioned as an archipelago of interconnected halo-like masses that mimic human sensations through subtle, coordinated movements. Across each floating island, densely interwoven structures and delicate canopies made of thousands of lightweight digitally-fabricated components are drawn together in nearly-synchronized breathing and whispers. Audiences walk into highly sensual, intimate sculptural spaces that support small clusters of activity interlinking into larger gathering areas. This experimental new work explores intersections between media art, interactive distributed mechatronics, and synthetic biology.

within a building façade system. Additionally, there are now many research groups and experimental practices engaged in the exploration and implementation of existing responsive materials such as shape memory polymers, shape memory alloys or thermochromic resin. In the context of the work of Skylar Tibbits and the Self-Assembly Lab at MIT, Manuel Kretzer[6] or Martina Decker of Material Dynamics Lab at NJIT, for example, their prototypes investigate the architectural potential of building materials and programmable matter that not only change, but also respond, self assemble, and adapt to environmental stimuli. Decker's speculative Homeostatic Façade System incorporates dielectric elastomers for dynamic shading in double-skin façade systems. A building's envelope must consider a number of important design parameters, including degrees of transparency, overall aesthetics, and performance against external conditions, such as sunlight levels, ventilation, and solar heat-gain. In addition to these existing examples of adaptive architecture, how might we meaningfully consider and embed the role of the human in response to changing conditions within the built environment?

Perhaps the closest example to this scenario is the work of Philip Beesley, whose sculptures and installations such as Hylozoic Ground, incorporate layers of chain responses and amplified effects that are the result of highly personal interactions. Feedback loops between these networked mesh systems respond,

Figure 1.4 Epiphyte Chamber, an immersive environment erected for the inauguration of the Museum of Contemporary and Modern Art, Seoul, 2014, demonstrates key organizations employed for Hylozoic Architecture group constructions, including lightweight resilient scaffolds, distributed interactive computational controls, and integrated protocell chemical metabolism.

adapt, and amplify user input, giving rise to emergent conditions that are the result of reciprocal loops between environment, code, and communication. In recent projects, Philip Beesley examines thermodynamics in order to, as he states, "seek a tangible exchange for the reality of an expanded physiology." Beesley's interest in a design process and form language rooted in what he calls *dissipative structures and diffusion* gives rise to adaptive architectures that are generated by the human body.[7] He further claims, "In turn, it suggests a craft of designing with materials conceived as filters that can expand our influence and expand the influence of the world on us, in an oscillating register: catching, harvesting, pulling and pushing." Beesley describes these constructions as, "a synthetic new kind of soil." These affordances, which are not features of organisms or the environment, actuate change through emergent forms. These architectural affordances act and they are also acted upon. Beesley's thermodynamic environments are in a perpetual state of formation and communication. In this sense, the new soil is both emergent and fully enmeshed in their environments and both of these attributes may be characterized as affordances. They are *emplaced* architectures that do not merely conserve energy, but rather exchange it.[8] His most recent work entitled, *Epiphyte Chamber*, which was erected for the inauguration of the Museum of Contemporary and Modern Art in Seoul, builds upon the periodic and aperiodic textile meshworks impregnated with interactive mechanisms that respond and adapt to the presence or absence of people and in turn engage in their type of learning or feedback. Additionally, this immersive environment is populated with what Beesley calls "protocell fields" i.e., glass flasks that add a stuttering and turbulent atmosphere through the aid of chemical reactions that affect, expand, amplify, and quiet the adaptive and responsive nature of what Sanford Kwinter calls a "hyper communicative

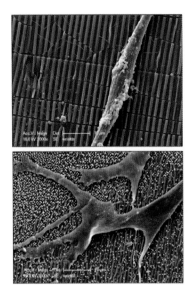 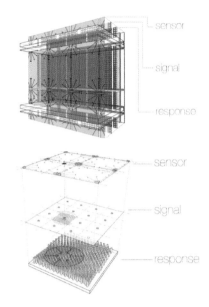

Figure 1.5 eSkin inputs: cell-matrix interface and architectural speculation as adaptive wall assembly

landscape."[9] Importantly, Beesley states, "These do not achieve high, efficient functions. Instead they offer a sketch of possibility."[10] Are there models in nature that exhibit similar reciprocity that we may mine?

The Sabin Design Lab at Cornell Architecture was launched in 2011 and is one of two current offspring of LabStudio. One of our driving questions is: How might architecture respond to issues of ecology and sustainability whereby buildings behave more like organisms in their built environments? We are interested in probing the human body for design models that give rise to new ways of thinking about issues of adaptation, change, and performance in architecture. As with LabStudio, our expertise and interests focus upon the study of natural and artificial ecology and design, especially in the realm of nonlinear biological systems and materials that use minimum energy with maximum effect.[11] Importantly, our practice and research offer another model for architectural affordance, one that is invested in developing an alternative material practice in architecture through the generative fabrication of the nonlinearities of material and form across disciplines. Together, the studio and lab investigate the intersections of architecture and science, and apply insights and theories from biology and mathematics to the design, fabrication, and production of material structures.[12] Seminal references for the work continue to include matrix biology, materials science, and mathematics through the filter of crafts-based media such as textiles and ceramics. In parallel, our work offers novel possibilities that question and redefine architecture within the greater scope of ecological design and digital fabrication.

Since the official public launch in the fall of 2010 of our National Science Foundation (NSF) Emerging Frontiers in Research and Innovation (EFRI) Science in Energy and Environmental Design (SEED) project entitled, *Energy Minimization via Multi-Scalar Architectures: From Cell Contractility to Sensing*

Figure 1.6 ColorFolds, a recent prototype by Sabin Design Lab, integrates eSkin material features with Kirigami principles and follows the concept of "Interact Locally, Fold Globally," necessary for deployable and scalable adaptive architectures. Using mathematical modeling, architectural elements, design computation, and controlled elastic response, ColorFolds showcases new techniques, algorithms, and processes for the assembly of open, deployable structural elements and architectural surface assemblies.

Materials to Adaptive Building Skins, we have led a team of architects, graduate architecture students and researchers in the investigation of biologically-informed design through the visualization of complex data sets, digital fabrication and the production of experimental material systems for prototype speculations of adaptive building skins, designated eSkin, at the macro-building scale.

The work of LabStudio and the Sabin Design Lab will be presented in the coming chapters, while the origins and intent of MEDstudio@JEFF, founded by Peter Lloyd Jones in 2013, are described, in part, via an interview with Mark Tykocinski within a separate chapter. For a comprehensive description of the eSkin project, see Part IV of this volume.

In parallel to the eSkin project, the work of the BIOMS group, directed by Maria Paz Gutierrez at UC Berkeley, takes direct inspiration from nature's skins. Gutierrez is also a recipient of and PI on one of the NSF EFRI SEED grants from 2010. As Gutierrez states, "Self-active matter is the new passive architecture."[13] Taking advantage of the textile as an important architectural element, the BIOMS multifunctional membrane features an integrative sensor and actuator system that is designed not only to answer to many functions through what Gutierrez calls the "synergistic optimization of heat, light and humidity transfer," it is also a closed loop system. Importantly, this system does not require energy input through mechanical actuators, sensors, and a mainframe. As the BIOMS group reports, "If the energy and material flows are synergistically optimized through a material programmed with self-regulation, the enclosure becomes, as in nature, a multifunctional skin."[14] Through an array of pores and apertures, the *breathing membrane* manages multiple functions through zero energy input.

In this sense, the material itself actuates and responds to multiple contextual inputs while optimizing for ideal conditions. The BIOMS group speculate that their breathing membrane, which is digitally fabricated through the integration

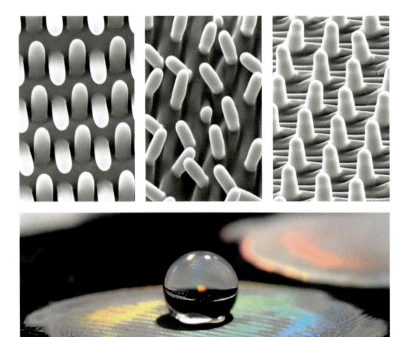

Figure 1.7 Yang's group at the University of Pennsylvania explores biomimetic concepts such as structural color, exhibited here

Figure 1.8 Multifunctional Building Membrane – Self-Active Cells, Not Blocks

of polymerization with 3-D printing extrusion, could be integrated with new construction such as in small deployable emergency housing or in public spaces in tropical zones such as markets and schools. Finally, Gutierrez and her BIOMS group articulate the importance of their research focus in the context of crisis. Rather than focus upon single solutions for conditions of crisis, as in the case of emergency relief housing, they are more concerned with how their research methodology and approach

> contributes to a paradigm shift in our understanding of how to approach resources (human and physical) in crisis and the transformations this entails from the design concept to the production framework from the nano or micro to the building scale.[15]

While Gutierrez and the BIOMS group focus upon the multifunctional capacity of self-actuated 3-D printed material membranes, the work of Achim Menges and his students at the Institute for Computational Design (ICD) at the University of Stuttgart operates at a larger scale through the explicit exploration of natural systems for novel structures in the context of computational matter.

Recently, the ICD and the Institute of Building Structures and Structural Design (ITKE) at the University of Stuttgart have constructed another bionic research pavilion, one of several in a series of research pavilions. Designed, fabricated, and constructed over one and a half years by students and researchers within a multi-disciplinary team of biologists, paleontologists, architects, and engineers, the focus of this project is upon the biomimetic investigation of natural fiber composite shells and the development of cutting-edge robotic fabrication methods for fiber reinforced polymer structures. As mentioned in the Foreword to this book, architects and structural engineers have historically looked to nature to design and build better shell and spatial structures. Cable nets have been inspired

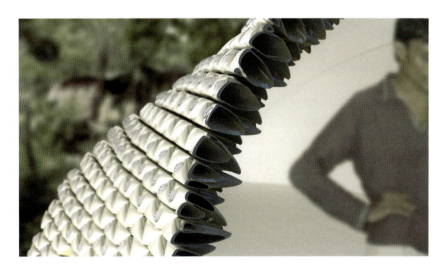

Figure 1.9 In contrast to many existing adaptive building assemblies and prototypes that require communication from a mainframe and electricity, the BIOMS breathing membrane operates on zero-energy input to self-regulate and optimize for heat, light and humidity.

Figure 1.10 The Institute for Computational Design (ICD) at the University of Stuttgart, operates at a larger scale through the explicit exploration of natural systems for novel structures in the context of computational matter. Recently, the ICD and the Institute of Building Structures and Structural Design (ITKE) at the University of Stuttgart have constructed another bionic research pavilion, one of several in a series of research pavilions.

by the high strength-to-weight ratio of the spider web; pneumatic structures after soap bubbles and films; and geodesics after radiolarian.

In the case of the new ICD/ITKE pavilion, the investigation of natural lightweight structures was conducted by an interdisciplinary team of architects and engineers from the University of Stuttgart and biologists from Tübingen University within the module "Bionics of Animal Constructions," led by Professor Oliver Betz (biology) and Professor James. H. Nebelsick (geosciences). With an interest in exploring material-efficient lightweight constructions, the Elytron, a protective shell for the wings and abdomen of beetles, proved to be an appropriate bionic model for the generation of innovative fiber composite construction methods through biological structural principles.

Through an analysis of SEM scans of the Elytra beetle, a biomimetic model of the trabeculae was extracted, synthesized, and redeployed through the aid of robotic fabrication. The trabecula is a matrix of column-like doubly curved support elements that is highly differentiated through the shell structure of the beetle. With an interest in working with this highly differentiated morphology as a model for a novel composite shell structure through the production of nonstandard unique elements, the robotic fabrication process involved two interacting 6-axis robots to produce doubly curved glass and carbon fiber reinforced polymers through a winding process. Through this simple process, which basically entails winding layers of fibers and strategically impregnating the hollow cores with resin, 36 unique components were generated for the lightweight pavilion. Overall, these lightweight structures rely upon the geometric morphology of a double-layered system inspired and informed by the Elytron beetle, then redeployed through the mechanical properties of the natural fiber composite.

Trans-disciplinary Research Practice　27

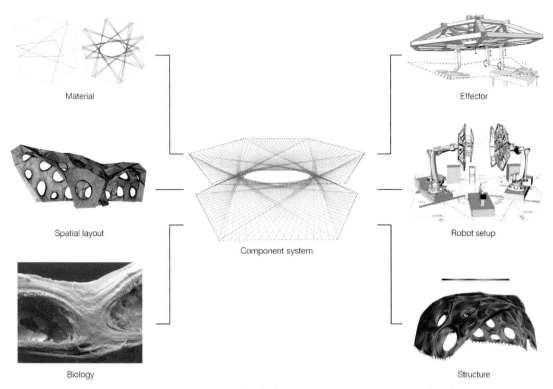

Figure 1.11　Integration of multiple process parameters into a component-based construction system

While nonlinear concepts have been widely applied in analysis and generative design, they have not yet been convincingly translated into the material realm of fabrication and construction, until recently. The ICD/ITKE Research Pavilion 2013–14 – Stuttgart 2014 showcases possible design routes and techniques that no longer privilege column, beam, and arch through a broadened definition of architectural tectonics successfully made with advances in computational design. How might these advances impact material practice in architecture, engineering, and construction at economic, technological, and cultural levels? Importantly, the ICD/ITKE is equally committed to the communication, documentation, and public dissemination of their advances in tooling and fabrication to encourage the design and production of nonlinear systems via complex geometries.

Central to all the work presented here is the integration of fields and industries outside of architecture in the practice of design research by multi-disciplinary teams composed of architects, engineers, scientists, and fabricators active in academia, practice, and industry. A primary thrust of the works is the evolution of digital complexity in the built environment. In parallel, this approach aims to make advances in material research and fabrication to effect pragmatic change in the economic and ecological production of complex built form and adaptive architecture.

Figure 1.12 The robotic fabrication process involved two interacting 6-axis robots to produce doubly curved glass and carbon fiber reinforced polymers through a winding process.

In all four cited design research practices, we are presented with architectural affordances that operate in counter-distinction to the solutionism of sanctioned and typical sustainable approaches to architecture, where models based on behavior are favored over the purely technological. In the case of Beesley's work, architectural affordances operate, affect, and interact as environments, entities, and beings. Beesley's thermodynamic environments are in a perpetual state of "catching, harvesting, pulling, and pushing." In this sense, his architectural interfaces are both emergent and fully enmeshed in their environments, exhibiting a dynamic reciprocity between context and form. In the case of eSkin or the work of BIOMS, programmable matter and self-actuating material systems operate as dynamic thresholds, interfaces that adapt, learn, and change in response to environmental cues with minimal to zero energy input. Here, geometry and matter are explored across multiple length scales and disciplines, where issues of sustainability are not merely about metrics and technology, but about new models for trans-disciplinary collaboration and design in the context of adaptive and ecological architectural matter. And, finally, in the work of the ICD, nonstandard tectonic elements emerge through the rigorous investigation of the behaviors of natural models and their corresponding translation into novel material systems where geometry, materiality, pattern, structure, and form are inextricably linked. Menges and his students resist the post-rationalization of complex form through an approach that engages materially directed generative design. Here, architectural affordances reveal themselves as evolving flows of force through geometry and matter that are computed, designed, and fabricated through robotic interfaces that dance, collaborate, wind, and weave. Perhaps the most important deliverable in the aforementioned examples to date are these new models for collaboration across

Figure 1.13 Thirty-six unique components were generated for the lightweight pavilion.

disciplines where architectural affordances generate transformative models that may in parallel provide potent contributions to issues of sustainability, ecology, and building energy in architecture. In the following pages, the work and research of the Sabin+Jones LabStudio will be described in depth, a new model for design research and practice between architecture and biology.

Notes
1 Adapted from Jenny Sabin, "Transformative Research Practice: Architectural Affordances and Crisis," *Journal of Architectural Education* 69(1) (2015): 63–71.

2   See James J. Gibson, *The Ecological Approach to Visual Perception*. (Hillsdale, NJ: Lawrence Erlbaum Associates, 1986). See also Andrew Lucia, *et al.*, "Memory, Difference, and Information: Generative Architectures Latent to Material and Perceptual Plasticity," paper presented at 15th Annual Conference on Information Visualization, London, July 2011. Finally, see Simone Ferracina, "Exaptive Architectures," in *Unconventional Computing: Design Methods for Adaptive Architecture*, ed. Rachel Armstrong and Simone Ferracina (Cambridge, ON: ACADIA and Riverside Press, 2014), pp. 62–65.
3   Rachel Armstrong, "Lawless Sustainability," *Architecture Norway*. Available at: www.architecturenorway.no/questions/cities-sustainability/armstrong/.
4   Jenny Sabin, "Sustainable Thinking," (review of the first installment of the Hans and Roger Strauch Symposium on Sustaining Sustainability: Alternative Approaches in Urban Ecology and Architecture), *The Architectural Review*, 1381 (2012): 88–89. Ithaca, New York: Department of Architecture, Cornell University, March 2012.
5   Maria Paz Gutierrez, "Reorienting Innovation: Transdisciplinary Research and Building Technology," *Architectural Research Quarterly*, 18(0)1 (March 2014): 69–82. Published online: 7 July 2014.
6   Manuel Kretzer, Jessica In, Joel Letkemann, and Tomasz Jaskiewicz. "Resinance: A (SMART) Material Ecology," in *ACADIA 2013 Adaptive Architecture*, eds. Phillip Beesley, Omar Kahn, and Michael Stacey (Cambridge, ON: Riverside Architectural Press, 2014), pp. 137–146.
7   Philip Beesley. "Diffusive Prototyping," paper presented at Alive International Symposium on Adaptive Architecture, Computer Aided Architectural Design, ETH Zurich, Switzerland, March 2013.
8   See Beesley's description of emplacement and architecture in his lecture on "Diffusive Prototyping."
9   Sanford Kwinter. "Creods," paper presented at Acadia 2008: Silicon + Skin, Biological Processes and Computation, Minneapolis, Minnesota, October 2008.
10 Beesley. "Diffusive Prototyping."
11 Marie-Ange Brayer and Frédéric Migayrou, "Naturalizing Architecture," in *Proceedings of the 9th ArchiLab*, Orleans, France, FRAC Centre, 2013, pp. 28–29, 142–145.
12 Ferda Kolatan and Jenny Sabin, *Meander: Variegating Architecture* (Exton, PA: Bentley Institute Press, 2010).
13 See project "Multifunctional Membrane: Self-Active Building Cells, Not Building Blocks," under BIOMS research initiative at UC Berkeley, Maria Paz Gutierrez (BIOMS director/lead) with L.P. Lee (BioPoets director), BIOMS team (Charles Irby, Katia Sobolski, Pablo Hernandez, David Campbell, and Peter Suen); B. Kim (BioPoets team).
14 Ibid.
15 Ibid.

2

Design Research in Practice: A New Model[1]

Jenny E. Sabin and Peter Lloyd Jones

You never change things by fighting the existing reality. To change something, build a new model that makes the existing model obsolete.

(Buckminster Fuller)

[T]he informal steps in easily, a sudden twist or turn, a branching, and the unexpected happens—the edge of chance shows its face. Delight, surprise, ambiguity are typical responses; ideas clash in the informal and strange juxtapositions take place. Overlaps occur. Instead of regular, formally controlled measures, there are varying rhythms and wayward impulses.

(Cecil Balmond)[2]

Nature as Muse

Cable nets based upon the high strength-to-weight ratio of the spider web; conditioning systems modeled after termite mounds; super-hydrophobic coatings based on lotus leaves; Pneumatic structures inspired by soap bubbles and foams; surface structures abstracted from the exoskeletons of radiolarians; structural color change engineered from the 3-D geometry and structure of butterfly wings, and towers generated and engineered from the distribution of stress forces observed in the femur bone. Architects, scientists, and structural engineers have always looked to nature and the sciences for inspiration, and each profession demands collaboration across a broad range of disciplines. Historically, architects, and biologists have borrowed from each other's disciplines in the pursuit of new ways of understanding and modeling living systems in context: their code, structure, geometry, and matter. Clearly, biologists and architects share similar concerns, and this is perhaps best reflected in the relationships that have emerged between their respective fields. Models borrowed from architects—such as tensegrity structures and geodesic domes—have led to radical new insights into how living systems, such as cells, tissues, and whole organisms, are assembled and function, as well as to a new understanding of how the exterior ecology of cells and tissues influences the interior of the cell, including signaling networks, the cytoskeleton, and the genome. Similarly, models borrowed from biology, particularly regarding self-organization, metadata structures, and the emergence of complex, nonlinear global systems from simple local rules of organization have led to radical new forms and structural organizations in architectural design. Examples such as these demonstrate how attentive architectural and scientific practices can be to each other—particularly within architecture and biology, which are constantly challenged to reinvent and question themselves in a manner similar to the historic avant-gardes, or in the face of new technologies.

Buckminster Fuller certainly appreciated this:

> About 1917, I decided that nature did not have separate, independently operating departments of physics, chemistry, biology, mathematics, ethics, etc. Nature did not call a department heads' meeting when I threw a green apple into the pond, with the department heads having to make a decision about how to handle this biological encounter with chemistry's water and the unauthorized use of the physics department's waves . . . nature probably had only one department and only one coordinate, omnirational, mensuration system.[3]

Fuller's structures and industrial designs repeatedly demonstrate these trans-disciplinary pursuits for a new architecture, with his geodesic domes perhaps representing the most memorable and copied examples. Many of his designs, including his Dymaxion Automobile from 1933, also turned to the natural and physical sciences for inspiration. Using the raindrop as a model system, Fuller attempted to understand the mathematical and geometrical systems behind the image of the raindrop, thus resisting the direct mimicry of its shape. Fuller understood that air resistance increases in ratio to velocity squared and applied these principles to his automobile design. The Dymaxion Automobile is designed through a careful study of the raindrop's chemical, physical, and morphological properties, giving rise to new ideas regarding form, function, and structure. However, as Detlef Mertins points out in his text entitled, "Bioconstructivisms" (see Chapter 1 in this volume), Fuller was after a universalizing geometry and mathematics, a "platonic oneness" with nature, linking his industrialized geodesic domes and structures with the exoskeleton surfaces of micro-scale creatures such as radiolarians.

Twenty-first-century designs that reflect the hidden and overt patterns and textures of nature are scattered throughout architectural practice, and Thomas Heatherwick's design for the British Pavilion or Seed Cathedral (nicknamed the dandelion) for the 2010 World Expo in Shanghai represents one example that also uses nature as muse. As described by the Heatherwick Studio:

> The Seed Cathedral is a box, 15 metres high and 10 metres tall. From every surface protrude silvery hairs, consisting of 60,000 identical rods of clear acrylic, 7.5 meters long, which extend through the walls of the box and lift it into the

Figures 2.1 and 2.2 Dome by Buckminster Fuller in collaboration with Thomas C. Howard of Charter Industries at the Vitra Museum, exterior and interior. The dome, called a Charter-Sphere Dome, is designed with fewer circles than a geodesic dome and is thus easier to erect and install. It was fabricated in 1975 and installed at Vitra in 2000.

air. Inside the pavilion, the geometry of the rods forms a space described by a curvaceous undulating surface. There are 250,000 seeds cast into the glassy tips of all the hairs.[4]

These powerful, yet more or less literal examples of design mimicking nature are paralleled by other more abstract and generative analogic approaches that

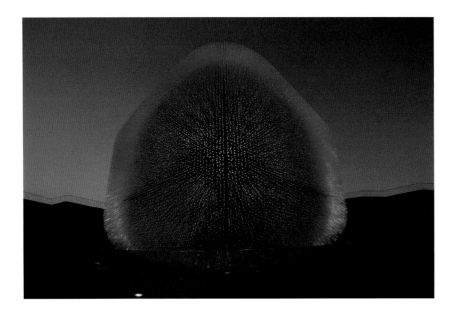

Figure 2.3 Seed Cathedral by Heatherwick Studio. Thomas Heatherwick's design for the British Pavilion or Seed Cathedral (nicknamed the dandelion) for the 2010 World Expo in Shanghai.

use the natural sciences as a methodological filter to incorporate and represent the basic principles and rules of life into design. The structural designer, Robert Le Ricolais, for example, studied the tension networks inherent to radiolarians in order to understand the dynamic properties and qualities of closed and open "skeletal" structures. He professed that he had "found no better discipline in this unpredictable problem of form than to observe the prodigies created by nature."[5] Particularly interesting is his brilliant observation that in nature, the art of structure and form is where to place holes, "all different in dimension and in distribution."[6] These discoveries inspired Le Ricolais' impossible desire to design and build with holes, to generate structures of "zero weight and infinite span." This seemingly contradictory statement shows us that in natural systems, we frequently find form that is extremely strong globally, yet is locally fragile. Le Ricolais argues for a higher level of (bio)synthesis: Why would we convert radiolarian structures into buildings? He exclaims, "Why should the Radiolarian help us to make money?"[7] Robert Le Ricolais worked to unfold and eventually discover more intelligent translations and deeper relationships between architecture and science, including pioneering the now ubiquitous space frame and corrugated sheet metal.

Also along these philosophical lines, and only one year after the description of the double helical structure of DNA by Watson and Crick (1953)[8] based on Rosalind Franklin's original discovery, Richard Neutra asked how nature-inspired design might contribute to the survival of the human race.[9] He states: "although no single deductive scientific method alone can yield all of the principles of design," that by understanding physiology (i.e. the scientific study of function in living systems), it may be possible to "begin to wield [design] tools which will enable us to

Figure 2.4 Space frame structures based on radiolarians; Robert Le Ricolais, a design engineer, studied the tension networks inherent to radiolarians in order to understand the dynamic properties and qualities of closed and open "skeletal" structures.

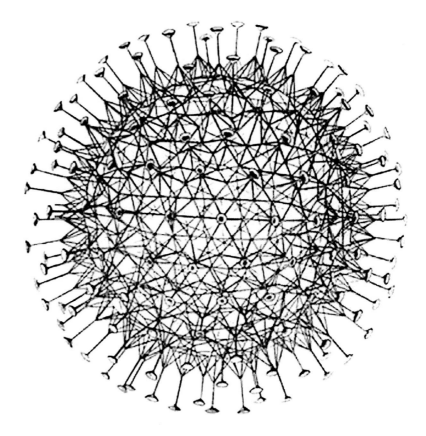

do the patient spadework which must be done. . . . It will be fascinating because it is so novel."

Fast forward to sixty years after the discovery of the double helix, and a central question in both the biology and architecture still remains as to what extent positional and functional information encoded in a blueprint versus the macro- and microenvironment influences how that code operates and is manifested. How does this code operate in a dynamic context within the whole organism in order to generate, maintain, or subvert diverse forms and their associated environments bearing multiple functions? To answer this, new tools are still needed. Given that the study of the code in context has gained tremendous ground in the past decade, both in biology and architecture, the time is now ripe to dig in.

Coding Architecture

Models informed by the study of code in context certainly exist within the field of design, especially in light of the advances in digital and parametric design modeling. While digital models and fabrication techniques are altering our formal landscape in the built environment, digital design tooling is revolutionizing how we think through a design process. Instead of composing or representing architectural elements through prescribed notations that relate one set of building instructions to the next, custom digital design tools enable the architect to think systemically through the design of relationships—be they geometric, material, programmatic, or otherwise. In some cases, a blueprint of instructions, or algorithms, that are then altered and informed through program and environmental inputs, defines sets of relationships across all scales. Pioneering examples include the transformable and deployable structures developed by Chuck Hoberman. These retractable roofs, chairs, tents, wall elements and even toys and medical tools, are capable of transforming at multiple scales, adapting to diverse environments with varied functions through the use of highly adaptable 3-D scissor mechanisms. Hoberman Associates' work is based upon the fundamental idea that a designed object can transform the way a natural organism does. Hoberman argues that while the smooth transformation of size and shape is ubiquitous in the natural world, it is rare among man-made objects.

Moving forward, the theoretical work of Karl Chu directly proclaims that the future of architecture and design is in genetic engineering, biotechnology, and universal computing. He argues that for the very first time, we are able to "think of a new kind of xenoarchitecture: an information labyrinth or, better still, a universal matrix that is self-generating and self-organizing with its own autonomy and will to being."[10] Chu references rule-based systems such as cellular automata, a discrete model invented by John Conway in Conway's *Game of Life*, and now championed by the scientist and inventor, Stephen Wolfram. This deterministic modeling system places code at the center of life and is studied in computation and theoretical biology. It consists of a regular grid of cells with a finite number of states, such as on and off. Each cell state is determined by a nearest neighbor relationship and is configured by an initial starting set of rules. Chu offers up the challenging notion that we may very well be growing buildings through the design and mutation of code in the near future.

In contrast to Chu's autonomous approach, François Roche of New Territories is interested in the phenomena of mutation, distortion, cloning, hybridization, grafting, and morphing as a means for generating new relationships with nature through their architecture, "as a vector of narration around the building."[11] Here, our personal engagement with nature and environment serve as generative approaches in the production of architecture, a machining of architecture through the politics of our own participation. The structural engineer and designer, Cecil Balmond, who is a contemporary pioneer of nonlinear thinking at the intersection of form and algorithm, and who is also the former head of the Advanced Geometry Unit at ARUP London, explores the use of mathematical, physical-geometric, and natural algorithms in relationship to an active datascape. For example, his proposed addition to the Victoria and Albert Museum's contemporary wing in collaboration with Daniel Libeskind, features an interlocking natural spiral structure and an external tiled façade generated from a mathematical model called the Fibonacci Sequence that forms fractal and branching figures. This mathematical mosaic moves from ornament to structure and back again. This example shows how part-to-whole relationships drive assembly where the resultant geometric figure emerges through contextual understanding. Mutations in the generative code and the environment both intersect and interact to alter the blueprint or algorithm, and so new understandings emerge. We ask, "What is it like to inhabit such a model?" It demands a willingness to engage actively in a generative design process, to let loose a simple set of instructions, and to see how they are made intelligent and functional through mutations of transformation, feedback, and context. It is a sleight-of-hand moment where an informed architecture is clarified and we are allowed to enter.

Formation of the Sabin+Jones LabStudio

In 2006, following a chance meeting at the inaugural conference for the Nonlinear Systems Organization (NSO)[12] initiated and led by Cecil Balmond at the University of Pennsylvania Graduate School of Design, we, i.e. Jenny Sabin (then a lecturer of Architecture and founding member of the NSO) and Peter Lloyd Jones (then a tenured Associate Professor of Pathology and Laboratory Medicine at UPenn and Core Member of the Institute for Medicine and Engineering [IME][13]), realized that in addition to the many approaches that use nature as a catalyst for design, that little to no work had been formally established that examined reciprocal intersections between our individual fields. Moreover, whereas the powerful and provocative examples cited above look to the natural sciences and code to deliver design tools and products, we were interested in establishing a new field that would also encompass process, new ways of seeing and doing, and a novel pedagogical model in which architects and scientists would not only collaborate in a shared space (a laboratory/studio), but one in which they would be involved in research projects as equals. Most importantly, we were looking to establish a hybrid space where information would be exchanged between and across disciplines to benefit not only a new field, but also one which would be of benefit to architects and scientists alike. To realize this, in 2006 we formally established the Sabin+Jones LabStudio at the University of Pennsylvania.

The common ground that allowed our initial conversation to proceed is rooted in the morphological, as well as molecular changes in cell structures forming functional 3-D tissue-like structures ex vivo, digital practices, computational design, and the subsequent ability to explore iterative, systems-based and complex formal strategies that have transformed architectural research and design. Key to our design research was understanding how the tissue microenvironment surrounding cells, made up predominantly of extracellular matrix (ECM) proteins, which includes collagens, laminins, and elastin, influences cell surfaces and their interconnected interior structure and function, including the genome/code itself. More specifically,

Figure 2.5 Graduate architecture students paired with scientists in the Jones lab, 2007. This photograph was taken on the first day of a course co-taught by Sabin and Jones in the Graduate Department of Architecture entitled, "Nonlinear Systems Biology and Design."

by studying models borrowed from biology, particularly regarding self-organization and the emergence of complex, nonlinear global systems from simple local rules of engagement, we have generated new forms and structural organizations in architectural design. In parallel, tools originating in generative architectural design have allowed scientists within LabStudio to posit new hypotheses, real and virtual models, and ways of evaluating and visualizing complex biological processes and datasets in unprecedented ways. Examples such as these demonstrate how attentive architectural and scientific practices can be to each other—particularly in architecture and biology, which are constantly reinventing and questioning themselves in a manner that is similar to the historic avant-gardes, or in the face of new technologies.

Before delving into specific examples of design research and practice conducted within LabStudio and beyond, however, it is essential to understand some basic principles in cell biology, including the concept of epigenesis. In a post-genomic world filled with genetically-altered "model" organisms that remain unchallenged by their controlled surroundings, together with sequencing of the human genome, this linear framework has come to overshadow more holistic, biological paradigms, including "epigenesis." This philosophy encompasses emergent behavior and network theory, and the notion basic to complexity theory more generally that the whole is greater than the sum of its parts.

A Brief History of Epigenesis

The model of the DNA double helix revealed by Watson, Crick, and Franklin continues to intrigue and to provide a key platform for understanding the chemical

Design Research in Practice: A New Model 39

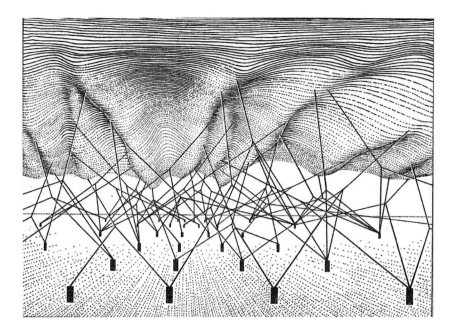

Figure 2.6 Drawing by John Piper illustrating the epigenetic landscape described by C.H. Waddington, a leading embryologist and geneticist from the 1930s–1950s. Piper's painting was first published in Waddington's book, *Organizers and Genes*.

nature, organization, and function of life.[14] Above all, this seminal finding ignited a gene-centered view of life revolving around the "Central Dogma" of molecular biology, which posits that information flows from DNA to RNA to protein. One caveat inherent with this linear model, however, is that DNA needs certain proteins to replicate, and RNA and protein need a blueprint to form; neither can operate without a dynamic, chemical, and physical structural framework. In other words, as C.H. Waddington noted in 1956, "genes are not only actors, but are acted upon."[15]

Why then did a more or less singular view of life come to dominate the contemporary life sciences, frequently ignoring the complex organizing principles that had already been appreciated in developmental biology, with its deep roots in paleontology, anatomy, and embryology? How can the fundamental, yet distinct approaches that differentiate molecular biology from developmental biology be reconciled? Could the principles that underscore these fields be combined into a unifying theory applicable to the study of other processes, such as emergent behavior and ecology, be it at the level of the genome, a cell, a limb, a skyscraper, or even a city?

The concept of epigenesis—which means developing progressively—was originally described by Aristotle who proposed that forms arising in developing organisms did so progressively, rather than being pre-formed. In 1893, Thomas Huxley stated "Evolution is not a speculation but a fact; and it takes place by epigenesis."[16] Today, although the word epigenesis is used also by molecular biologists to encompass chemical, heritable modifications to genes other than changes in the basic DNA sequence itself,[17] an older biological definition exists whereby environmental factors act on the genome to generate a multi-dimensional, dynamic organism.[18] In other words, the physical and chemical microenvironment of cells within tissues represents an epigenetic entity in that it modifies the genome in a context-dependent fashion. The idea that external forces can mold the developing organism, however, has faced considerable challenge from the seventeenth century to the present, exemplified by the pre-formationist idea that the sperm not only contained the embryo, but even a miniscule human—termed a homunculus—and the predominant reductionist idea that the genome harbors the secret of life.

Architecture and Dynamic Reciprocity

The idea that cells within tissues function as integrated architectural units that include their surrounding microenvironment was elegantly described by the developmental and cell biologist Paul Weiss in 1945: "the living units enmeshed in [the microenvironment—which includes the extracellular matrix (ECM)] . . . bind them to the substratum. It thus confers upon what otherwise would be isolated units, the character of a coherent tissue."[19] Scientific descendants who advocate this type of model include Mina J. Bissell, who has refined this idea to suggest that a state of "Dynamic Reciprocity" exists between cells and their immediate microenvironment:

> A dynamic reciprocity exists between the extracellular matrix on the one hand and the cytoskeleton [which supports translation of messenger RNA into protein] and the nuclear matrix [which associates with chromatin, the site of transcription of genes into messenger RNA] on the other hand. The extracellular matrix is postulated to exert physical and chemical influences on the geometry and the biochemistry of the cells via trans-membrane receptors so as to alter the pattern of gene expression by changing the association of cytoskeleton with the mRNA and the interaction of chromatin with the nuclear matrix. This, in turn would affect the extracellular matrix, which would affect the cell. and so on.[20]

What is the evidence that cell and tissue architecture, specified by the microenvironment, forms part of a dynamic chemical-physical loop that signals to cells and their genomes and back again? Mostly inspired by the works of Buckminster Fuller and Kenneth Snelson, tensegrity has been successfully transposed from architecture and sculpture to cell biology. Significantly, as early as 1935 in his article entitled "Le Toles composées et leurs applications aux constructions métalliques légères," Le Ricolais imagined a rapport of relationships in opposition, leading to the conclusion that "there is a correlation between a mechanical principle and a geometric pattern."[21, 22] Interestingly, Fuller's concept of tensegrity, which he understood as a system of energy where space is not static,

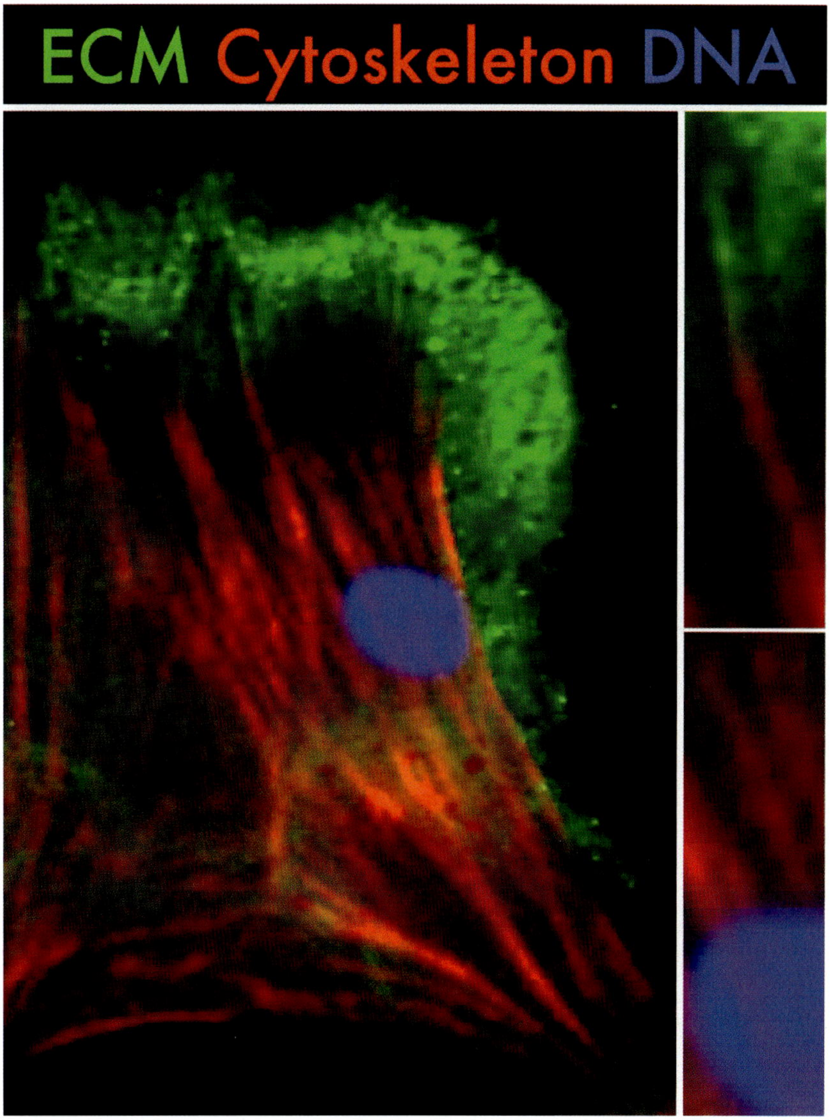

Figure 2.7 Cells can be viewed as "hard-wired" networks of molecular struts, which extend from the extracellular space to the DNA via the cytoskeleton.

has subsequently been adopted by another pioneering cell biologist, Donald Ingber, as a model for understanding how cells are structured at the nanometer scale. As with models of architectural tensegrity, tension in cellular tensegrity is continuously transmitted across all structures within the cell so that tension, in one of the members, results in increased tension in members throughout the structure.[23] How does this relationship relate to environmental influences on gene expression and cell behavior? Inside cells, a network of filaments extend throughout the cell that is exerting tension. In turn, this structure is linked to the extracellular matrix and to the nucleus via filaments that comprise the nuclear matrix. Thus, the cell can be viewed as a "hard-wired" parametric network of molecular struts, which extend from the extracellular space to the DNA via the cytoskeleton.

> If the cell and nucleus are physically connected by tensile filaments and not solely by a fluid cytoplasm, then chemical or physical stimulation of receptors [which interact with the matrix] at the cell surface should produce immediate structural changes deep inside the cell.[24]

Indeed, both actual and simulation models of tensegrity reveal how mechanical forces applied to the cell surface lead to the realignment of cytoskeletal fibers/filaments and structures within the nucleus (where the genome is located). What is more, soluble biochemical reactions are known to take place on the solid-state cytoskeletal fiber bundles, indicating that changing extracellular matrix-dependent cytoskeletal geometry can modulate signaling to and from the genome. At the physical level, this model is remarkably similar to Le Ricolais' Trihex network structures, and to his Funicular Polygon of Revolution system, which is described as the "connectivity of the compression system, and the chain action of the tension cables, acting as bundles of fibers."[25] In reality, form shaped by the environment is essential to unleash the functionality of the genome. In fact, "geodesic forms similar to those found in viruses, enzymes and cells existed in the inorganic world of crystals and minerals long before DNA came into existence."[26] In other words, architecture and the environment can be considered major evolutionary forces that actually generated and shaped the genome.

As written previously,

> Perhaps limited by technology and a modernist bias towards the reproduction of single parts, Fuller was not able to realize his theories of variation between parts and whole systems in his functional geometries. Recent advances in digital and fabrication technologies allow for the exploration of interconnected parts and material behavior where forces external to three-dimensional tensegrity structures may influence and alter the continuity of the tensional forces and the discontinuity of the compressive forces. Ingber's link between the mechanics of the cell cytoskeleton with the dynamics of tensegrity is perhaps the closest translation of Snelson's and Fuller's structural concept as it forms a bridge between the purity of mathematics and geometry with the instability, resilience, and complexity of nature. Imagine these structures growing, contracting, and expanding in response to the presence of people, light, or temperature![27]

And further, imagine a conceptual design space enmeshed in this dynamic reciprocity where architecture and science meet through shared and co-evolving relationships. At root, this book is about co-produced methodologies, tools, materials, prototypes, spatial systems, and most importantly new models for thinking and teaching in design across disciplinary boundaries in both architecture and science.

Notes

1. Portions of this text were adapted from: Jenny Sabin and Peter Lloyd Jones, "Nonlinear Systems Biology and Design: Surface Design," in *Proceedings of the 28th Annual Conference of Acadia 2008: Silicon Skin, Biological Processes and Computation*, ed. Andrew Kudless *et al.* (2008), pp. 54–65; Peter Lloyd Jones, "Content Messaging: Modelling Biological Form," in *Models: 306090*, eds. Jonathan D. Solomon and Emily Abruzzo (New York: Princeton Architectural Press, 2008), pp. 33–38.
2. Cecil Balmond, *Informal* (New York: Prestel, 2002), p. 111.
3. Buckminster Fuller,
4. Heatherwick Studio,
5. Robert Le Ricolais, "Things themselves are lying, and so are their images," *VIA 2: Structures Implicit and Explicit* (University of Pennsylvania, 1973).
6. Ibid.
7. Ibid.
8. James D. Watson and Francis H.C. Crick, "Molecular Structure of Nucleic Acids: A Structure for Deoxyribose Nucleic Acid," *Nature* 171(4356) (1953): 737–738.
9. Rosalind E. Franklin and R.G. Gosling, "Molecular Configuration in Sodium Thymonucleate," *Nature* 171(4356) (1953): 740–741.
10. Karl Chu,
11. François Roche, "New Constellations/New Ecologies," presentation, ACSA 101, California College of the Arts, San Francisco, 2013.
12. See the Nonlinear Systems Organization (NSO) (2005–2011): The NSO was launched in 2005 at the School of Design within the Graduate Department of Architecture, University of Pennsylvania. The research unit was founded by internationally renowned structural engineer and designer, Cecil Balmond, under the chairmanship of Detlef Mertins, and directed by David Ruy and later Jenny E. Sabin. The mission of the NSO was to explore ways in which architecture can demonstrate, test, and apply insights and theories from mathematics and the sciences—nonlinear, algorithmic, and complex—in the design of material structures across an open-ended range of scales, materials, and design disciplines. By transferring theoretical scientific knowledge to the applied design arts, it sought to expand the horizons of design and, at the same time, promote a broader appreciation of these theories by the general public. The NSO conducted think tanks and design workshops that brought together researchers in the sciences with architects and other design professionals to identify scientific models that could be developed into new design techniques and processes. It offered research fellowships to support work that advanced the mission of the organization. NSO fellows and Senior Researchers included: Ben Aranda, Daniel Bosia, Peter Lloyd Jones, Ferda Kolatan, Chris Lasch, Philip Ording, Jenny E. Sabin, and Roland Snooks.
13. See the Institute for Medicine and Engineering (IME). The mission of the Institute for Medicine and Engineering, directed since its inception by Peter F. Davies, is to stimulate fundamental research at the interface between biomedicine and engineering/physical/computational sciences, leading to innovative applications in biomedical research and clinical practice. The IME was created in 1996 by a mandate from the Trustees of the University to bring together the Schools of Medicine (SOM) and Engineering and Applied Science (SEAS) to pursue opportunities for collaborative research. This collaboration extends the mission to encompass the School of Design.
14. Watson and Crick, "Molecular Structure of Nucleic Acids."
15. C.H. Waddington, Scott Gilbert, and Sahotra Sarkar (eds.) "Embracing Complexity: Organicism for the 21st Century," *Developmental Dynamics* 219(1) (2000): 1–9.
16. Thomas Huxley, *Darwiniana: Collected Essays* (London: Macmillan and Co., 1893), vol. II, p. 202.
17. Christine B. Yoo and Peter A. Jones, "Epigenetic Therapy of Cancer: Past, Present and Future," *Nature Reviews: Drug Discovery* 27 (2006).
18. See F.S. Jones and Peter Lloyd Jones, "The Tenascin Family of ECM Glycoproteins: Structure, Function, and Regulation During Embryonic Development and Tissue Remodeling," *Developmental Dynamics* 218 (2000): 235–259.
19. Paul Weiss, "Experiments on Cell and Axon Orientation In Vivo: The Role of Colloidal Exudates in Tissue Organization," *Journal of Experimental Zoology* 100 (1945): 337.
20. Mina J. Bissell, H.G. Hall, and Gordon G. Parry, "How Does the Extracellular Matrix Direct Gene Expression?" *Journal of Theoretical Biology* 99(1) (1982): 31–68; and Celeste M. Nelson and Mina J. Bissell. "Of Extracellular Matrix, Scaffolds, and Signaling: Tissue Architecture Regulates Development, Homeostasis, and Cancer" *Annual Review of Cell and Developmental Biology* 22(1) (2006): 287.
21. Peter McCleary, "Visions and Paradox: An Exhibition of the Work of Robert Le Ricolais," exhibition, University of Pennsylvania, Philadelphia, PA, 1996.

22 Robert Le Ricolais, "Things themselves are lying, and so are their images," pp. 89–91.
23 Donald Ingber, "The Architecture of Life," *Scientific American* 278 (1998): 48–57. Ingber imported tensegrity into cell biology via his studies in sculpture at Yale. Tensegrity is a holistic and systemic theory of structure made famous in the postwar period by Buckminster Fuller. Fuller's architectural work pre-dates the discovery, re-design and application of C60 fullerenes, named after his geodesic dome structures. Of course, carbon fullerenes, in their many forms, are now widely used in nanotechnology, electronics, optics, and other fields, including material and medical science.
24 Ibid.
25 McCleary, "Visions and Paradox."
26 Ingber, "The Architecture of Life."
27 See Jenny Sabin, "Geometry of Pure Motion: Buckminster Fuller's Search for a Coordinate System Employed by Nature", in *The Language of Architecture: 26 Principles Every Architect Should Know*, ed. Andrea Simitch and Val Warke (Beverly, MA: Rockport Publishers, 2014), p. 190.

PART II

Design Computation Tools for Architecture and Science

New Tools and Forms

Introduction to Design Computation Tools for Architecture and Science: New Tools and Forms

Jenny E. Sabin and Peter Lloyd Jones

The objective of Systems Biology [can be] defined as understanding network behavior, and in particular their dynamic aspects, which requires the utilization of modeling tightly linked to experiment.

(Cassman)[1]

This research collaboration puts into operation a renewed dialog between architecture and biology that has been gaining momentum in recent years and holds great potential for the co-evolution of both disciplines.

(Detlef Mertins, Chair, Department of Architecture, University of Pennsylvania, 2007)

The IME-LabStudio collaboration is a unique approach to "Nature's design" of biological systems and their regulation by the local environment. It promotes the flow of new ideas across disciplines, identifying and developing biological principles directed to architecture and an understanding of design approaches in biomedicine and potentially personalized medicine. Facilitating these important interactions through the IME is an enormous pleasure.

(Peter F. Davies, Director, Institute for Medicine and Engineering, Department of Pathology and Laboratory Medicine, University of Pennsylvania, 2007)

How would an architect provided with as many as 30,000 individual building blocks of different shapes and sizes, which interact in multiple ways with one another and with their surrounding environment in time and space, design a final form that is unique on its exterior, yet is relatively uniform at its core, at certain scales at least? As an additional part of the brief, the architect is instructed that the fully self-assembled, modular structure needs to have a personality, be intelligent, regenerative and appealing, while retaining a memory of the intermediate processes that gave rise to the ultimate form. What if the client dictated that this design goal always had to be met on time, with minimal cost and energy, yet with a high degree of reproducibility and fidelity, using a slightly different version of the original blueprint for each and every project? What if many parts of the structure had to execute more than one function at a specific moment in time, even at the same or a different location within the developing and final configuration? What if the rules of engagement between the emergent and final form, and the immediate and larger environment, continually changed at all phases of building, and at every possible scale and time-point? How would the designer and engineer manage a structure that is continually relocating from one city block to another, as well as to one that is constantly being remodeled and rewired from within? How would the form appear if the client decided to selectively remove or modify one or more of the building blocks during construction, or even after completion of the structure, without the designer's input? What if every form and structure represented in the designer's portfolio always had to influence those of subsequent generations? These seemingly unattainable design directives are some of those required for the genesis and maintenance of each and every living being, and so the aforementioned nonlinear rules, problems and questions represent many of those that natural scientists are continually trying to test and model, *in vitro, in vivo* and *in silico*. The aim of the work featured in the following sections is to foster new and ongoing dialogues between the disciplines of architecture and biology, and to jointly investigate fundamental processes in living systems and their potential application in structures and buildings and vice versa.[2]

Following our chance meeting at the inaugural conference for the Nonlinear Systems Organization at the Graduate School of Design, University of Pennsylvania, in the fall of 2005, we, Sabin and Jones, launched a year-long exchange to articulate and formulate a means of collaboration through hybrid venues and discursive convergence; and to initiate possible research collaborations and areas of shared research. Although we had no set goals, our initial aim was to learn how to effectively communicate across disciplinary boundaries. We shared in joint studio reviews, weekly lab meetings, invitations to seminars and conferences, and informal gatherings. We learned that biomedicine and architecture share a common language and direction. Both disciplines share a fundamental concern with ordering principles that have structural and organizational consequences, and with the formal principles and possibilities of morphogenesis and epigenesis. In addition, the connection between biomedicine and architecture has not often evolved beyond the literal: a reduction of biology to formal pattern making and "organic" aesthetics, on the one hand, or of architecture to simplistic structural models, on the other. In hindsight, this investment of shared and open-ended discussion was perhaps the most important ingredient in the success of our exchange and to our current collaborative endeavors in architecture, science, medicine, and engineering. Over the course of this initial exchange, highlights included a joint, day-long event at the Institute for Medicine and Engineering and PennDesign with Matilda McQuaid, Deputy Curatorial Director at Cooper Hewitt National Design Museum, and organized by Peter Lloyd Jones and Jenny Sabin. This spring event was followed by two fall events:

The annual NSO conference, entitled *Architecture Bits*, and an evening discussion at the Slought Foundation entitled "Cross-Catalytic Architectures", a conversation between Cecil Balmond and Peter Lloyd Jones, moderated by Peter Davies, David Ruy, and Jenny Sabin, and organized by Eric Ellingsen and Aaron Levy. We successfully ignited an ongoing conversation and it was time to further articulate our next steps through research, design, and application.

During the summer of 2007, we officially launched the Sabin+Jones LabStudio, jointly housed within the Institute for Medicine and Engineering and PennDesign. Three graduate architecture students, Erica Swesey Savig, Wei Wang, and Allison Schue were hired to work alongside post-doctoral fellows and MD-PhD students, Jan Baranski, Agne Taraseviciuete, and Mathieu Tamby in the Jones Lab. The research agenda for the inaugural year of LabStudio was entitled "Nonlinear Biosynthesis." In "Nonlinear Biosynthesis," we argue that through the analysis of biological design problems in specialized 3-D designer microenvironments, the biologist and architect are afforded new ways of thinking about design through an understanding of how dynamic and environmental feedback specifies structure, form, and function. Through the design of digital and algorithmic tools, we aim to escape the direct mimicking of these biological structures (popularly known as biomimicry) in favor of biosynthesis, where new models for architecture and biomedicine are generated. Overall, the mission of LabStudio is to foster new and ongoing dialogs between the disciplines of architecture and biology, and to jointly investigate fundamental processes in living systems, connect their historical and contemporary relationships to generative design and fabrication in architecture, and innovate their potential application in architecture and biomedicine.

Figure 0.1 First Day of the Sabin+Jones LabStudio, summer of 2007. Pictured left to right in the Jones Laboratory at the Institute for Medicine and Engineering, University of Pennsylvania: Dr. Peter Lloyd Jones, Allison Schue, Wei Wang, Erica Swesey Savig, and Jenny E. Sabin (taking photo).

This first summer session of LabStudio also provided the necessary foundation for our first jointly taught seminar, "Nonlinear Systems Biology and Design." This graduate seminar (2007–2010) was housed jointly between the Graduate School of Design within the department of architecture and the Institute for Medicine and Engineering at the University of Pennsylvania. Later, this seminar was followed with a revamped fifth installment, taught by Sabin at Cornell University within the Department of Architecture. The title of this seminar was "Special Topics in Construction: Bio-Inspired Materials and Design." The following body of work highlights six years of collaborative research in LabStudio and in the Sabin Design Lab at Cornell University, including design research methods and teaching across three specific areas of inquiry: *Networking*, *Motility*, and *Surface Design*. We will start by presenting and describing the methods and structure for this design research. Each subject area and biological system will be explained in detail, preceding the project descriptions. Overall, the following section explores novel methodologies in computational design and for discovering new, formerly unseen relationships in dynamic biological systems. The chapters demonstrate how the production of catalogs of visualization and simulation tools are used to discover new behaviors in geometry and matter. Project work from LabStudio and our co-taught graduate seminar will form the basis for discussion and exploration. Through the analysis of biological design problems, students in both venues were exposed to new modes of thinking about design ecology through an understanding of how dynamic and environmental feedback specifies structure, function, and form. Specific physical (including 3-D print and various digital fabrication outputs) and visual exercises (generative and algorithmic studies) will be explained to familiarize the reader with the topics in architectural and scientific terms. A short summary of the three areas of inquiry is as follows:

- *Networking*: The first project topic investigates part-to-whole relationships found during the generation of branching structures formed by interacting vascular cells. The study and quantification of this network allow for a greater understanding of how the design of variable components give rise to structured networks. Fabrication experiments in 3-D printing technologies explore these variable components as they structure network design.
- *Motility*: This project investigates the role of the microenvironment in controlling cellular movement and the shifting geometries inherent to this movement. In short, diseased cells move differently from healthy cells and we seek to derive 4-dimensional motility signatures for these movements. These projects aim to develop metrics to quantify these differences in motility and thus enable a novel understanding of the space and structure of movement and change in architecture. 3-D prints, models, and simulations of these signatures enable a deeper understanding of the frame of reference that articulates the discrete space and structure of movement.

Figure 0.2 LabStudio field trip to NYC and the Cooper Hewitt Smithsonian Design Museum, summer 2007

- *Surface Design*: This project topic seeks to quantify and spatialize cellular contour information through the design of surface architecture in dynamic space. The projects look at the role of personal shape change as it relates to surface design. Here, the study of relationships found within the closed and open structure of neighboring cells gives rise to an abstract understanding of cellular form as it relates to dynamic boundary conditions and cell mechanics. Through the use of algorithms such as the Delaunay Tessellation and its Voronoi Diagram, geometric abstraction generates dynamic shell and spatial structures capable of shape shifting in environmental context. Fabrication experiments in rapid prototyping (RP) technologies are used to study part-to-whole and whole-to-whole relationships within the context of packing performance, component design, the sub-division of space and environmental response.

Criteria and Goals

Based upon our investigations, we posit that any future investigation between architecture and biology should require a consideration of models that capture and cultivate the dynamic reciprocity of the less obvious organic systems of architecture and the more obvious living complexities of biological systems. To address this, we ask whether architecture can take a cue from biology in matching the complexity of its generative design models to the very dynamic features of the living environment and organic milieu in which the architecture is a part. We may then begin to move toward a more dynamic and volumetric model where architecture acquires connectivity, performance, and time as embedded features and thus responds dynamically to both environment (context) and to deeper interior systems.

This approach requires us to establish bi-directional dialogues and a favoring of process-driven research over goal-driven research. The vitality of this research depends upon a collective intuition, or knowing where to look, along the way. This clarity comes forth from doing, making, collaborating, and failing. "Knowing where to look" requires a constant refinement of one's intuition. There are no fields within this space of refinement; rather, it undulates within a multi-set of honed intuitions. It embraces, teaches, and explores abstract concepts that reflect across and over perceived pedagogical and disciplinary boundaries. To 'see' demands collaboration. It is neither about breaking down walls, nor sampling across, but rather it is about continuity and connectedness. Specificity, specialty, and subjectivity, then, are in the 'seeing.'

Systems biology represents a relatively new and major emerging field that focuses on the systematic study of complex interactions in biological systems, thereby using a new perspective (integration instead of reduction) to study them. Importantly, its relationship to architecture and vice versa have not yet been probed. As with architectural design and more specifically generative and computational design, systems biology examines the nature of nonlinearities, emergent properties, and loosely coupled modules that are cardinal features of complexity. New models for research and design in architecture have grown in response to radical breakthroughs in technology and an increasing interest in algorithmic tools used in architectural form generation and sophisticated responses to complex problems, such as those that negotiate issues of feedback, sustainability, and performance in buildings. Algorithmic imaging, applied mathematics, and molecular tools found useful in analyzing nonlinear biological systems may therefore prove to be of value to new directions in design in architecture.

The aim of the following project work is to foster and expand the emerging dialog between architecture and nonlinear systems biology to gain new insights into how dynamic living systems might operate, to develop techniques for digital modeling, and to create experimental designs with rigor at multiple length scales, ranging from the nanoscopic to the human. The work of LabStudio and its current formations in the Sabin Design Lab at Cornell University and MedStudio@JEFF at Sidney Kimmel Medical College serve to deepen knowledge of biological complexity from a radically different viewpoint, and to develop design thinking and tooling through the study of complex biological systems, alongside an introduction to advanced scripting logics in parametric and associative software, and the production of hybrid material systems, including 3-D printing fabrication.

The pedagogical framework for the featured projects was structured through a sequence of discussions and workshops based upon a detailed understanding of systems biology, and corresponding explorations in generative design and

Figure 0.3 LabStudio methodological approach across scales, materials, systems, and disciplines

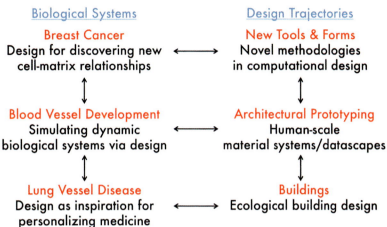

experimental fabrication in architecture. This was augmented with lab meetings, readings, and field trips to inter-disciplinary research laboratories at the University of Pennsylvania and now at Cornell University and MedStudio@JEFF. The project work follows three methodological trajectories including:

▶ *Visualization and simulation*: The generation of digital design tools, whereby cellular-mediated changes in pattern, geometry, material, and environment are simulated in 3-D digital environments via custom-written architectural algorithms. These models and simulations visually describe the dynamic and nonlinear human cell behaviors and processes being researched in 3-D and 4-D space/time.
▶ *Experimental material systems*: The abstraction and application of nonlinear and dynamic cell behaviors to the experimental design of materials and geometries at the human scale with maximum response to environment leading to a catalog of surface effects (e.g., color or pattern change).
▶ *Generative fabrication*: Transformation and translation of the design ecology developed in the first two phases into the design and fabrication of a series of analogic prototypes that are materially directed. These may include experimental and responsive systems that function at the human and architectural scales. The final fabricated physical models are composed of hybrid material systems that may include 3-D printed components.

Matrix Architecture

A novel ecological model for architecture: Embedding reciprocity into material tectonic organizations through the study of code in context.

While the first phase of our design work resides within the spirit of research and discovery, the current phase engages design-oriented applications in contemporary architecture practice ranging from new concepts of materiality to adaptive structures and complex geometries. Here, material technology and design ecology are informed by the visualization of complex biological systems through the generation of new design tools. The introduction and design of many advanced tools, such as parametric and algorithmic software into the design process, not only facilitate the work process, but also more importantly allow for a far more intelligent design ecology in which various factors are interlinked in responsive feedback loops. Next to issues of research, technique, and performance, we seek to explore the material and ecological qualities produced by these models. This design-based research spans across design studios, courses in visual studies, fabrication, and parametric modeling. Here, design research exists as process and search, as means for exploration and discovery, as an end in itself and as a bridge and catalyst toward application in architecture. For example, several of the featured projects place emphasis upon the production of *hybrid material systems* inspired by understanding the force-generating behavior of human cells as they interact with a geometrically and chemically-defined dynamic surface, which might ultimately be abstracted and translated into a responsive skin at the architectural scale. By immersing oneself in complex biological design problems, and abstracting the inherent relationships of these models into code-driven parametric and associative models, it is possible to gain new insights into how nature deals with design issues that feature pattern formation, part-to-whole relationships, complexity, and emergent behavior. The following experimental projects engage these topics across three areas of inquiry: networking, motility, and surface design, and primarily aim to unfold a robust catalog of tools and methods for modeling and simulating complex biological behavior in context. Threaded throughout the projects, are edited quotes and discussion strings captured during final reviews and critiques over the course of four years of teaching and production, 2007–2011.

Notes
1 Cassman (2005).
2 Adapted from course syllabus by Sabin and Jones for "Nonlinear Systems Biology and Design" (syllabus, Graduate Department of Architecture, School of Design, University of Pennsylvania, 2007–2010) and Peter Lloyd Jones, "Context Messaging: Modelling Biological Form," in *Models: 306090*, eds. Jonathan D. Solomon and Emily Abruzzo, 11 (New York: Princeton Architectural Press, 2008), pp. 33–38.

3

Networking: Elasticity and Branching Morphogenesis

Jenny E. Sabin and Peter Lloyd Jones

> Because every cell in an individual's body harbors the same genome, it is reasonable to hypothesize that it is the nonlinear coordination of protein-based events, superimposed upon the linear genome, that leads to morphological complexity, patterning, and differentiation of cells within tissues. This "epigenetic" viewpoint of development suggests that embryonic structures arise as a consequence of environmental influences—both physical and chemical—acting on the genome, rather than an unfolding of a completely genetically specified and pre-existing invisible pattern.
>
> (Peter Lloyd Jones)[EPIX]

Introduction

In support of the aforementioned concepts and models, this chapter provides specific examples of parallel design thinking between architectural research and cell biology that has been garnered over six years of research and education within LabStudio and now in its current formations. This section highlights specific project work on the topic of cellular *networking* behavior through examples that examine the nature of nonlinearities, emergent properties, and loosely coupled modules that are cardinal features of complexity. Specific physical (including 3-D print and various digital fabrication outputs) and visual exercises (generative and algorithmic studies) are explained to familiarize the reader with the topics in architectural and scientific terms.

Specifically, Part II explores one research track within LabStudio centered upon cellular networking behavior. Six case study examples are presented from student work. The aim of this section is to report upon one stream of research and to demonstrate to natural scientists and architects alike how high-risk, nonlinear, design-driven philosophies and practices emanating from 3-D spatial biology and generative design in architecture can result in radical advances in both scientific and design research and applied architectural practice.

The primary goal of the Jones Lab research group is to determine how the extracellular matrix (ECM), a cell-derived woven and globular protein network that envelopes or contacts most cells within the body—an architectural textile of sorts—changes throughout development and disease, and how alterations in this 3-D ECM environment feed back to control cell and tissue behavior at the level of the genome and beyond in real time. Toward these goals, and in the context of the Sabin+Jones LabStudio collaboration, we began by studying the influence of different extracellular matrix components, such as laminin and type I collagen, on a number of nonlinear biological processes, including tumor formation, cell motility, and networking behavior. Ultimately, at the patho-physiological level, the main aim of our collaborative research is to derive new structural and functional information from each of these dynamic systems in an attempt to diagnose, prognosticate, and treat human diseases.

Over the past three decades, it has been firmly established that 3-D cell and tissue architecture, at the level of the ECM, exerts a dominant influence over the genetic makeup of a cell or individual. In this sense, the ECM represents a key phenotypic determinant of cells within tissues. In essence, much of the secret of life resides outside the cell within the extracellular matrix. Importantly, the ECM not only acts as a physical entity, which lends flexibility and physical support to cells within tissues, but it also behaves as an informational entity, an ongoing and historical document that records what transpires in and around a cell during the lifetime of an organism.

Clearly, these biological models are underscored at the cellular and molecular levels by complex nonlinear responses to complex and dynamic scenarios. Thus, we posit that a rigorous understanding and analysis of these types of models will allow architects to retool and reevaluate how we may negotiate topics such as nonlinear fabrication, feedback, and performance in architecture. On the architectural side, rather than seeking direct translation of the science to architecture and vice versa, this research collaboration is about working through biological design problems that give rise to new modes of thinking and working in design and biomedicine.

On the design studio side, our aim is to introduce biologists to digital and algorithmic architectural tools that may be used to reveal new complexities within the biological systems being studied. Additionally, by specifically examining dynamic cellular systems, our intent is to discover new ways of understanding, revealing, and abstracting how these biological systems negotiate issues of auto- and artificial-fabrication. Here structure and material are inextricably linked. Cells and tissues not only produce their own underlying, self-produced fabric, but they modify and respond to this woven, felted, or embellished environment. A digital code begets a physical code that informs the digital code, and so on.

Background

For the networking projects, we are investigating part-to-whole relationships during capillary formation by lung endothelial cells. At the biological level, coordinated

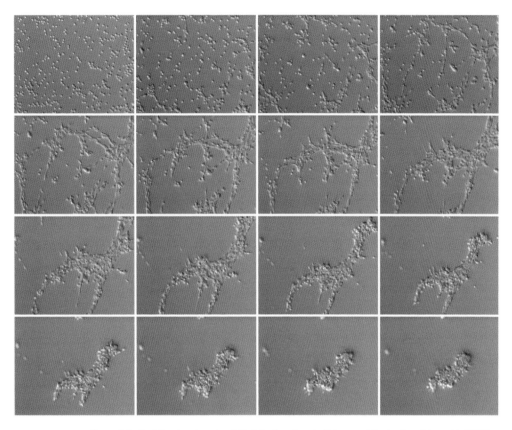

Figure 3.1 Real-time imaging of endothelial cells cultured within a specialized extracellular matrix (ECM) microenvironment

cellular networking, a component of angiogenesis, is required to form the exquisite fractal network that emerges in the developing and mature lung to facilitate efficient gas exchange from birth onwards. Real-time imaging of endothelial cells cultured within a specialized extracellular matrix (ECM) microenvironment, designated the basement membrane, that either suppresses or promotes networking, formed the basis for these projects.

The study and quantification of this vascular network allow for a greater understanding of how variable components give rise to structured networks in both biology and architecture.

Blood Vessel Development within the Lung

At birth, the entire airway network within our lungs fills with air as we take our first breath. Like the vascular network, the airways start off as a large space in the upper respiratory tract region, and become smaller and smaller in caliber. These distal airways are called alveoli, and adjacent to these structures resides a network of blood vessels called capillaries, that contain blood. Both the airway sacs and capillary tubes are each surrounded by a specialized ECM, called the basement membrane. The purpose of the airway-capillary interface is to provide oxygen from the air to red blood cells within the capillaries. To accomplish this, oxygen within the airways diffuses from the alveoli to hemoglobin molecules within the red blood cells. Next, the heart pumps this oxygenated blood away from the lung to tissues and organs, all of which require oxygen and nourishment provided by other components in the bloodstream. Once they reach these sites, oxygen and nutrients are released, and deoxygenated blood is returned to the lung via the heart. And the cycle begins again.

The interface between the airways and the blood vessels is carefully controlled prior to birth during fetal development, and in fact, the development of both of these systems is dependent upon the other. Our research, as well as work by others, has shown that if the vessels do not develop, then the airways also fail to thrive, and vice versa. So this means that vessel and airway development go hand in hand, which means that patterning of these parallel and inter-dependent systems must be carefully controlled and coordinated, which is where the ECM comes into play.

Before the final branching pattern of vessels and airways is established, the primitive airways and vessels have no recognizable structure: they simply exist as an outcrop or a lump. What the extracellular matrix does, together with numerous other factors, is to form a template ahead of this outcrop, upon which the primitive cells can divide, thereby increasing in number, and upon which they migrate into the surrounding connective tissue to form the final interconnected fractal networks that structurally define the lungs. So who makes the ECM? Both primitive airways and blood vessels can do this, but only when they are talking to one another in proximity. Take one element away, and the other one will fail to make an ECM and vice versa.

So this seemingly simple set of experimental facts tells us a number of things about the ECM in lung vascular development. First, that it can control cell growth, as well as cell movement, and both of these processes depend upon the appropriate use of the DNA code. Second, it tells us that the ECM associated with one cell type can have a major influence on a different cell type, which can be located at some distance from the other cell, at least at a biological scale. Third, it indicates that the existence of one matrix can impact the production of another matrix in a different cell type, once again showing that matrix can control code usage, both locally and more globally.

Whether the cells or the ECM templates are acting as hubs or nodes is left for the students to decide, but the point is to highlight the requirement for a structural network provided by the ECM, and to point out that the ECM itself can even be regarded as a type of code. Thus, a DNA code begets a protein code (the ECM), which informs the DNA code, and so on. The following projects explore, visualize, simulate, and abstract these fundamental dynamic processes.

Case Study: Modeling a Complex Multi-Variable System: Nonlinear Cellular Networking

Jonathan Asher, A.J. Chan, Kenta Fukunishi, Christopher Lee, and Andrew Lucia with Jan Baranski, IME, B.S.

Nonlinear Biosynthesis
Instructors: Jenny E. Sabin and Peter Lloyd Jones
Sabin+Jones LabStudio

As previously mentioned, the primary function of the lung is to allow for efficient gas exchange between the airways and blood vessels in post-natal life onwards. However, determining how networks of blood vessels (the lining of which is composed of interconnected endothelial cells [ECs] forming tubes) are generated and maintained during development represents a major challenge in contemporary lung biology. The aim of this project is to sequentially model the process of tube formation *in vitro* and *in silico*, and then to abstract this process into a set of potential design tools for architectural application. To approach this, we studied the parameters that govern EC morphology in response to the underlying extracellular matrix (ECM), and how this alters cell-cell and cell-ECM interactions during networking. By comparing the behavior of human lung ECs cultured either on polystyrene (i.e. a non-networking branching environment) with those cultivated on reconstituted ECM (i.e. a condition that promotes network formation), we suggest that there are more than merely visible parameters governing the emergent structures and networks that eventually give rise to EC networks. Through our research, we have explored potential parameters that potentiate or prohibit networking behavior, including intercellular communication, environmental instigators, and cellular geometry.

This project is situated within a larger research framework that aims to garner new ways of understanding design as a complex multi-variable system. Given the models developed thus far, we are capable of measuring relative forces, rates of change among those forces, angles of attraction among cells, and rules of proximity. Furthermore, these variables can be adjusted to create stable or unstable networks.

Our initial investigations allowed us to consider the following questions: Given the environment, in this case, the ECM, what makeup and density may allow for proper cell aggregation? Similarly, what are the forces that are at play in and among the cells and their environment? Certain behaviors are evident within the *in vitro* test systems that we investigated. By abstracting primitive parameters that govern these behaviors (cell-cell interactions, cellular geometry, and the influence of the surrounding ECM), simplified computer models were created. Studying these parameters within a parametric environment allowed us to explore multiple configurations with dynamic variables, increasing the accuracy of our computer models, and continually defining, and redefining potential parameters. In addition, these investigations allowed us to reformulate *in vitro* models of EC networking that focused on the geometry and movement of the ECM during tube formation. Yet, given the conditions within the models, how can forces and relative densities be measured both within the simulation and the actual samples to determine accuracy?

If the models prove to be accurate, what conditions must be met within the models to achieve simulated networking? What are the non-visible variables (physical and chemical) that instigate aggregation and formation of ECs into blood vessels? What are the temporal/spatial elements at play? What role does the ECM play in proper vessel formation (i.e., *in vivo* vs *in vitro*)? What roles (if any) do forces generated by the cells within the ECM play in the establishment of proper vessel formation? Does this play a role beyond cell geometry or initial cell organization alone? What role does cell geometry play in angiogenesis? Above all, how can these novel studies lead to development of new rules of engagement in architecture at different time and length scales?

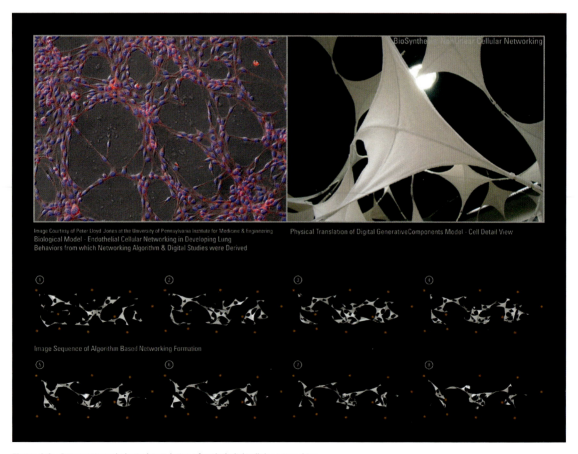

Figure 3.2   Parametric and physical translation of endothelial cellular networking

Cellular Networking Algorithm Principles
1. Matrix points act as anchors and find nearest cells
2. Each cell identifies all neighboring cells according to proximity
3. Cells individually calculate their angle to adjacent cells & identify towards which cell they should move based on an Angle/Proximity Evaluation
4. The cell moves specified unit in direction of neighboring cell, then the next cell begins an identical calculation. This continues until every cell has moved one unit while the matrix points remain stationary, at which point the process loops again for a user-defined number of steps or until the network stabalizes

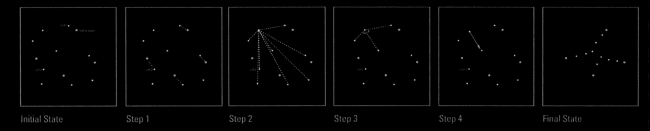

Initial State   Step 1   Step 2   Step 3   Step 4   Final State

Reflected Ceiling Plan from GenerativeComponents Digital Model Used for Installation
Grid Represents 4"x 4" Ceiling Grid in Installation Space - Numbers Represent Distances of Cell Centroids from Ceiling - Green Boxes Represent Suspension Points

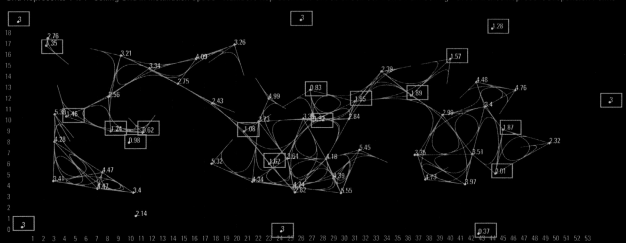

Figure 3.3 Cellular networking algorithm principles and integration of scaled contextual parameters of a room for physical prototyping

As we develop these new tools, we are continually asking how this could influence the role of the designer in an ever evolving and complex world. In this regard, we have explored how this system can materialize as an architectural proposition. As spearheaded by Asher and Chan, the networking algorithms have integrated real-world environmental inputs, including floors, walls, and ceilings that have led to large-scale physical models to study the effect of nonlinear networking in three dimensions. This was followed by more in-depth research by Jonathan Asher in the context of an independent study with Jenny E. Sabin and LabStudio.

In the following suite of simulations by Lucia and Lee, fundamental questions were posed as to the role communication and connectivity play through a structured or non-structured ECM environment during the process of branching morphogenesis. Specifically, parameters were limited to cells in attraction with each other whose linkages were tethered to a networked or non-networked material

62 Design Computation Tools

Figure 3.4 Physical prototype on view at the *Fiber* exhibition, Philadelphia, PA, 2008. Organized by Jenny E. Sabin with Peter Lloyd Jones and Philip Beesley.

substrate (simulated ECM) through time. The principles of the simulation sets were quite simple, with one major difference between them—whether or not a structured environment was in place through which the cells could communicate and distribute force.

The rule sets demonstrate a simple condition, while the actual simulations were carried out with much greater density of both cells and environmental matrix points (virtual material ECM). A series of these denser simulations was carried out with exemplary cases called out here (Figures 3.7 and 3.8). Of interest are the images in the right panel of both Figures 3.7 and 3.8. These images depict the movement of simulated ECM points through time. In both simulations, all rules for cell behavior and attraction remain the same, save the networking and communication of their forces through the underlying ECM. In the absence of communication, no discernible pattern emerges (Figure 3.7). What does appear is a disorganized agglomeration of movement without any

Figure 3.5 Cell attraction with force exchange through matrix. Matrix deformation diagram over time with multiple attractors

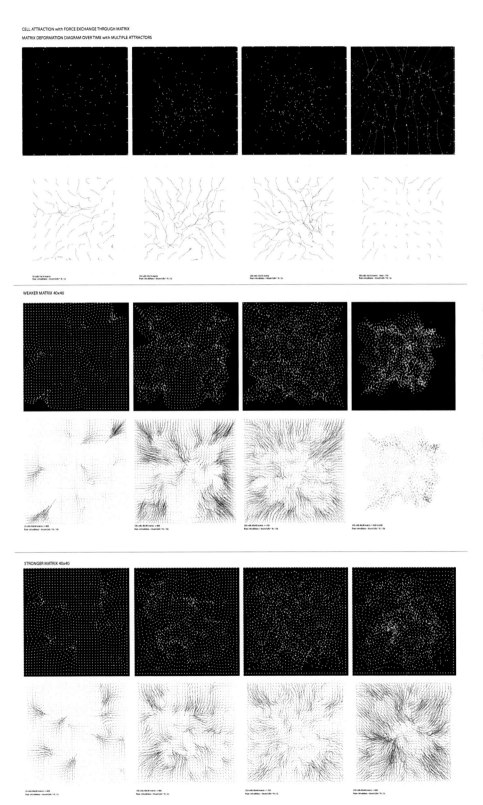

64 Design Computation Tools

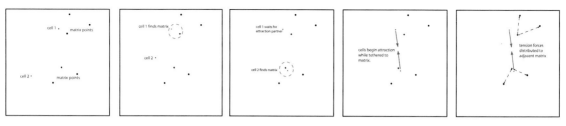

Figure 3.6 Rules:
1. Cells walk with Brownian motion looking for matrix.
2. If cell finds matrix, initialize attraction and wait for another cell connected to the matrix at any other anchor.
3. When next cell finds matrix, begin attraction to previous cell that located matrix.
4. While cells attract, tension forces distribute through adjacent matrix connections.

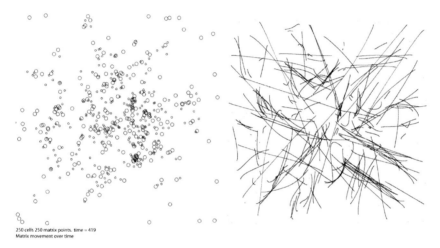

Figure 3.7 Networking simulation with no environmental ECM matrix communication. This simulation is played out according to the rules outlined above, however, the material substrate to which the cells adhere is not networked. Left, cell and substrate image. Right, path of the substrate matrix points over time.

intuitively meaningful structure. In the presence of a networked ECM in which forces are communicated and distributed, an entirely different image emerges (Figure 3.8). In the latter case, an underlying formal pattern emerges in which the relational distribution of force is made legible as the matrix points are tethered to cells in mutual attraction over time. This ultimately suggests that meaningful organization, at least in this simulation, is facilitated not by the behavior of the cells themselves but rather through the environments in which they reside. Again, the cells in both simulations behave according to exactly the same rules with the only difference being the presence or absence of communication within the environment.

The initial physical prototypes informed another catalog of advanced networking simulations, this time exploring a dynamic ECM with programmed components featuring variable surface densities.

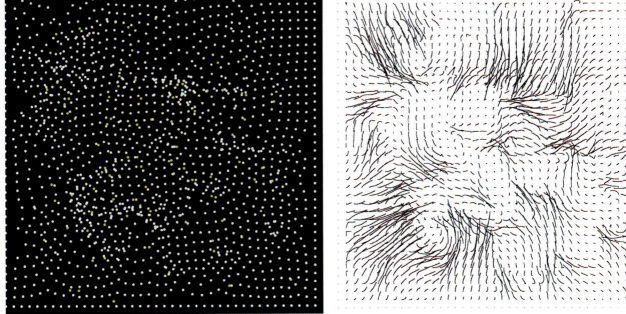

100 cells 40x40 matrix. t=400
float criticalMass = ((numCells * 9) / 5);

Figure 3.8 Networking simulation with environmental ECM matrix communication. This simulation is played out according to the rules outlined above. In this case the material substrate to which the cells adhere is networked. Left, cell and substrate image. Right, path of the substrate matrix points over time.

Figure 3.9 Networking simulation over time

Figure 3.10 Surface connections formed during networking process, shown in simulation

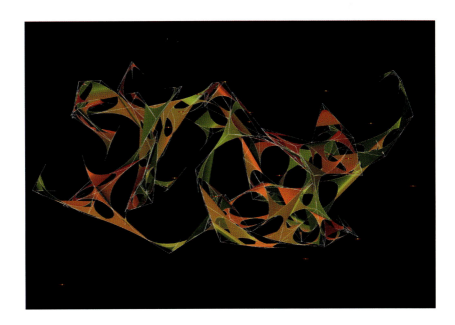

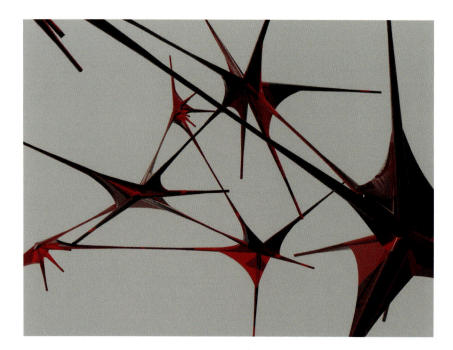

Figure 3.11 Rendering of larger network

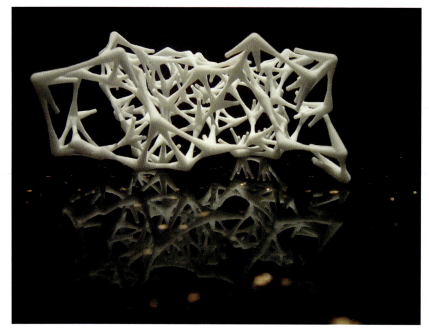

Figure 3.12 3-D print of a static moment within the overall simulated network

Case Study: Scale-Free Networks
Joshua Freese, Jeffrey Nesbit, and Shuni Feng

Nonlinear Systems Biology and Design
Instructors: Jenny E. Sabin and Peter Lloyd Jones
Sabin+Jones LabStudio

Rules and Logics of Scale-free Networks Derived from the Study of Angiogenesis

The angiogenic networking behavior of lung mesenchymal cells, designated RFL-6 or MFLM-4 cells, which have the ability to transform themselves into networking endothelial cells in response to basement membrane proteins, was evaluated in contrasting conditions: with or without Prx-1, a gene that enhances networking.[1] When Prx-1 is absent, cells form clusters on basement membrane material. On the other hand, when Prx-1 is present, cells form branched networks on this matrix. The network formations are also influenced by the topological geometries found within the underlying ECM. With the help of metalloproteinase, cell behaviors could either dissolve the existing extracellular matrix or regenerate and attach new structures to the matrix. Thus, the roles played by Prx-1 (code) and the ECM (environment) are highly influential in creating these network morphologies by endothelial cells (components).

This collaboration between code, environment, and component has to operate on a series of scales, which start with the single cells pairing, grouping, and clustering, and eventually reconfiguring the clusters into larger networks, which create the lung vasculature. By studying the relationship between these effectors, we can observe how inter-scalar relationships occur and move from the micro to the macro. Our research examines and explores a variety of organizational operations in biology and architecture by studying the scale-free networks that define order and relationships in global systems, including the internet and air travel. These network systems, like those in the biological model, rely on

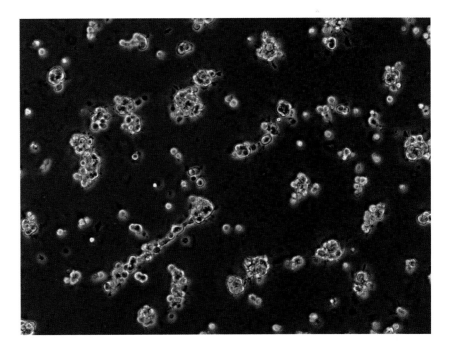

Figure 3.13 Networking without Prx-1

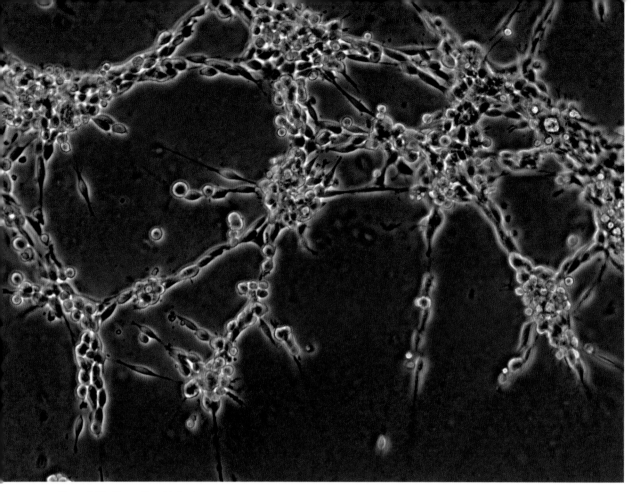

Figure 3.14 Networking with Prx-1

managing complexity by sharing simple organizing principles that govern behavior at all scales.

Scale-free networks, as opposed to random networks, represent complex systems in which some nodes have a tremendous number of connections to other nodes. The popular nodes are called "hubs."[2] One characteristic of scale-free networks is that they are robust against accidental failures, yet are vulnerable to coordinated perturbations or attacks. A growing network with preferential attachment will become scale-free. Our objective is to understand the code as a switch that can amplify the generation of cellular structures in response to the environment, to study the relationships between component, code, environment, and the formations produced, and to understand networking behavior in a scale-free system.

The first phase of development included the establishment of a series of organizing principles derived from the biological models so that we could apply these logics to the computer model. The topics, which governed our operations, were connectivity and clustering. These terms imply a set of particular observed behaviors that are abstracted and controlled through the design and production of custom algorithms and digital tools. To understand connectivity, we first examined

Figure 3.15 Cell to hub clustering establishing initial hub communities and neighborhoods

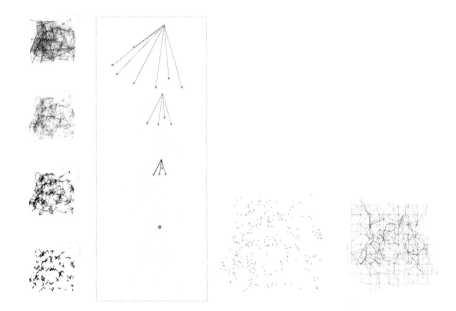

several modes of principles for generating connectivity using the criterion of proximity as an initial guide for establishing network connections. These inputs were based on neighborhood and community conditions as opposed to a hierarchical model. This scale-free approach, like the biological systems of study, gives rise to an organization that can add, shift, or translate its hubs and nodes over time and space without having fixed roles similar to top-down systems.

The next step entailed activating these connections through a web-like matrix, which could then pull and engage cells, thus converting a community of cells connected to a hub into a hub of cells. This hub could then attract other hub communities. This step exemplifies a basic concept of scale-free networking where behaviors and conditions remain the same as one navigates through various scales. Thus, the behavior of a node, hub, community, neighborhood, or region retains its governing principles through any relative scalar transition. This clustering effect defined which cells were hub cells, and which neighborhoods or communities could also become hubs at new scales. The definition of the hubs is significant because organization is established via their density and distribution, such that an eventual assemblage of hub communities results. These hub communities exhibit formations similar to those observed in the angiogenesis phase.

A critical element in this project entailed the design and generation of a translational tool for behavioral modeling, such that we were not merely constructing mimetic or simulation models. This behavioral approach opened up new strategies for abstract component typologies, which also behaved in a scale-free system in response to their input cells (points), locations and behaviors. This pursuit of a component typology culminated in a method that could translate the relationship of hub communities and individual cells, and use them as inputs

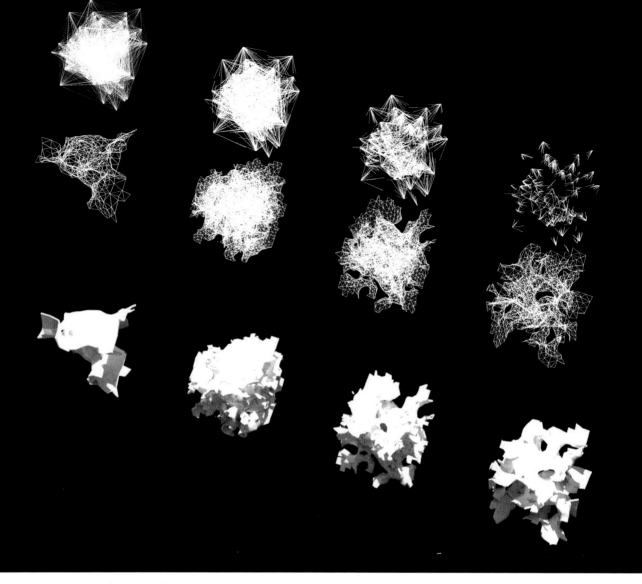

Figure 3.16 A four-phased clustering model and the resultant mesh components produced

weighted by their community density. This in turn, defined a perimeter and interior geometry that provided a more tangible means of studying, analyzing, and iteratively testing the temporal progressions as the networks developed and clustered.

     This process also established another scalar leap as hub communities network like the individual cells had in the previous scale/step. This is significant because as one moves from the micro-scale to the macro-scale in these fractal-like structures, a transitional method is necessary for shifting resolutions, so that there are no disjunctions between all steps at all scales. Thus, the system exhibits cohesive and fluid transitions between scales where consistent behaviors of connectivity and clustering occur at all phases and scales.

Networking 71

Figure 3.17 Secondary scale clustering and transitional mesh component, detail

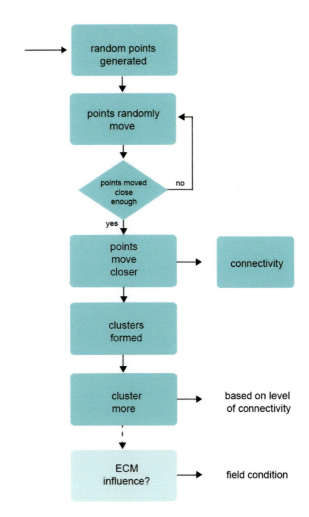

Figure 3.18 Schematic diagram of model development through conditional phases

Once a working model for connectivity and clustering was established within a random field environment, the next step included the integration of context. In the case of the biological model, context is defined as the extracellular matrix (ECM). This matrix establishes a base field for cellular movement and networking, which also has the capacity to influence and reform the ways in which the cells navigate, connect, and cluster, based upon the geometrical and directional properties of the matrix. This step is very relevant as it integrates and instrumentalizes the ability of cells to behave in different environmental conditions, again with and without the Prx-1 gene as a networking enhancer.

This last step includes the exploration of feedback relationships that are behavioral and anamorphic rather than static and fixed. We are seeking to produce a dynamic system that can be analyzed through the production of static temporal or phase-based models, which may further inform and reform the dynamic model. The extracellular matrix could shift over time while the cells are connecting and clustering, causing a perpetual catalytic feedback loop, which could generate infinite possibilities. The inclusion of the ECM does not alter the scale-free network in the direction of a hierarchical model; rather, it reinforces the scale-free network and becomes a component for maturing and evolving the cellular formations and their mesh outputs as part of the established system. The difficult task of establishing qualifications for defining what is a complete model is dependent upon at what scale and system we seek to apply these tools. As previously stated, we do not seek to merely replicate or mimic the biological model; we are in pursuit of various inter-scalar applications ranging from the cellular to the architectural to the urban, and eventually the global.

As we discovered early on with scale-free networks, the same governing principles that define the network inside of a computer or mobile phone also define the entire internet, the telephone network and other systems such as intercontinental air travel. This capacity for operations at any scale also points to the significance of biological models in architecture. Certain aspects have to reform themselves at least in scalar terms, but the behavioral logic of how these systems, buildings, and environments come together may rely on the same fundamental rules, elements, and tools that together work to establish, develop, and execute their production.

Through this behavioral and conditional modeling, we can establish and discover new relationships for designers to observe and build upon, not just in architectural terms but in many different professions and practices which also range from the micro to the macro in terms of phases, processes, and products. These relationships and dependencies will inform both biology and architecture as we seek to cross-fertilize and inform new modes and methods for modeling, representing, and solving conditions and problems that exist in various cases and at a myriad of scales.

In conclusion, our research is based on biological models and how they can be translated into computational models for biological and architectural research into conditional modeling and behavioral analysis. The basis for our studies is cellular networking in lung tissue formation and how that can become a model for designing

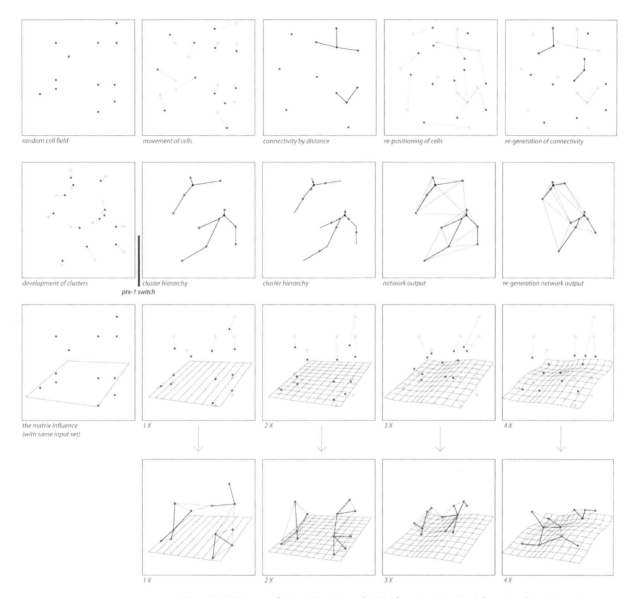

Figure 3.19 Diagram of potential variations of cellular formation given the deformation of an orthogonal or regular ECM and diagram of the scale-free networking process

responses to subtle variations in conditions present within the environment. Custom tools in GenerativeComponents (GC) were generated to process the biological information by operating via a scale-free network to create a more flexible computational networking model that is defined by conditional relationships between cells and their environment. This led to a series of advanced dynamic networking models where components respond locally and globally to simulated environmental cues. These models are inherently open where more advanced

Figure 3.20   3-D printed mesh component family showing surface deformations based on networking density

contextual constraints, such as those that would come through building façade design and optimization, may be integrated and layered readily. Here, advanced adaptive capacity to negotiate sunshine, moisture, and heat is made possible. More importantly, these models and simulations present a new way of thinking about dynamics and form in the context of the shifts occurring in architectural representation, notation, and digital fabrication. As Greg Lynn states:

> This shift from the dynamics of the plan and the statics of the section to a pliable multi-dimensional space of curved surfaces is one of the most significant shifts in the way we think about architectural space since the invention of perspective.[3]

The models presented here are inherently pliable and multi-dimensional where surface is not just a product of formal invention, but of hyper-communication between environment and code.

Networking 75

Figure 3.21 Rules for networking behavior

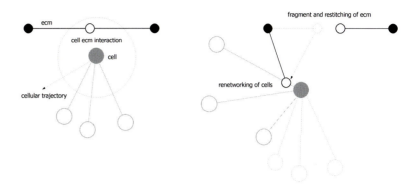

Figure 3.22 Pseudo code for networking conditions

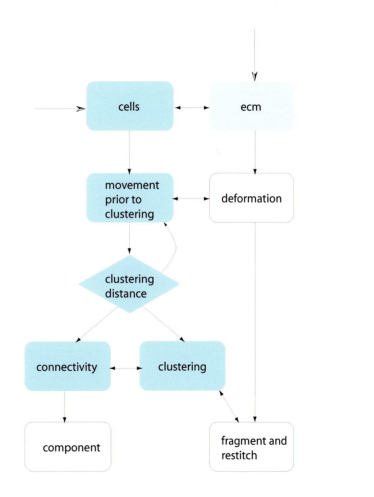

initial connections
component guide map polygons
component by area of guide polygons
most area = highest tension
mid area = mid tension
least area = least tension

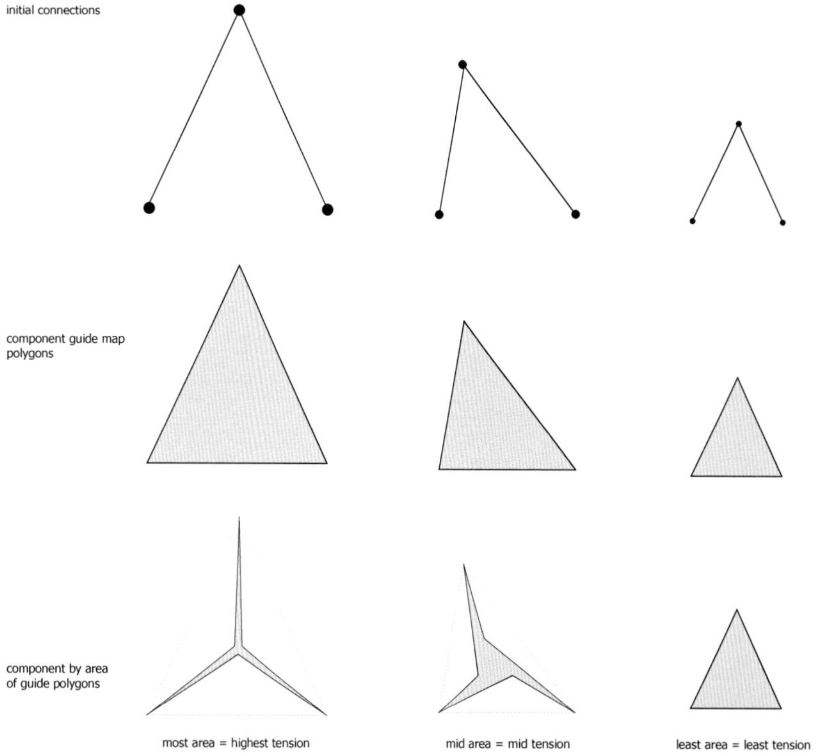

Figure 3.23 Component generation based on tension distribution

.5- 2
1- 2
2- 2
4- 2
.5- 2.1
1- 2.1
2- 2.1
4- 2.1
.5- 2.2
1- 2.2
2- 2.2
4- 2.2
.5- 2.4
1- 2.4
2- 2.4
4- 2.4
NETWORK DENSITY/ CELL TO COMPONENT TRANSLATION

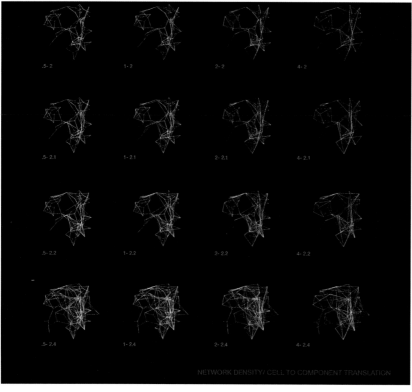

Figure 3.24 Component translation with increasing cell network density

Figure 3.25 Dynamic networking of components with the presence of an active matrix

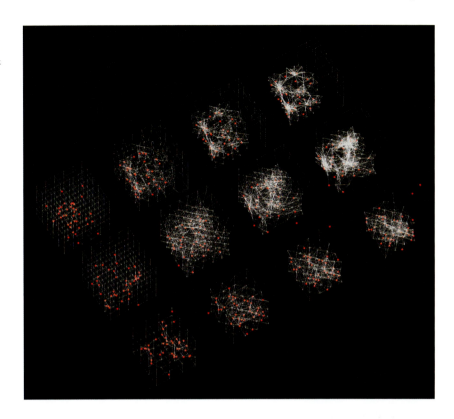

Figure 3.26 Adaptive component system with rendered surface distribution

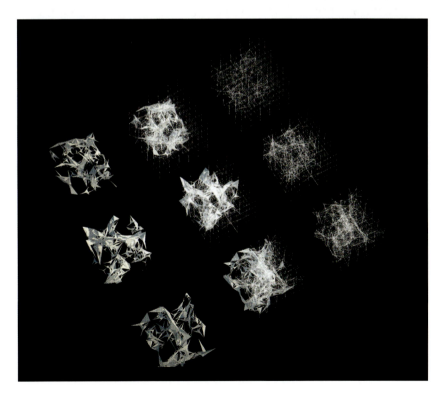

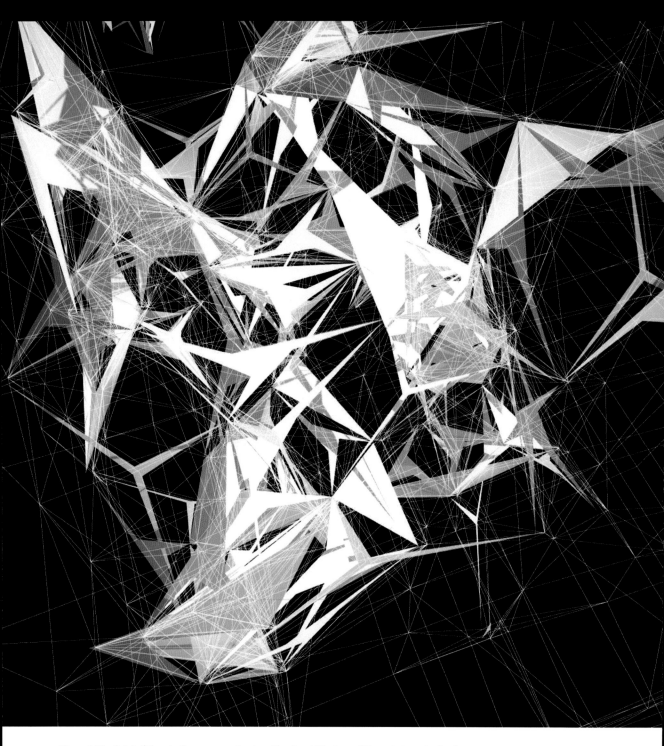

Figure 3.27 Detail of the adaptive component system. The aim of this phase of the project was to develop a family of adaptive components that respond to changes in tension, external environment, and internal code.

Notes
1. Kaori Ihida-Stansbury *et al.*, "Paired-Related Homeobox Gene Prx1 Is Required for Pulmonary Vascular Development," *Circulation Research* 94(11) (2004): 1507–1514.
2. Albert-Laszlo Barabasi and Eric Bonabeau, "Scale-Free Networks," *Scientific American* (May 2003): 60–69.
3. Greg Lynn and Mark Foster Gage, "From Tectonics to Cooking in a Bag," in *Composites, Surfaces, and Software: High Performance Architecture* (New Haven, CT: Yale School of Architecture, 2010), p. 22.

Comments on the Role of the Matrix
Sanford Kwinter
Invited critic and Professor of Theory and Criticism, Pratt Institute, New York, 2008
Jenny E. Sabin

Kwinter: I'm not sure where the dream lies right now but I know one thing, when you ask the question to the architects, the answer is that you needed the biological constraints here to act both as a limit, but also, to ensure that the default scope would actually include the important concept of the matrix. Or the concept of the regulator or the dial, whether Prx-1 is there or not there, or even the critical degree to which it is there. These principles don't naturally occur to architects when creating computer models. They don't seek the ecology of multiple pressures acting upon form. I hold this to be really important because without it, you aren't really doing anything new. But with it, there is the potential for a significant breakthrough. It's important for architects to include the role of the matrix as an active environment; a networked, hooked-up, interactive, correlated, version of the complex. I personally think this is the next step in thinking about structure. Not that it's new historically—for there was incredible intuition in 1950s and 1960s engineering to move out of three- or even four-dimensional descriptions of force and structure and to carry it to the next level. That can only be done by knowing that you are dealing with moving boundaries or parameters. This for me has been the promise and disappointment in the use of contemporary software. The current consensus is that it does reprogram the user when one gets used to working inside a parametric environment, but not if one is constantly bringing it back to pure formal familiarities. That's why I have always seen biology as a kind of "art of structure." That's why the present type of investigation is rich with potential for architects as well. What you brought to them is the concept (which I hope is going to be in their minds, reflexes, and design methods for the rest of their careers) that there is always a matrix, always an active matrix. That's a pretty new spatial idea in architecture.

Sabin: The presence of the matrix, its intricate relationship with form and function, is what I think has the most potential to inform and inspire an alternative mode of making and designing in architecture. It provides a set of criteria to respond to in terms of successive models where we ask: Which models are more behavioral? They are models that exhibit a series of feedback loops between matrix, code, component, material, and so on. These behavioral models have incredible potential for architectural application if they start to move beyond the filter of information or data and toward materialized information through analogic experiments.

Critic: But see, it's a really, really interesting point and I absolutely agree but I can't help but be perplexed by it. When I listen to what you say or when I listen to what Sanford says—and Sanford says we have to move away from spatializing as architects—and I admit that I am very guilty of that. I always try to find the form in the space, which brings me to my question. When we talk about materializing an idea based on an ecology, we talk about structure, organization, form, where everything at one scale or another is connected to each other, but we can differentiate from space. I guess what I'm not quite understanding is at what point does that information in a valid way translate to an architectural condition as knowledge, like for us as architects to talk about its material?

Kwinter: Let's assume there is a blind watchmaker, a non-theological cause responsible for what nature presents to us. With a lot of design processes, the questions are: did that decision or this issue hamper the outcome?—should I make spatial things or should I make temporal things? But nature's reply is that this is not a valid set of questions. Nature says, "Make things that support and manifest behaviors and interactions." Of course, this may one day be seen as a fad, but it's clear today that many of the last vestiges of gravity-based limitations on the conception of architectural form are falling away fast, setting it free—architectures, if you like, for the emerging world. The economic, political and, I would say, epistemological contexts—the whole relationship to knowledge in our world and the relationship that architects have to it—have changed. So, in one sense, when you first asked the question, you pivoted around a stable term—the architecture that we know of from the past—churches, buildings, things. You then talk about the architect in the future, setting free a lot of the parameters. Now you are a parameter person. You are used to the flexible relationship to terms and positions, even to addresses and coordinate points in the world, used to moving in a fluid environment, and understanding how an input here or there might affect an output elsewhere. Why, generally speaking, wouldn't this be applicable to the world you see projected for you in the future: displaying the trust that architectural expertise will connect to that world? I find the conservatism or the doubt that you harbor does not belong in your universe, and I don't expect it to last long . . .

Critic: I totally agree. The question concerns the notion of ecology. But I also find myself becoming perplexed because I have to make things.

Kwinter: But "a thing" is not limited to something you can stub your toe on, a thing rigidly bounded in space and time. You might more profitably see it, ecologically as you say, as embedded, just as we witness it here. You must accept it as part of the matrix concept, as an active entity embedded in conversation with its environment.

Case Study: Dynamic Surface Modeling
Kara Medow, Kirsten Shinnamon, and Young-Suk Choi

Nonlinear Systems Biology and Design
Instructors: Jenny E. Sabin and Peter Lloyd Jones
Sabin+Jones LabStudio

Real-time imaging of endothelial cells cultured within specialized extracellular matrix microenvironments may either suppress or promote cell networking, i.e. angiogenesis. The relationship between cell-to-cell communications during the process of angiogenesis formed the basis of this project. Critical to our study was the impact of a gene designated Prx-1 on the process of angiogenesis. As was discussed previously, Prx-1 enhances endothelial cell networking on a specialized extracellular matrix called a basement membrane. At a functional level, expression of Prx-1 allows for the extension of cell "tips" on the basement membrane outwards toward other cells.[1] *In vivo*, when Prx-1 is absent, blood vessels fail to form in the developing lung, leading to demise at birth when the first breath is taken.

The ultimate aim of this project was to determine how Prx-1 facilitates networking between endothelial cells. An abstraction of those relationships would

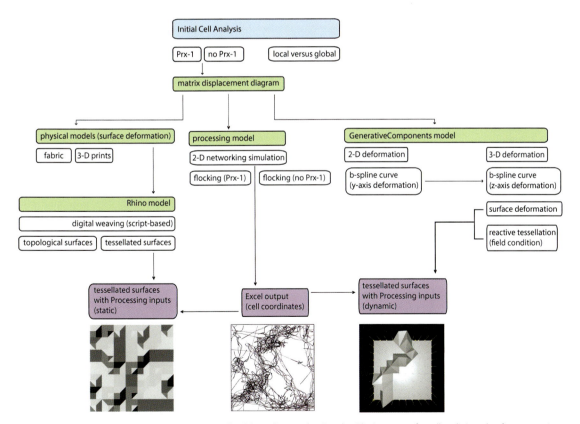

Figure 3.28 Feedback loop diagram showing algorithmic strategy for cell analysis and surface generation

then fuel new investigations related to, but not exclusively about, cell networks. Resulting physical and digital investigations all centered on the notion that as cells network and converge in space, they distort the matrix they are plated on. This unique relationship between object, environment, and code laid the groundwork for the breadth of our studies, as all concurrent investigations took each of these three variables into account. The feedback between these variables played a vital role in our investigations; as variables changed in level or degree, so did the resulting outcome.

Primary analyses of cell cultures included studying two movies of cells exhibiting or not exhibiting networking behavior. From these movies, snapshots of cell behavior were extracted in order to extrapolate the rules by which cells behave under these distinct conditions. Through three concurrent digital investigations, we were able to study the local condition of cell-to-cell networking in distinct ways. Output from each of these studies ultimately created input for the other studies, so that the project itself exhibited a strong feedback loop condition. In all of the investigations, the basement membrane is abstracted and reduced to a surface condition that reacts to corresponding points or cells. The hyper-local condition of cell-to-cell communication and the environment seemed inextricably linked, and through these studies we hoped to prove that connection.

Primary comparisons of these two different cultures of endothelial cells on basement membrane revealed that in the presence of Prx-1, cells at the periphery of clusters possessed a more elongated phenotype. In the cultures of cells plated in the absence of Prx-1, this elongated phenotype never occurred and the cells never seemed to reach any level of overall organization. In contrast, the cultures in the presence of Prx-1 seemed to form increasingly denser and more organized aggregations over time where these elongated cells appeared to be instrumental in facilitating the greater network of cells. Their length was perhaps a result of continuous interaction with the extracellular matrix as these longer cells were closer to the basement membrane due to their position at the edge of the cell clusters. At the local condition, these longer cells seemed to operate in particular directions in relationship to overall cell clusters: they trended toward being parallel or perpendicular to the existing aggregations of cells. As these cells networked and moved toward each other, they remained in contact with the matrix, thus allowing for a distortion of the basement membrane.

The idea that points attached to a surface would distort that medium led to initial physical models where a flexible surface of canvas was distorted via the precise movement of a string or wire threaded through it. These physical models prompted the creation of a GenerativeComponents model that utilized sets of B-spline curves organized in a field. These curves, reacting to points, or cells, would move between the top and bottom curve. As a surface was ultimately attached to sets of curves in the model, one or more points moving through the field would elicit an accurate distortion of those surfaces.

Figure 3.29 3-D prints of surface deformation studies based on changes in networking density

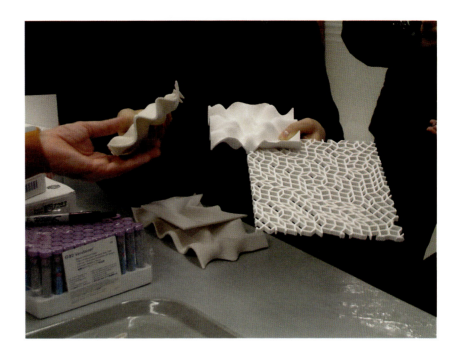

    This notion of spline curves attached to a deformable surface led to the development of precise rule sets intended to create static deformed surfaces utilizing RhinoScript. *StringWeave2*, written by Jenny E. Sabin and Raymond Kettner, formerly of CabinStudio+, generates a series of digitally woven curves using binary inputs. In this script, sets of binary information acts as a digital on/off switch for the code, as 1 or 0 signifies an up or down movement of the curves. Resulting output from the script creates a set of x-direction and y-direction curves that are woven together. Initial tests utilized sets of x- and y-direction curves from randomly generated sets of binary data to create topological surface models from Patch, Surface from Curve Network, Sweep, and Surface from Edge Curves in Rhino. Ultimately, information was separated so that either the x- or the y-direction curves would serve as an input to generate the surfaces. As well, the topological surface models were reduced to tessellated triangulated models (by ignoring the interpolated curve information and utilizing only the 1 or 0 up/down data) so that information could be more easily "read" from these increasingly complex surfaces.

    To return to the notion of networking, it was necessary to infuse the information in our digital studies with some content related to cell-to-cell communication. Concurrent to the static Rhino surface model investigation, processing models were developed in order to create data sets that simulated the difference between Prx-1 and non-Prx-1 cell samples. Cohesion and alignment of points in a network determined the variables and parameters to create digitally scripted models of cells moving and flocking to ultimately cluster, which would imitate the condition of networking. Within the code, cohesion allowed cells to

Figure 3.30 3-D printed catalog of surface deformation studies

steer toward other local cells within a defined distance, while alignment allowed cells to navigate within a field of continuously moving and changing locations. The alignment parameter allowed the overall cell clusters to aggregate on a more global scale, due to the fact that the cohesion factor only happens on a hyper-local scale. The visualization of the raw data from the processing movies indicated (in composite or frame-by-frame over time) the distinct behavioral difference between the varied cell cultures. The precise locations of point-to-point movement in both the Prx-1 and non-Prx-1 simulations in processing were extracted as x, y, z coordinate sets so that the data could be imported back into the GenerativeComponents and Rhino investigations.

In GenerativeComponents, we input the point coordinates from the Processing simulations. The movement and convergence of the points as they change their x- and y-location allowed the surface to react in the z-direction. The final GenerativeComponents models abandoned the B-spline condition for a triangulated tessellated representation so that ultimate visual comparisons could be drawn between the GenerativeComponents and the Rhino surfaces, which used identical processing inputs. The reading of this surface condition, while

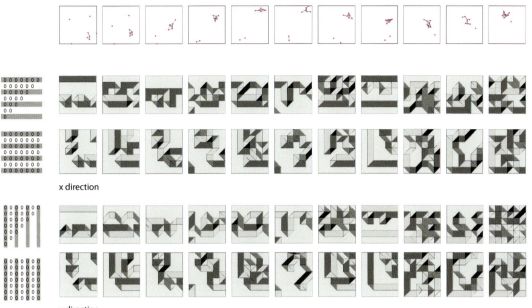

Figure 3.31 The notion of spline curves attached to a deformable surface led to the development of precise rule sets intended to create static deformed surfaces utilizing RhinoScript. Resulting output from the script creates a set of x-direction and y-direction curves that are woven together.

literal, is an accurate representation of the convergence or non-convergence of cells over time.

The final Rhino models used the same point coordinates, separated out into frames so that the networking state of each set of cells could be analyzed as time progressed. As networks develop, connectivity between nodes increases. The point-to-point connections between discrete nodes in a network are, in effect, a binary condition: at any given point in time, each node is connected or is not connected to another node. Given varied length parameters, the cell points from the processing studies were analyzed in order to determine network connectivity and thus were able to generate sets of binary data that were directly related to the state of the network and its connections for that time frame.

This binary code created from point-to-point convergence information was fed back into the *StringWeave2* script. The resulting x- and y-direction curves provided information to generate tessellated surfaces that then conveyed the increasing or decreasing degree of connectivity in both the Prx-1 and non-Prx-1 simulations.

Ultimately, the resulting models in Processing, GenerativeComponents and Rhino have implications beyond the tooling and methodology used to construct these investigations. The tessellated surface models as well as the

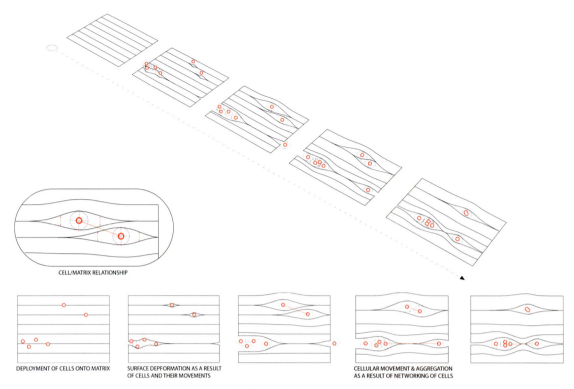

Figure 3.32 Cellular movement and aggregation as a result of networking of cells

Processing data sets and their visual representations have the potential for diagnostic and simulation tools for biologists in order to see simple part-to-whole relationships within a highly dense and complicated developing cellular network.

These investigations also capitalize on the visualization of information in a network through resulting surface conditions. The architectural potential for the use of these systems is vast: to be able to generate increasingly complex surfaces from the informational content of networks is a step beyond the notion of simply materializing a static diagram. The potential of these studies to move beyond the surficial and into the spatial dimension holds great possibilities for the process of creating new and different architectural typologies utilizing the ideas and data embedded in network conditions.

Note
1   Kaori Ihida-Stansbury *et al.*, "Paired-Related Homeobox Gene Prx1 Is Required for Pulmonary Vascular Development," *Circulation Research* 94(11) (2004): 1507–1514.

Scientific Representation
Jamer Hunt
Invited critic and Associate Professor of Transdisciplinary Design, School of Design Strategies, Parsons School of Design, New York, 2007

My sense of the entire history of scientific representation is that it's full of design in the sense that it was rarely and possibly never an issue of seeing something in a less than empirical level, but about translation to a scale that we could understand. It's always a series of visual translations in that the history of scientific representation is one of creative leaps and bounds where the tools don't necessarily determine how empirical the outcome is if you represent it differently. So it seems like you almost have to assume that whatever tool you are bringing in to model or see, is inherently a kind of design tool. It's adding something that is not necessarily there but there's no way of grasping what's there because it doesn't have a pictorial quality. We always have to give it a color, a boundary, a feature, in order so that we may see it conceptually. And so, I'm not sure if that's what we're getting at in terms of the importation of these kinds of architectural modeling systems, that they kind of bring with them their own baggage.

Case Study: Percolative Transition and Network Formation
Chi Dang, Alexander Lee, and Annabelle Su

Nonlinear Systems Biology and Design
Instructors: Jenny E. Sabin and Peter Lloyd Jones
Sabin+Jones LabStudio

To distribute nutrients throughout the body, vertebrates have evolved a hierarchical branching blood vascular system that terminates in a network of invariant units, or capillaries. The development of capillary networks characterized by typical intercapillary distances ranging from 50–300 μm is instrumental for optimal metabolic exchange. The ability to form networking capillary tubes is an autonomous property of the endothelial cells (ECs), which need permissive but not instructive signals from the extracellular environment. In the article "Modeling the early stages of vascular network assembly," by Guido Serini, Davide Ambrosi, Enrico Giraudo, Andrea Gamba, Luigi Preziosi, and Federico Bussolino, the authors present a model where chemo-attraction serves as the fundamental mechanism for cell-to-cell communication in order to identify key parameters in the complexity of the formation of vascular patterns.[1] With biological experiments, theoretical insights and numerical simulations, the model shows the correlation between endothelial cell number and the range of activity of a chemo-attractant and how this factor regulates vascular network formation and size. A mechanism linking the scale of formed endothelial structures to the

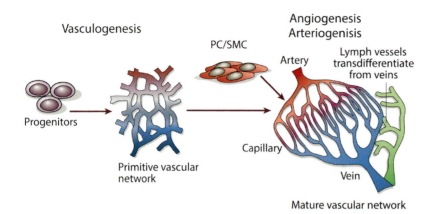

Figure 3.33 Vasculogenesis: Endothelial progenitors give rise to a primitive vascular labyrinth of arteries and veins. Angiogenesis: Primitive embryonic vasculature is remodeled into a mature vascular bed comprising arteries, capillary networks, and veins.

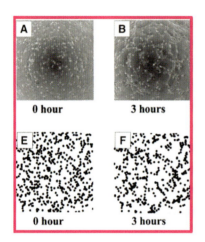
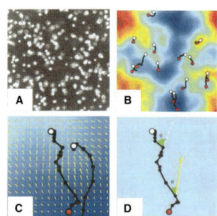

Figure 3.34 Movement in different directions, interaction, and adhesion with neighboring ECs, spreading and eventually forming a continuous multicellular network (left). Highly directional movement toward zones of higher cell concentration (right).

range of cell-to-cell interaction mediated by the release of chemo-attractants is proposed.

Upon reviewing the article, we were drawn to the vital role of cell density in the formation of capillary networks. The research from the article shows in the first 3–6 hours, human ECs start moving in different directions, interact and adhere with their neighbors, spread, and eventually form a continuous multicellular network. Tracking of individual trajectories shows a highly directional movement toward zones of higher cell concentration. A certain cell density has to be reached for a cellular network to form; or the vascular network fails to complete, resulting in only many discrete partially formed networks instead of one complete one.

Such an abrupt change in the connectivity properties of a randomly formed structure through the variation of a control parameter is known in physics as percolative transition. Once the percolative transition is reached, the cells complete the network very suddenly. We hope our inquiry will uncover methods for understanding the structural resilience of completed vascular networks, and perhaps a relationship to the percolative transition, whether networks that partially

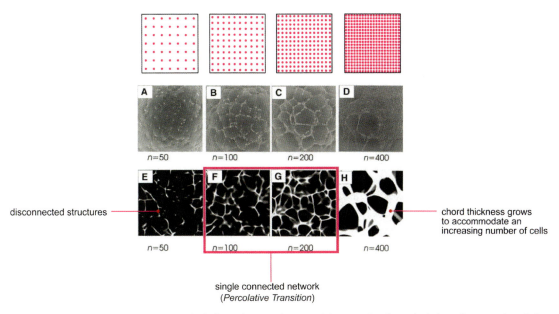

Figure 3.35 Such abrupt change in the connectivity properties of a randomly formed structure through the variation of a control parameter is known in physics as percolative transition. As can be seen between image F to image G, once the percolative transition is reached, the cells complete the network very suddenly.

form have any structural integrity, or whether structural performance of a network is proportionate to the density of cells that produced it. We did not seek to model a copy of the vascular network itself, rather, we sought to model a meta-network of the potentials that exist across a sample population of points in space based on density and distance relationships originally uncovered in a biological context.

Percolative transition occurs at many scales. At a micro-scale, percolative transition may be described as the moment when any given cell is within close enough proximity to another cell to make a connection. At a somewhat larger scale, percolative transition is the moment that fragmented clusters of cells suddenly form a network. In this respect, a cell is a "part" of the network as a "whole." However, the cell by itself can also be viewed as a whole and therefore actually has a whole-to-whole relationship with the network. This has important implications for understanding the significant role of the cell within a network. A cell without context has inherent structural characteristics in order for that cell to exist. A cell within the context of other cells has the potential to network and create larger structures, whose resiliency is affected by the behavior of the individual cells.

We can begin to study percolative transition across different scales by starting with a point. When there are many points, there is the possibility for making connections, which can be abstracted as lines. And when there are many lines, we can take their vertices and intersections to create surfaces. And similarly, with many surfaces which share edges, volumes can be formed. So, from points, we can create

NETWORKING: Percolative Transition

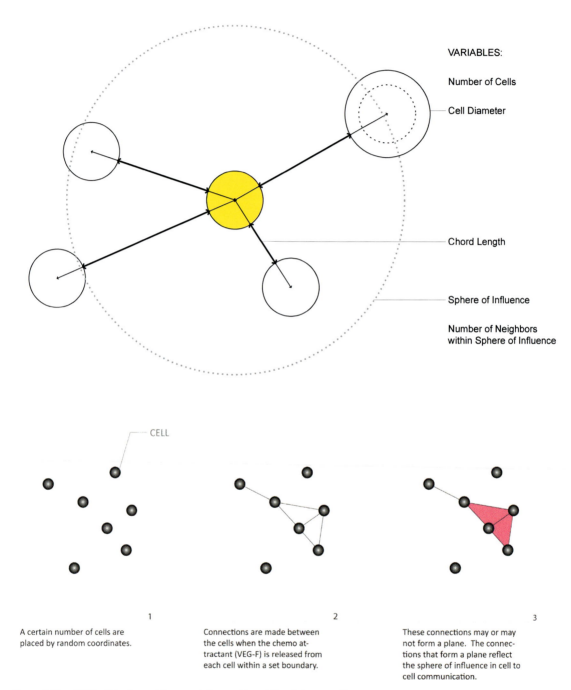

Figures 3.36 and 3.37 Rules for modeling percolative transition within cellular networking

lines; from lines, we can create surfaces; and from surfaces, we can create volumes. The evolution of a point into a volume has three phase changes or moments of percolative transition. Like the cell to the network, the point plays an important role in forming volume.

For our own analysis, we focused on key variables from the biological model, which demonstrates different formal outcomes over the same period of time given different initial populations of cells. Three distinct forms are: fragmented quasi networks, complete network, and overly redundant porous mass. In our digital model, working with a fixed volume, we made two adjustable variables: the initial population of randomly generated points, and the threshold distance that determines the potential connection between those points. By adjusting those two values, the outcomes in our simulation produce forms ranging from fragmented to networked to dense and tangled. By taking snapshots of these moments, the collected images formed a context for each of the individual images. Each is arranged as a matrix organized along the lines of the two variables (initial population and distance threshold).

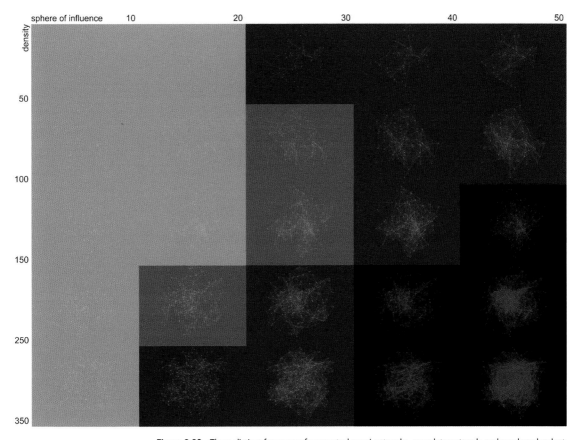

Figure 3.38 Three distinct forms are: fragmented quasi networks, complete network, and overly redundant porous mass.

In our work, the percolative transition occurs at a moment when the same system produces one formal expression within one range of parameters, and a different formal expression at the next range of parameters. Drawing an analogy to water helps—having one state until either heated to its boiling point or cooled until it freezes, respectively transforming into vapor or solid. A similar phenomenon arguably occurs in the systemic process that drives vascular network formation, however, rather than adjust the temperature, the varying forms result from changes to the population of cells that typically behave to form arteries.

Pairing that concept with the intent to use a component-based method to model the system, we sought to allow a component to emerge from the process of evolving point information into lines, and from those lines, determining the conditions to create surfaces, which would create volumes. The conditions generated both distinct volumes and points and lines within the networking system. Perhaps the biggest challenge was to repeat the model under a thorough variety of

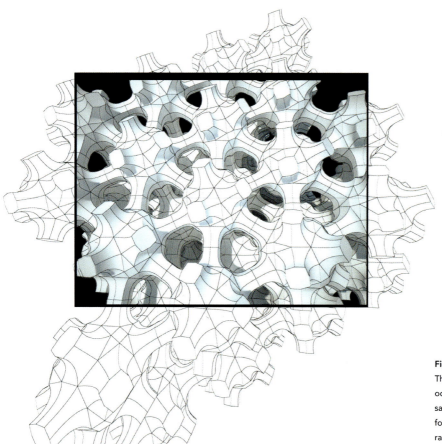

Figure 3.39 Dense network. The percolative transition occurs at a moment when the same system produces one formal expression within one range of parameters, and a different formal expression at the next range of parameters.

conditions in order to recognize the patterns that emerged across the spectrum of outputs, in order to identify the moments between our snapshots that keyed the transition from one formal expression to the next.

In our pursuit of a physical model, our team decided to work within the parameter of a kit of parts to reproduce the logic of the networking system. In a digital 3-D model, we were able to produce a sample of densities of nodes, but our components were restricted to 12 connections per point at fixed angles. Using only one type of node and only two lengths of connectors, we arranged the components into a limited number of arrangements. Node-to-node structure produced the densest configuration, but connecting fewer nodes in the same volume required the use of long connectors and produced a lighter contiguous network structure. Reducing the number of nodes further, however, resulted in many small fragments that failed to join, and in a physical space that would fail to stand or hold its shape. The ideal model may be one that changes very gradually

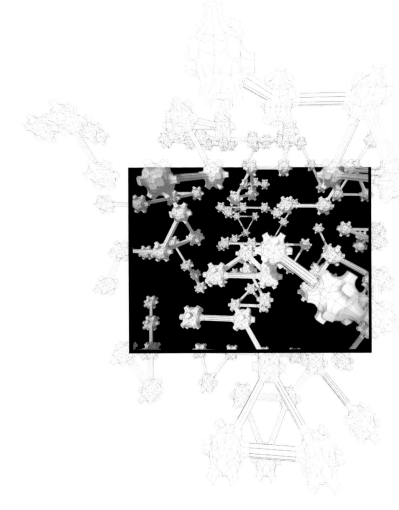

Figure 3.40 Fragmented network

throughout, punctuated only once or twice by radical shifts in form, in how the networks are organized.

Note
1   G. Serini, D. Ambrosi, E. Giraudo, A. Gamba, L. Preziosi, and F. Bussolino, "Modeling the Early Stages of Vascular Network Assembly," *The EMBO Journal* 22(8) (2003): 1771–1779.

Case Study: Flexible Scaffolds and Emergent Space
Dale Suttle, Qiao Song, and Ziyue Wei

Nonlinear Systems Biology and Design
Instructors: Jenny E. Sabin and Peter Lloyd Jones
Sabin+Jones LabStudio

As we have seen in the previous projects, the movement of endothelial cells (ECs) within the pulmonary arterial system is largely dependent on and also influenced by the extracellular matrix (ECM). The influence is largely possible through the connection of the cytoskeleton within each cell to the matrix. In the case of the pulmonary arterial cells, on which we will be focusing, the ECM is more specifically known as a basement membrane.

Based on the definitive relationship of ECM and EC, we explore the movement and networking capabilities of the ECs through the definition of

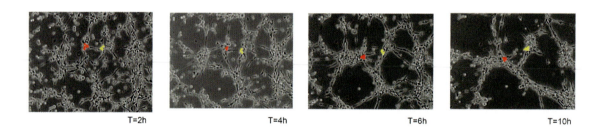

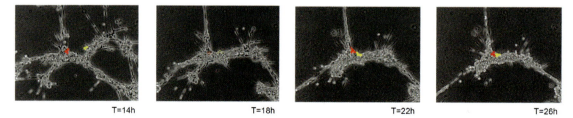

Figure 3.41   Mapping of networking cells over a 24-hour period

the ECM around the cells. Rather than defining the relationship of the cells to one another directly, we strive to capture a dynamic map or understanding of the networking of cells over time through the modeling of the construction and destruction of the ECM. In the way that a general trend can be derived from the mapping of a series of points in a graph, we believe that by looking at the void space rather than the cell networks themselves, we gain a greater understanding of the networks and their useful properties from the plotted record of their paths in the ECM.

Our understanding of the dynamic network of cells and the ECM can be used to digitally model the resultant matrix that surrounds the cells. This model is then used to derive an analog model in which the objects in the interior of the network are expressed through the exterior support system. By presenting the model in this manner we provide a better understanding of the entire system that involves not only the endothelial cells, but also equally important the extracellular matrix, which ultimately provides greater potential for architectural discoveries.

The extracellular matrix (ECM) is the extracellular part of animal tissue that usually provides structural support to the animal cells in addition to performing various other important functions. The extracellular matrix is the defining feature

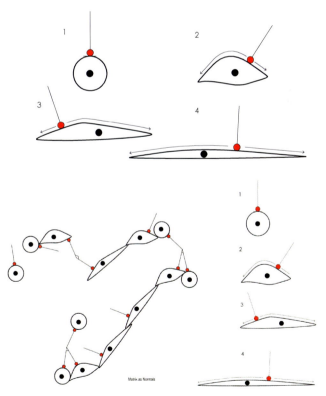

Figures 3.42 and 3.43 Simple translational rule set describing matrix as normal to the ECs

of connective tissue in animals. Due to its diverse nature and composition, the ECM can serve many functions, such as providing support and anchorage for cells, segregating tissues from one another, and regulating intercellular communication. The ECM also regulates a cell's dynamic behavior. Components of the ECM are produced internally by resident cells, and secreted into the ECM via exocytosis. Once secreted, they then aggregate with the existing matrix. The ECM is composed of an interlocking mesh of fibrous proteins and glycosaminoglycans. Many cells bind to components of the extracellular matrix. Cell adhesion can occur in two ways: by focal adhesions, connecting the ECM to actin filaments of the cell, and hemidesmosomes, connecting the ECM to intermediate filaments such as keratin. This cell-to-ECM adhesion is regulated by specific cell surface cellular adhesion molecules (CAM) known as integrins. Integrins are cell surface proteins that bind cells to ECM structures, such as fibronectin and laminin, and also to integrin proteins on the surface of other cells. Fibronectins bind to ECM macromolecules and facilitate their binding to transmembrane integrins. The attachment of fibronectin to the extracellular domain initiates intracellular signaling pathways as well as an association with the cellular cytoskeleton via a set of adaptor molecules such as actin.

In order to better understand the biological systems involved in the movement and networking of endothelial cells and in order to generate an algorithmic digital model of the system, we began our research by creating a simple translational rule set. The biological networking system is simplified to a series of steps beginning with the generation of extracellular matrix by the cell followed

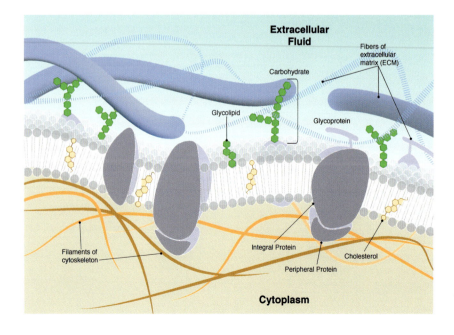

Figure 3.44 Components of the extracellular matrix at the interface of the cell membrane

Networking 97

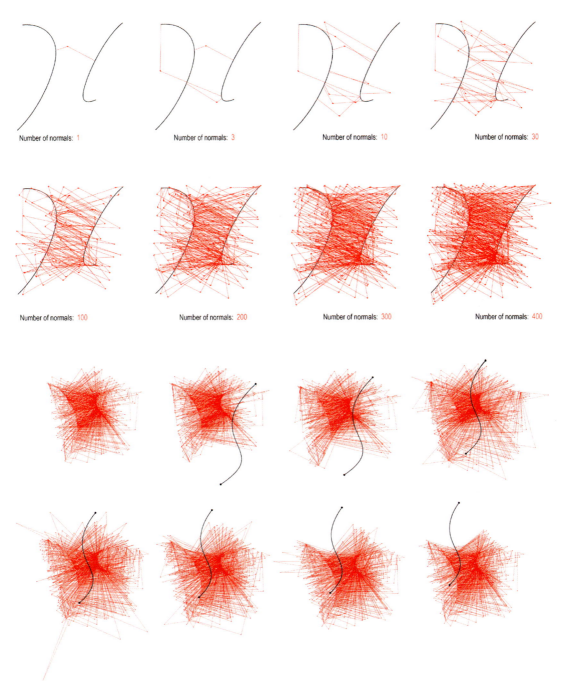

Figures 3.45 and 3.46 Simulation of branching system. The first frame of the series begins with the random placement of x number of points representing each cell's initial point of interaction with the other cells and the whole system. Each of these cells or points generates a matrix of six normals radiating in both directions on each axis to a specific length.

by the conglomeration of these pieces into a matrix. Next, the matrix is solidified and capable of having a reciprocal influence upon the cell. Finally, the cell moves forward as directed or influenced by the matrix and in turn is able to influence the overall organization of the matrix surrounding it. This simplified rule set is translated into a scripted algorithm that accurately simulates the movement and interaction of the cells within the pulmonary arterial system.

The algorithm derived from the rule set is formatted such that the processes of cellular networking are simulated through a series of frames that are compiled into a final resultant sequence that display coordinated cellular movement. The first frame of the series begins with the random placement of x number of points representing each cell's initial point of interaction with its neighbors and the whole system. Each of these cells or points then secretes or generates a matrix of six normals radiating in both directions on each axis to a specific length. Each of these lines checks the proximity of its endpoint to the other lines of matrix within the system. If the endpoints are within a specific distance of each other, they meet at the midpoint between the two endpoints. This movement of the matrix then elicits the rotation of the two cells to realign their axis with the piece of matrix. This rotation of the center point of the cell affects the secretion of matrix in the next frame of movement of the cell because the normal lines are secreted along the new rotated axis. After rotation, the cell moves to the next frame.

The ECM may affect the movement of the cells to the next frame. Three rules are generated for this movement based upon the development of the motility (movement) algorithm:

(1) *Influence*. The cells are pushed away by the ECM previously secreted by the other cells. In this way, all the other cells within a certain area can affect one cell. As the cell moves, the force could be different at different moments, so the path could be an unexpected curve. Then we apply this rule to all the cells so that they can interact with each other.

(2) *Elastic connection*. Although there is no boundary for the whole system, the cells cannot disassemble infinitely. There are two reasons for this: one is that the shape of the cells may not be deformed to an extreme; the other is that ECM, which cannot be extruded to extraneous lengths, connects them. As a result, a spring is added to the cells so that the cells resist moving too far away from each other.

(3) *Obstruction*. The movement of the cells can be affected by the existing ECM, which means the network of ECM will impact to the movement. So the cells will move slowly or bypass the ECM based upon near proximity.

We use vectors to translate the rules to a coordinated algorithm. Through the combination of all vectors, a motion path of the cells is generated, which serves as the foundation for the movement of the cell within the algorithm of matrix

Networking 99

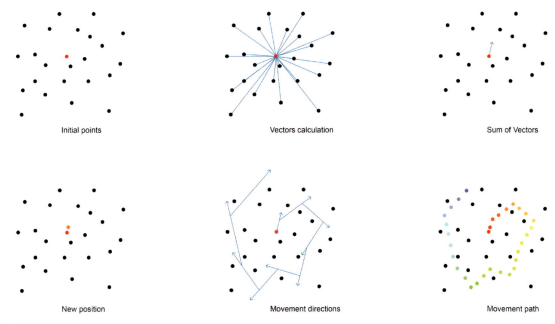

Figure 3.47 Cell influence

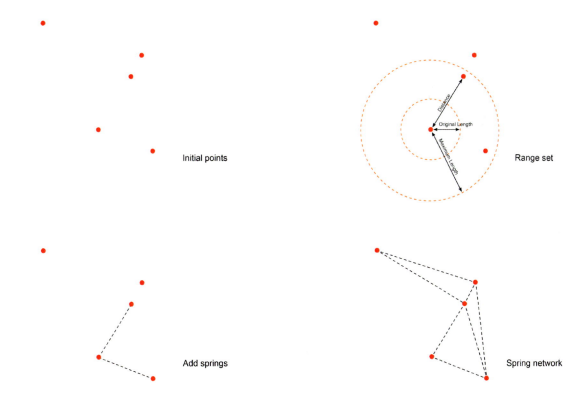

Figure 3.48 Elastic connection

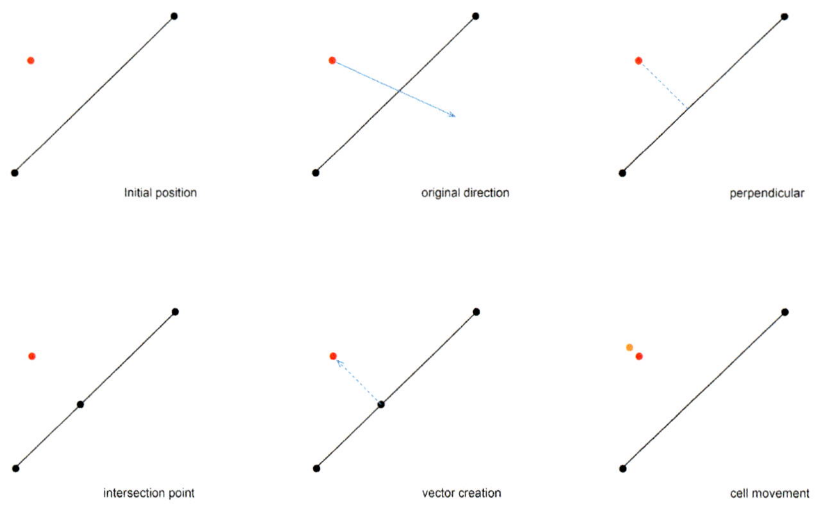

Figure 3.49 Obstruction generated by existing ECM

Initial position | original direction | perpendicular

intersection point | vector creation | cell movement

generation. The vector paths of the surrounding cells are also used to determine the form of the cell within any given frame of the system. The form of the cell increases or decreases in size based on the proximity of the cell to its neighbors. If the cell is far from its neighbors, it will expand to a much larger size in search of other cells. If it is in close proximity to other cells, it will shrink in size and continue to move towards the other cells. The resultant form provides a map of the movement and interaction of the various cells within the network, but also provides an interesting potential for architectural application.

The process provides a generative structural system where large open spaces (the outline of the cell) emerge surrounded by the structure necessary to support and define the void space. The notion of open space surrounded by support structure is a common one to architecture, however, this system provides a unique definition of the idea through the dynamic reciprocity that exists between the surrounding flexible scaffold, the resultant void and overall form. This system is ambiguous and redundant and therefore has the potential to create flexible spaces that have the ability to self-organize based upon required program and attached support structure.

The representation of cellular networking where void is privileged over solid to model cellular movement and interaction, provides not only an understanding of the relationship of cells over time, but a greater understanding of the structural and largely influential matrix in which the cells move. The back and forth relationship or biological reciprocity between the cells and the matrix—the cells generating the matrix, the matrix altering the cell movement, and so on—provides a great potential for architectural application in the creation of large open spaces and the necessary support system surrounding them. It also provides a deep structural transparency through endless expansion and contraction.

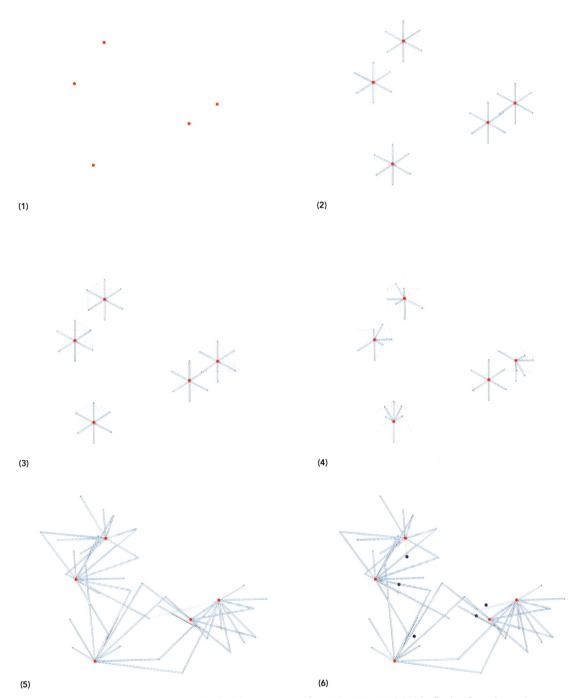

Figure 3.50 Final diagram sequence for matrix generation. 1: Initial cell points; 2: matrix secretion or generation; 3: proximity check; 4: matrix adjustment; 5: matrix variation; 6: cell movement.

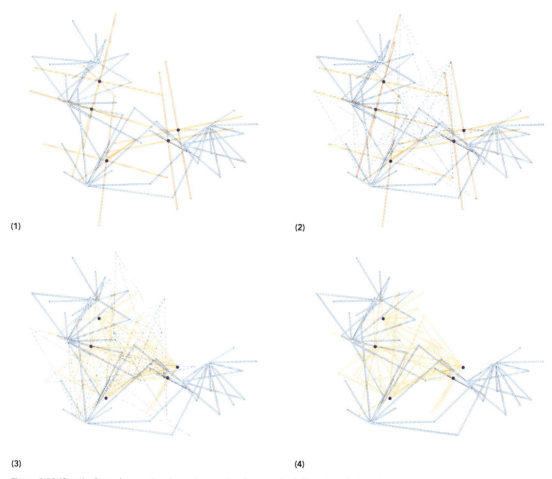

(1)　　　　　　　　　　　　　　　　(2)

(3)　　　　　　　　　　　　　　　　(4)

Figure 3.50 (Cont.)　Second generation; 1: matrix secretion; 2: range check; 3: matrix adjustment; 4: resultant matrix.

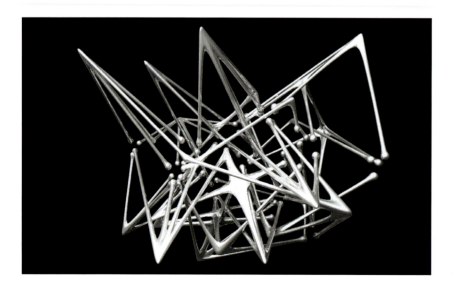

Figure 3.51　Final generative system

Case Study: Adaptive and Resilient Networks: Dynamic Material Systems

Winglam Kwan, Mark Nicol, and Yohei Yamada

Nonlinear Systems Biology and Design
Instructors: Jenny E. Sabin, Shawn Sweeney, and Peter Lloyd Jones
Sabin+Jones LabStudio

In a native environment cells gain stability from tensile forces, which are exerted on them by the extracellular matrix and are expressed through their cytoskeleton. When stressed, these tensile connections transmit contextual information throughout the cell and in so doing drive cell behavior. If there is a shift in the extracellular matrix, those forces are distributed through the integrin receptors to the cytoskeleton. The resulting realignment of the actin network within the cell's cytoplasm triggers a response from the cell. Therefore, information about connectivity can drive the condition of the cell, prompting it to respond by either growing, differentiating, moving, or dying. Just as chemical gradients in the environment stimulate behavior, so too can gradients in the organization of the extracellular matrix. The interrelationship between gradients of chemo-attractants and the composition of the extracellular matrix can inform the design of nonlinear systems, which have a similar dynamic relationship to their environment. Gradients of contextual triggers will spark local reactions, which will reverberate through the tensile network. Our research and simulations are the beginning of

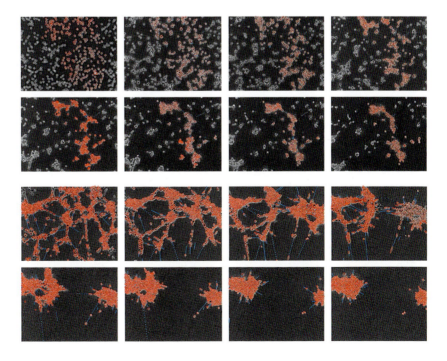

Figure 3.52 Analysis of endothelial cells networking with variegated extracellular matrix conditions

Rules of Behavior

Figure 3.53 Rules of networking behavior

1. cells give birth to chemoattractants
2. chemoattractants disperse from cell
3. chemoattractants degrade over time

1. cells are attracted to regions with high density of chemoattractant
2. field of gradent develops
3. field fluctuates in time

Rules of Behavior

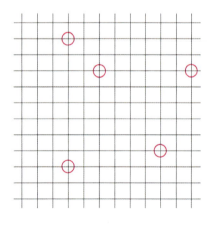
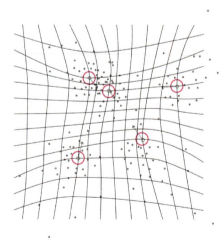

Figure 3.54 Rules of behavior within the matrix environment

1. cells are interconnected by extracellular matrix
2. tensile forces can be communicated throughout network

1. cells are attracted to regions with high density of chemoattractant
2. gradient develops in both the chemoattractants and the matrix
3. matrix restrict movement but also translates forces

Networking 105

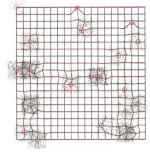 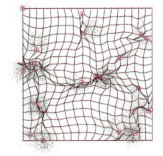 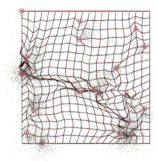

Figure 3.55–3.57 Matrix with chemo-attractant emitters under deformation

Figures 3.58 and 3.59 4-D metamaterialistic system; final responsive material prototype based on analog rules of behavior in resilient networks; composed of shape memory polymer and 3-D printed nodes

a framework for considering dynamically reactive building systems and material assemblies. Nicol with Sabin later extended this work through an independent study in the context of the eSkin adaptive building skins project, which will be discussed in Part IV.

New Architectural Concerns
Roland Snooks
Founding partner of Kokkugia and director of Studio Roland Snooks; Senior Lecturer, Architecture and Design, RMIT University, 2010

One thing that I really like about this method is that it is not biomimetic. You aren't imitating the forms of nature but instead are engaging architectural concerns and processes within the biological systems. But I'm interested in what other types of forms, tectonics, and aesthetics that emerge at the intersection. I'm not suggesting that you make architecture that looks like the cell videos that you have analyzed, but I'm interested in the dynamic self-organization that is occurring in the data; not only in how you apply that within an architectural context, but how that may generate entirely new architectural concerns, where new tectonics and aesthetics are developed. I think that is probably what this seminar is pushing towards and that's something that I'm really interested to see.

4

Motility: Adaptive Architecture and Personalized Medicine

Jenny E. Sabin and Peter Lloyd Jones

Mapping techniques, used with care, can offer fresh insights into the data about the world around us . . . The flood of data now coming online and the emergence of new forms . . . mean that visualization will be increasingly important for scientists. Such diverse windows on data should also strengthen civil society by giving scientists and citizens alike the power . . . to challenge their own preconceptions of the world.[1]

Introduction to the Biological System

We will now turn to our second research track on the topic of cellular motility or how cells move *in context*. The aim of the motility projects was to discover new ways of describing, measuring, and visualizing the behavior of human cells being studied either at the individual level, or as communities within a tissue culture dish. Ultimately, the goal was to generate behavioral "signatures" that are specific to each cell type, and/or the condition under which these cells exist and function. In their own way, these projects aim to use these signatures to begin to personalize individual behaviors, a task that has yet to be established within the medical field at the cellular level, at least using design research. By offering new ways of describing cell behavior, the highly unconventional methods and metrics outlined herein provide a platform for generating new ideas and hypotheses. Further, these studies begin to provide an alternative view on what constitutes personalized medicine and architecture, burgeoning fields for which new methods are constantly being sought and refined.

Personalizing Cellular Behavioral Changes Associated with Pulmonary Hypertension

Personalized or precision medicine involves the systematic use of information about each individual patient to select or optimize the patient's preventative and therapeutic care. A great deal of activity in this field involves consideration of genetic factors and other molecular parameters. However, these approaches may be considered as falling short of capturing the range of influences on a person's health, health risks, and ideal choice of treatment, including within the pre-clinical screening of therapeutic drugs and compounds that are routinely tested in cultivated cells. The main obstacle to this latter phase, however, is finding the means to measure global differences in cell behavior between control and test samples.

As proof-of-principle, and to support this alternative line of inquiry, we created tissue culture conditions that either represented the normal environment of a lung blood vessel wall, or one that mimicked changes associated with development of pulmonary hypertension (PH), a disease associated with increased motility of vascular smooth muscle cells (SMCs) in response to cell-mediated structural damage to the vessel wall.[2-5] Before describing the preliminary impetus underlying the development of this system, it is crucial to understand the macro details of pulmonary hypertension (PH).

The Pathophysiology of Pulmonary Hypertension (PH)

PH is a devastating, female-biased disease, defined as increased blood pressure within the lung vasculature, causing progressive shortness of breath, exercise limitation and, in the absence of effective therapy, progressive right heart failure and early death—the median survival time without treatment is 2.8 years from diagnosis.[6] In addition, PH may be caused by other defects or diseases, as in PH associated with connective tissue or liver disease, HIV infection, anorectic agent use, genetic mutations, or without any identifiable risk factor. The presence of PH also portends a bad prognosis for patients with prevalent ailments, including chronic obstructive pulmonary disease (COPD) and scleroderma. Simply put, the death of millions of patients worldwide is hastened by PH, including >15m patients with schistosomiasis, making the potential impact of this project highly significant to human health at the national and global scales.[7] The clinical complexity associated with PH may help to explain the current lack of diagnostic and therapeutic options, and why this disease remains under-diagnosed, difficult to treat, and incurable. Compounding this, current diagnostic methods, such as highly invasive catheter-based PA pressure measurements or lung biopsies, are at once both insufficient and/or are frequently contraindicated for PH patients. Therefore, there exists a dire need to develop rapid, yet personalized approaches for PH diagnosis and treatment.

Pulmonary arteries, which paradoxically transport deoxygenated blood from the heart to the lung, are composed of multiple concentric layers of different cell types. In the simplest sense, these vessels are lined in their interior and sealed by endothelial cells (ECs), which act as a sensor system that responds to the constant changes in the chemical and physical makeup of the blood vessel lumen. ECs also act as a selective barrier to invasion, and serve as a surface that prevents blood clots. Moving outwards, underlying these cells is a gelatinous membrane made of extracellular matrix (ECM) proteins, which melds into one or more concentric

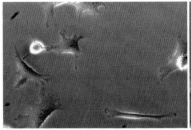  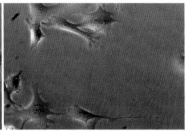

Figure 4.1 Motile smooth muscle cells over three distinct time states

layers of vascular smooth muscle cells (SMCs), which are themselves enveloped by a different type of ECM that is rich in type I collagen and elastin, depending upon the size of the vessel. Using the surrounding ECM as a foothold, and because of the connection of these cell-ECM feet to an internal tension system, called the cytoskeleton, SMCs have the capacity to relax or contract upon the ECM in response to external stimuli, including signals from ECs, thereby allowing the vessels to control the pressure of blood within the lumen as they constrict or dilate. Pulmonary artery SMCs produce and secrete enzymes that degrade their surrounding ECM, which includes type I collagen. This remodeling has two consequences: It promotes the contraction of the vessel wall by SMCs (Chapados *et al.*),[8] and it induces SMCs to produce an entirely new repertoire of ECM proteins that supports SMC proliferation and migration (McKean *et al.*, JCB).[9] Collectively, these events conspire to occlude the vessel wall, causing the usually low-pressure system of pulmonary arteries to increase as vascular tone and lumen diameter alter. This increase in pressure makes the heart work faster, putting a strain on this vital organ. There is no cure for PH, and so determining how SMC behavior changes in response to ECM degrading enzymes is a highly worthwhile pursuit.

Why Study Cell Motility as the Basis of the System?

As mentioned above, a cardinal feature of PH is the inward and outward remodeling of the PA wall by resident vascular SMCs that actively move through their existing type I collagen-rich ECM in an attempt to harness increases in lung blood pressure. However, over-exuberant SMC motility associated with PH often leaves PAs occluded and functionally deficient. Current methods for assessing motility are unable to discern highly discrete differences in cell motility at multiple length scales and in real time. In addition, most motility assays do not account for cell-ECM interactions. As well, existing methods to measure cell motility cannot be transferred to an automated platform due to problems in standardization, handling, and visualization. Given these deficits, generating accurate, cost-effective, and personalized multi-dimensional readouts of SMC motility requires paradigm-shifting approaches that will reveal both subtle and overt differences in motility behavior occurring within an appropriate 4-D context, a task that will be accomplished through the lens of these experimental projects.

The Space and Structure of Movement

Early inspiration for our approach to the motility project was inspired by architectural research previously conducted at PennDesign that looked at the space and structure of ice dancing.[10] The project, led by Jackie Wong, entitled *Dance and Space*, was an investigation into the spatial properties of figure skating, with a specific look at compulsory ice dancing. The choice of figure skating as the subject of investigation stemmed from an interest in how the ephemeral quality of body movements in a dance can shape space and leave behind impressions or signatures of form in the mind of the viewer. As Wong points out, "Like ballroom dancing, compulsory ice dancing is algorithmic by nature. Each dance has its own specific tempo, steps, and movement patterns." In the study, various methods to capture the spatial essence of dance were proposed. In order for the movements of the dancers to be mapped, certain parts of the body were tracked to map and extract their spatial qualities as a set of signatures and to analyze their relationship to one another. Six points of the body were chosen to abstract the skater's tracings in space as he/she traverses the dance pattern.

The unique signatures that emerged for each dance type especially intrigued

Figure 4.2 Choreographies of ice dances were mapped using points on dancers' bodies to produce unique spatial signatures of each dance type.

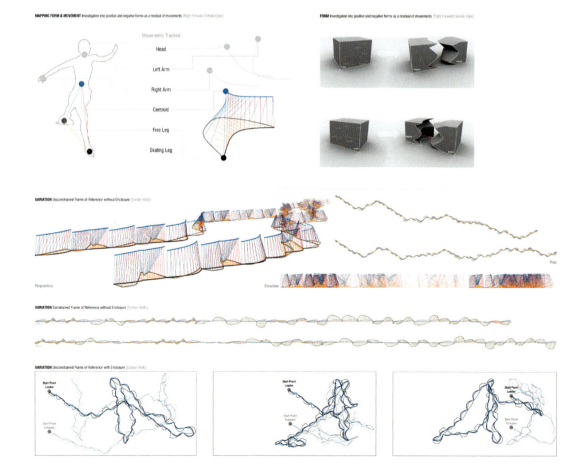

Motility **111**

Figure 4.3 Categorizations of 42 different dance patterns of a range of difficulties and tempos. The different spatial qualities of the dance patterns are shown as unique signatures where complex changes in movement, symmetry, and directional inclination are diagrammed over time and space.

us.[11] We speculated that it could be possible to extract unique signatures for cellular movement through similar methodologies and algorithmic thinking, but in the reverse. In Wong's project, the choreographies are defined and are used as input to generate the spatial signatures for each dance type. In contrast and with the cellular motility project, our task was to extract a unique set of choreographies or "cell dances," which could then potentially reveal and characterize personal cell motile signatures. But what are these choreographies? The following projects engage this task.

By examining and abstracting the cellular edge behaviors of cells within a normal or diseased tissue microenvironment, we are able to generate novel ways of quantifying, visualizing, and characterizing cell motility on a patient-to-patient basis. Through computation and abstraction of edge behaviors, we aim to derive personalized signatures for patients with pulmonary vascular disease.

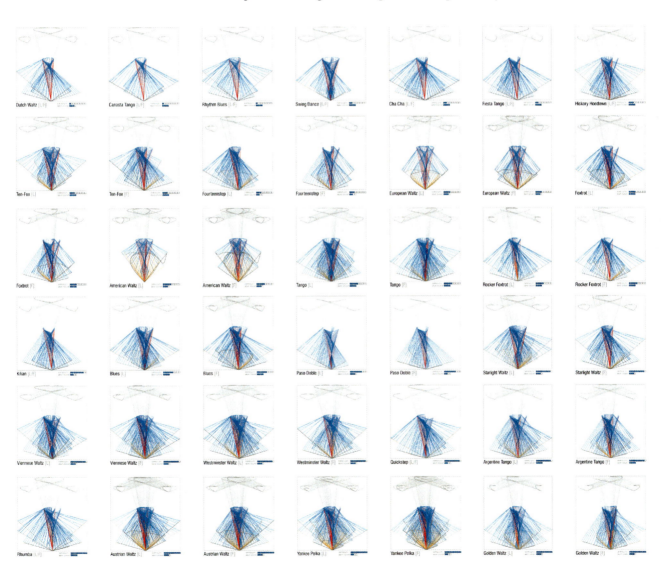

Architecturally, this project has resulted in a more complex understanding of a dynamic cellular structure that negotiates and integrates changes in external forces with internal cellular mechanics. In Part IV, we will see how some of these tools and concepts for modeling motile behavior on geometrically defined substrates are transferred to the design and engineering of highly functional adaptive building skins. Together, the following projects showcase a robust catalog of methods and tools for modeling and simulating cellular motility, each of which informs adaptive architectural prototypes and material assemblies in later scaled applications.

Notes
1 Editorial, "Virtues of Visualization," *Nature*, 455(7211) (2008): 264.
2 M. Oka, K.A. Fagan, P.L. Jones, and I.F. McMurtry, "Therapeutic Potential of RhoA/Rho Kinase Inhibitors in Pulmonary Hypertension," *British Journal of Pharmacology*, 155(4) (2009): 444–454.
3 D.D. Ivy, I.F. McMurty, K. Colin *et al.*, "Development of Occlusive Neointimal Lesions in Distal Pulmonary Arteries of Endothelin B Receptor-Deficient Rats: A New Model of Severe Pulmonary Arterial Hypertension," *Circulation*, 111(22) (2005): 2988–2996.
4 P.L. Jones, K.N. Cowan, and M. Rabinovitch, "Tenascin-C, Proliferation and Subendothelial Fibronectin in Progressive Pulmonary Vascular Disease," *American Journal of Pathology*, 150 (1997): 1349–1360.
5 V. Karamanian *et al.* "Novel Molecular Signatures for Human Pulmonary Arterial Hypertension," *Molecular Biology of the Cell*, 19 (2008): 432.
6 J.L. Snow, P. Lloyd Jones, and D.B. Taichman, "Pathobiologic Mechanisms of Pulmonary Arterial Hypertension," in *Pulmonary Vascular Disease*, eds. J. Mandel and D.B. Taichman (Oxford: Elsevier, 2006), pp. 33–49.
7 M. Rabinovitch, "Pulmonary Hypertension: Updating a Mysterious Disease," *Cardiovascular Research*, 34 (1997): 268–272.
8 R. Chapados, K. Abe, K. Ihida-Stansbury, D. McKean, A.T. Gates, M. Kern, S. Merklinger, J. Elliott, A. Plant, H. Shimokawa, and P.L. Jones, "ROCK Controls Matrix Synthesis in Vascular Smooth Muscle Cells: Coupling Vasoconstriction to Vascular Remodeling," *Circulation Research*, 99(8) (2006): 837–44.
9 D.M. McKean, L. Sisbarro, D. Ilic, N. Kaplan-Alburquerque, R. Nemenoff, M. Weiser-Evans, M.J. Kern, and P.L. Jones, "FAK Induces Expression of Prx1 to Promote Tenascin-C-Dependent Fibroblast Migration," *Journal of Cell Biology*, 161(2) (2003): 393–402.
10 J. Wong, "Dance and Space," advanced research seminar, *Form and Algorithm*, C. Balmond and J.E. Sabin, PennDesign, Philadelphia, PA: University of Pennsylvania, 2006.
11 Jenny Sabin and Peter Lloyd Jones, in conversation on J. Wong, "Dance and Space," advanced research seminar, *Form and Algorithm*, C. Balmond and J.E. Sabin, PennDesign, Philadelphia, PA: University Pennsylvania, 2006.

Case Study: Capturing the Fleeting and Fractile Nature of Cell Motility
Megan Born and Andrew Ruggles
with Mathieu Tamby, Keith Neeves, and Vanesa Karamanian

Nonlinear Biosynthesis
Instructors: Jenny E. Sabin and Peter Lloyd Jones
Sabin+Jones LabStudio

In this project, we evaluated human pulmonary artery smooth muscle cell motility *in vitro*, with the specific aim of understanding the edge behavior of these cells. Seeking to understand and possibly quantify the complexity of these conditions,

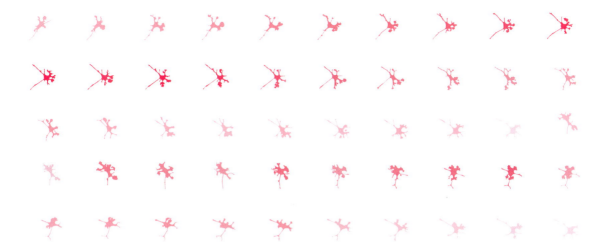

Figure 4.4 Smooth muscle cell edge conditions and motility

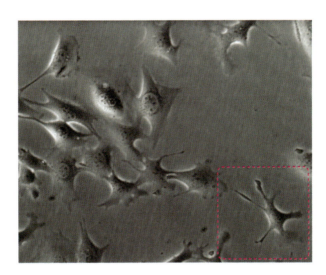

Figure 4.5 Smooth muscle cells and highlighted cell of interest

we developed a technique of geometric filtering that allowed us to convert available data into rigorous mappings. These mappings both located static and dynamic points on the cell edge and traced these points through time.

Beginning with an interest in mapping the morphological complexity occurring at the cell edge over time, our data set consisted of 50 frames extracted from a recorded video of smooth muscle cells cultivated on a plastic substrate taken with a phase contrast microscope. Initially, one cell was studied for the following reasons: the cell remained almost entirely within the frame of the video and it had minimal physical contact with other cells on the tissue culture dish.

The edge of the cell was traced manually to create the mapping using CAD software. In short, the filtering of the cell edge tracings was accomplished through a technique of locally refining mesh structures to approximate the fractal nature of the cell edge.[1] More specifically, we employed a recursive system of subdividing a

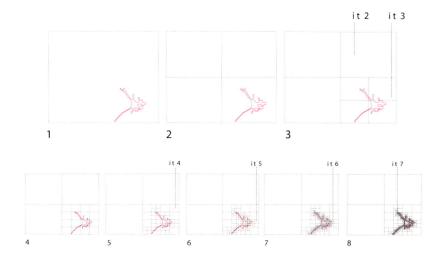

Figure 4.6 Edge tracing of smooth muscle cells

grid on a local level responding to the changing edge shape and length; this appears to change through a system of fractal branching. Beginning with a single rectangle, an algorithm was used to detect the presence of the cell tracing within the rectangle. If this condition was met, the rectangle subdivided into four nested rectangles of equal proportion. Each new rectangle was similarly checked for the presence of the cell edge tracing. To approach a more refined map, the process was repeated recursively.

In each frame of the dataset, we obtained a unique, yet consistently systematic map of a discrete and temporally isolated condition. Due to this consistency, we are able to determine points between the chronologically sequential maps that were either static or dynamic in nature. For example, an area from one frame that may feature a very refined subdivision of a highly complex edge condition would, in the next frame, be less subdivided. From this we can conclude that this region of the cell was dynamic in the space between the two frames being compared, i.e. cellular motility increases. Conversely, an area of a frame that remains the same between sequential frames shows that this area is static at that moment in time, i.e. cellular motility decreases. In order to graphically represent this, we stacked the 50 mappings in the Z dimension through time and used parametric software (GenerativeComponents [GC] by Bentley) to map the static and dynamic points for the entire dataset.

This work has led to both a visual and quantifiable understanding of cell edge motility. The GC model defined a set of conditions and parameters to describe the motility of a given cell's edge, where each segment of the cell edge is visualized to show areas of increasing or decreasing movement at a given moment in time as well as what scale the movement is occurring. Further study of the GC model could reveal direction and velocity of specific lengths of a cell's edge, as well as how these parts relate to the movement of the entire edge. If a neighboring cell were to be mapped in the same way, relationships between cells could be established.

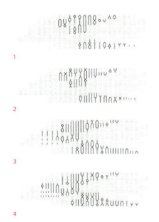

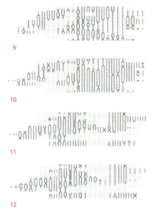

Figure 4.7 Movement through time of traced cells

Figure 4.8 Exploded Axon; static and dynamic edge conditions over time

1 filter 2 static points 3 dynamic points

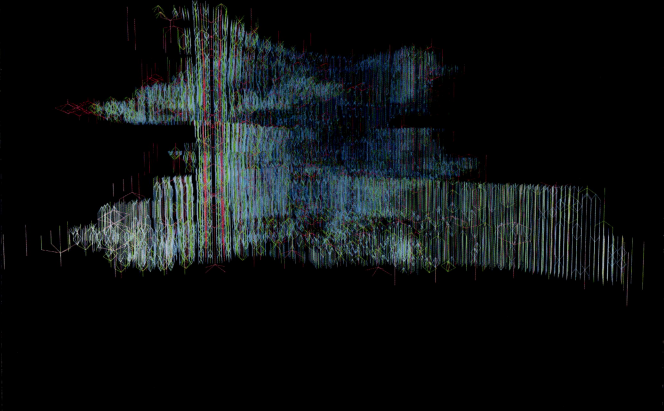

Figure 4.9 Combined layers of filter: static points and dynamic points

A challenge in studying movement of any kind is the difficulty of quantifying something that is continuous, fleeting, and controlled by a complex set of conditions. We have worked to develop a useful tool to output both a discrete dataset corresponding to a cell's edge in a specific moment in time as well as mapping the interstitial moments unable to be captured otherwise. Abstracting this complexity through a geometric filter allowed us to understand the edge behavior without oversimplifying the original condition. For us, the geometric filtering afforded us the ability to quantify conditions through a rigorous system of measurement.

Note

1 See Benoit B. Mandelbrot, *The Fractal Geometry of Nature* (San Francisco: W.H. Freeman, 1982). He presents a new geometry of nature, a family of shapes that he calls fractals, to describe "amorphous" forms such as clouds and coastlines that are difficult to describe through traditional Euclidean geometries.

Case Study: Mapping Motility through Dynamic Penrose Tiling
Jared Bledsoe, Vahit K. Muskara, and Gillian Stoneback

Nonlinear Systems Biology and Design
Instructors: Jenny E. Sabin and Peter Lloyd Jones
Sabin+Jones LabStudio

As previously described, pulmonary hypertension is an incurable disease characterized by increased blood pressure within the lung, leading to heart failure. At the level of the blood vessel wall and with pulmonary hypertension, smooth muscle cells that line and reinforce these structures move from their usual resting place into the lumen of the vessel, leading to vessel occlusion. In order to do this, the cells use enzymes to break down their surrounding collagen extracellular matrix, and respond to this degraded microenvironment by changing their motile behavior as an attempt to repair the blood vessel wall. Understanding how smooth muscle cells respond to normal or degraded collagen may therefore provide new ways of understanding and ultimately treating this disease. To model these behaviors, we have examined real-time movies of pulmonary artery vascular smooth muscle cells cultivated on native or denatured collagen.

Thereafter, we abstracted these behaviors into digital and fabricated design environments, based on newly devised rule sets. Overall, the furthest reach and

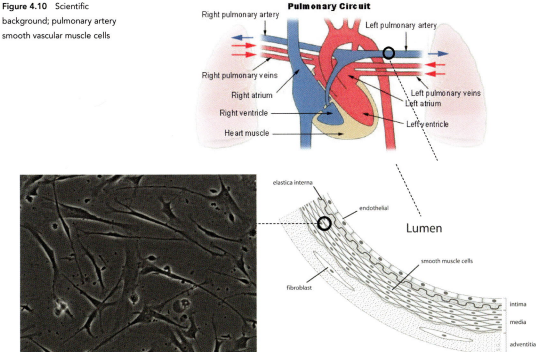

Figure 4.10   Scientific background; pulmonary artery smooth vascular muscle cells

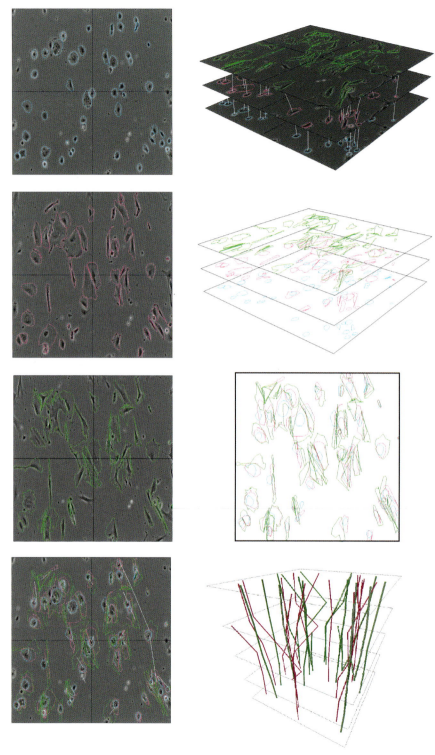

Figure 4.11 Initial investigation and analysis

Motility 119

Figure 4.12 Binary system tracking movement: A grid is imposed across the ECM field to document cellular motility in terms of x and y. The cellular movement is documented in addition or subtraction from its original placement within the grid. When a cell is present in a square portion of the grid, the grid is turned on, which creates a tiling pattern and in turn, a geometric record of each cell movement emerges.

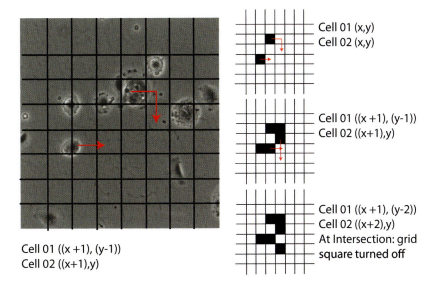

Figure 4.13 Binary system tracking movement combined

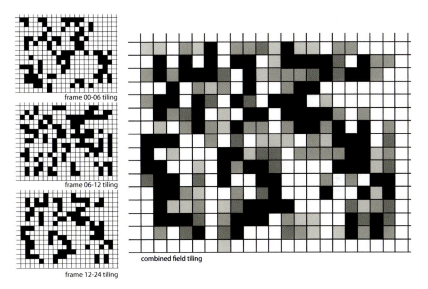

direction of motility differed between native and denatured collagen, as did cell growth. One analysis looked at the overall scale of the cell by examining the relationship between the nucleus and the furthest extension of lamellipodia or filopodia.

Cells on denatured collagen also appeared to have increased growth rates. To describe and simplify these behaviors, a simulation of various cell behaviors can be used to mimic cell behavior using a series of circles with a center at the nucleus and radius to the furthest reach. Initial variables considered included: small movements, no movement, growth, consistent growth, fluctuating growth, and no growth. In combination, these parameters were used to depict differentiated

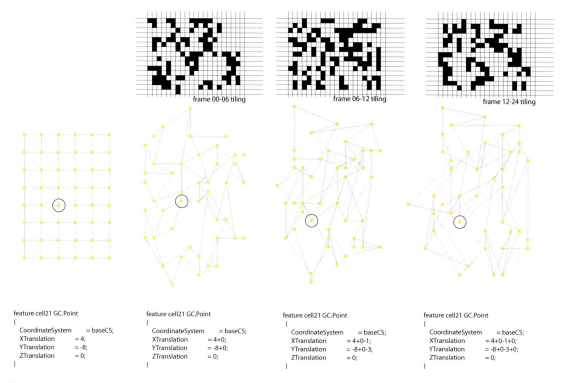

Figure 4.14 Field study: Using the coordinated changes from the previous exploration, a study was conducted to determine the effects of movement on a matrix visualized as a generic grid populated with cells at points of intersection.

cell behavior. In a supplementary study, connections between the cells were made illustrating association and gathering as well as to represent the extracellular matrix. Considerations for the connections include fixed connections, connections between same-cells, and flexible connections between different corresponding cells.

The Penrose, a geometric filter denoting orientation and position within a field, allowed an initial exploration into the mapping of motility through intentional patterning. Closed configurations of Penrose tiles were used to locate the position within the given cell dataset frame as well as the cell's movement from frame to frame. This technique has proven to yield simple mappings of directional movement and orientation of local cells. The infill, even though it changed according to the cell's movement and orientation, became the dominating feature within the patterns.

Continuing with this useful simulation method, a grid of Penrose tiles was set up to document the tracking as well as the motility effects upon the ECM. The first round of mapping proved to be unsuccessful as a single cell had too much influence on the field, thus indicating an incorrect simulation of the cell's local interaction with the ECM. Nevertheless, the movement of the cell is clearly seen by the grid of Penrose tiles subdividing its local geometry as the cell comes within

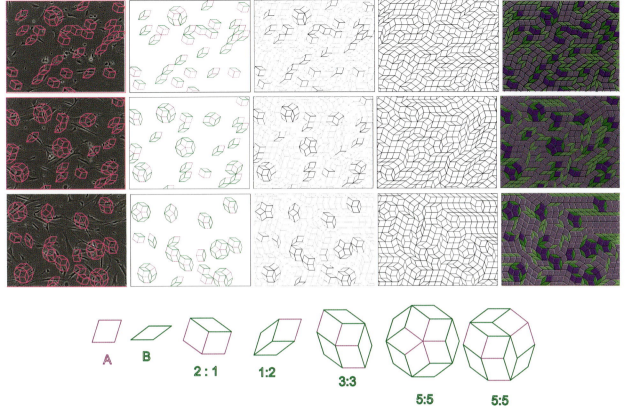

Figure 4.15 Penrose tiling analysis: The goal of the Penrose tiling system is to demonstrate visual patterns and relationships between cells correlating to a moment in time. The variegated tiling patterns visualize changes in cellular motility as a field condition.

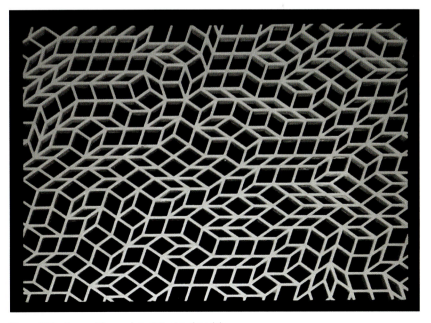

Figure 4.16 Penrose tiling analysis, 3-D printed model

a range of each portion of the grid. The next level within this investigation proved to be more successful. With the cell having a more localized influence on the grid, the documentation of multiple cells became possible, changing the reading of the pattern as a field condition of motility. A moving cell was assigned the most subdivided geometry. With the rate of change or the cell movement over time, the path became less subdivided. This enabled two readings: cells frozen within a single time frame, and as a tracking device from one cell dataset to the next. This technique allowed for the entire tiled field to be influenced by the movement of the cells, producing a more dynamic simulation of patterning through motility. These field conditions also enabled analysis of a given cell's possible effect on the ECM. With this technique, a cell would not replace another cell's position within the grid until the original cell's influence was no longer present in the grid.

An alternate trajectory looked at using an aperiodic Penrose tile configuration to represent and analyze cell motility in the native and denatured samples. The two tiles chosen were both four-sided, whose angles were: 36, 144, 36, and 144 degrees; and 72, 108, 72, and 108 degrees. This tiling system was chosen because it is a system composed of two tiles easily arranged in many configurations.

Figure 4.17 Alternative Penrose tiling procedures: native smooth muscle cell. Building upon the previous study, this Penrose tile pattern documents the path of movement and the rate of change through the field. The presence of a cell within the generic Penrose tile grid produces denser sub-divided areas. Thus, the Penrose tiling pattern acts as a filter to visualize cell motility over time in an alternate way.

The cells were mapped by closed configurations of arrangements of three or more tiles. The term *closed* refers to the configuration containing no acute exterior angle, creating a closed or round shape when the tiles are combined. Three-tile configurations were used for individual cell mapping, while larger groupings of cells were mapped with the six or ten tiling closed systems. The groupings containing more of the 72-108-72-108 degree tiles were used to represent cells and cell groups with more lamellipodia armatures, whereas the groups containing more of the 36-144-36-144 angled tiles were used to represent cells with more filopodia extensions. Individual tiles then fill in the extracellular matrix, completing the aperiodic tiling pattern.

With the use of 3-D parametric applications, we were able to abstract several variables from the cells within the ECM. Displacement, rotation, scalar shift, and connection were some of the characteristics derived from the native smooth muscle cell films. Initial studies and diagrams abstract the morphological and directional changes through the series of frames in the native ECM. Through these studies, we were able to generate simple geometric components to represent a cluster of cells. The models display various differences between the native and denatured cultures. The native ECM illustrates the most radical variation in scale and rotation of the cells.

Case Study: Cell Motility: Spatial Signatures
Christopher Allen, Benjamin Callam, and Katherine Mandel

Nonlinear Systems Biology and Design
Instructors: Jenny E. Sabin and Peter Lloyd Jones
Sabin+Jones LabStudio

Data extracted from scientific papers and cellular motility videos served as a basis for the study of the relationship between cells and the extracellular matrix (ECM) during motility. More specifically, this research aimed to explore the relationship of ECM compliance and actin/cytoskeleton structural changes within the cell. Research has shown that the ECM can influence a cell's structure and therefore its geometry, motility, and proficiency to adhere to the ECM; this relationship between cellular changes and ECM can have reciprocal effects.

This relationship was explored through the development of a tool that analyzes an object moving within an environment over time. While this research was directed at information learned from the biological systems studied, it was intended to be applicable to external data sets of various kinds. Of particular interest was the opportunity to look at how human movement is influenced by architectural space.

Cell motility is an ongoing process in the body critical to tissue development and wound healing. During the process of cell movement, changes within the cell

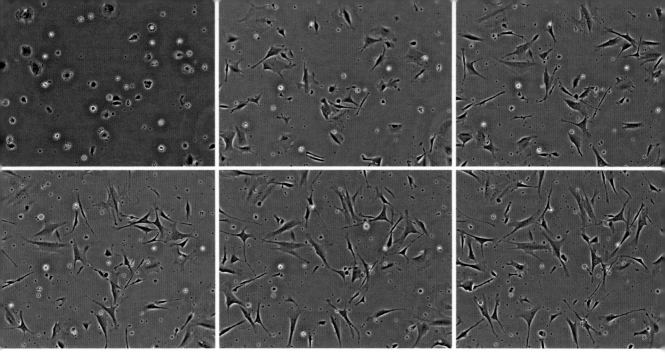

Figure 4.18 Denatured pulmonary artery vascular smooth muscle cells over six time states

are concentrated in its cytoskeleton system. Movement is achieved through changes in cell geometry that are a result of changes in cytoskeletal structures within the cell. These structural changes result in the extension of filopodia, which attach to the surrounding environment. The forces developed by the structural changes to the cell and its interaction with the surrounding environment translate into motion. In this way, the environment of the cell takes on a more regulatory role. The extracellular matrix (or ECM) is the interstitial space between cells that provides structural support and biochemical information as well as promoting certain cell behaviors. Thus, the ECM is critical for proper cell migration, proliferation, and interaction.

Video recordings of cells moving on native and denatured collagen were the first set of data used in analyzing the relationship between the ECM and cellular behavior. Collagen is a prolific extracellular matrix tissue that many of the body's systems rely on to function properly. Native collagen is structured in a way typically found in a healthy body whereas denatured collagen is in an unraveled state. In this state, it does not maintain the necessary geometry or structural integrity to operate appropriately in the body. Patients with pulmonary hypertension suffer from this deformation of collagen. Changes to the collagen composition cause smooth muscle cells lining the pulmonary arteries to multiply excessively, creating blockages in the lung's arteries.

In addition to the video data sets, a series of scientific papers detailing behavioral changes in cells with respect to ECM compliance aided in our exploration. In particular, the following two papers contributed important results that influenced the direction of our project.

Figure 4.19 Cell motility behavior analysis

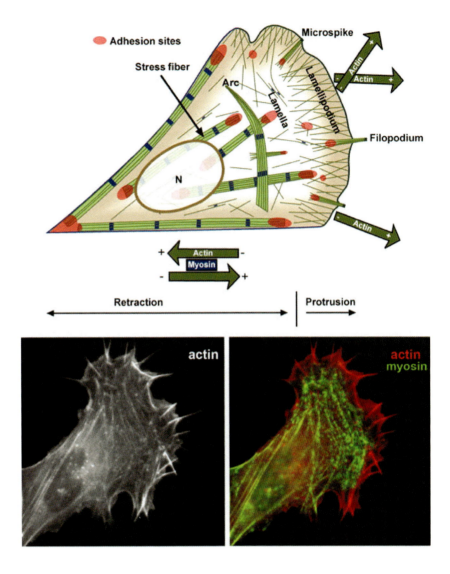

In the paper entitled, "Geometric Determinants of Directional Cell Motility Revealed Using Microcontact Printing,"[1] the authors state that cells "pull themselves toward stiffer environments." At a global scale, the environment's stiffness has some influence on the directionality of cell movements. Within the cell, this means that actin developments are promoted by stiffer substrates.

In the paper entitled, "Fibroblast Adaptation and Stiffness Matching to Soft Elastic Substrates,"[2] the authors document internal cell stiffness change as a function of the substrate stiffness and test if cell area is proportional to substrate stiffness. At a global scale, this means the environment directly influences the shape and area covered by a cell. Within the cell, cytoskeleton organization changes due to changes in the environment's compliance.

In addition, Brock *et al.* and Solon *et al.*,[3] were helpful when studying the differences in the two cell motility videos. The speed at which the cells move across the environment, the average area the cells encompassed, and differences in the probing edge of the cell were all evident when comparing the two videos. The videos coincide with the arguments made in both articles and provide insight into the gradient of movements possible on environments of different compliance.

After this initial exercise where we used videos to track cell movement over time as a measure of cell to ECM interactivity, examination of the changing cell behaviors of smooth muscle cells on native and denatured collagen provided a set of behaviors seen in both normal and extreme conditions. The motion capture algorithm established a method of abstraction through which we could observe cell behavior within a context. It extracted the choreography of a cell in its environment over time using color as a metric. The intensity of the color in the field, its lifespan over time, and the repetition of color geometries indicated behavior variables such as speed, area, geometry, and filopodial adhesion points.

By studying cells in various environments, the repetition of local and global geometries and color intensity could help establish the composition and compliance of an unknown context. The scientific implications of the tool for a disease like pulmonary hypertension include the possibility of extracting a patient's collagen around the pulmonary arteries, applying smooth muscle cells to it, and recording the cell behavior using our algorithm to provide insight into the collagen defect and the stage of the patient's illness.

More specifically, questions arise as to whether or not new behaviors or nuances in cell behavior can be detected on various ECM compliances using our algorithm. If so, these nuances might be used to examine the state of disease within a patient's ECM tissue. These nuances could even be traced to a specific problem in the ECM such as composition or signaling.

The analogic investigations began to record the geometric data gathered from the motion capture program in a physical process. We struggled to establish variables that would not create a "column" of data and reduce the algorithm to a finite structure. Our process demanded that time is used as a measure of the change in the field for the duration of the video.[5] We embraced time as a variable for recording the changes and repetitions of the algorithm. Additionally, the finite structure created by physically modeling the geometric data established another means by which to analyze the geometric differences of objects moving in an environment over time.

The physical model archives the history of the interactions. It provides a singular representation of all the moments processed by the motion capture algorithm. It is the first attempt to identify a geometric signature to the cellular behavior of motility. Although we established this modeling process with the cellular data, we designed it to be applicable to any scale of data the algorithm analyzed.

Figure 4.20 Denatured environment showing history of change with physical artifacts; cell motility 3-D signature over time

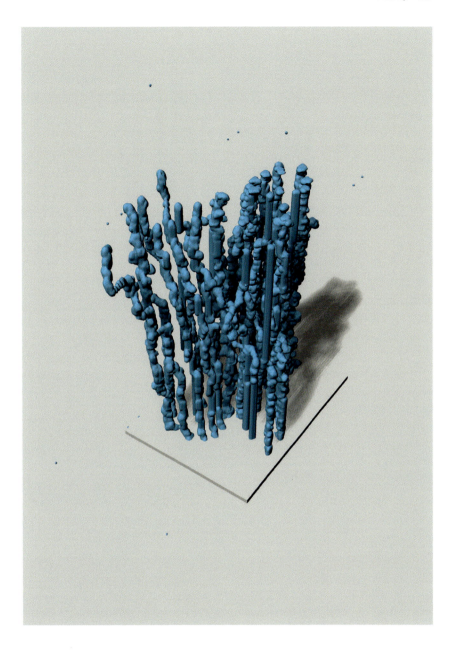

The motion capture algorithm explored the relationship of ECM compliance and actin/cytoskeleton structural changes within the cell. More specifically, it explored the implicit changes that ECM compliance has on cell migration and differentiation. The algorithm demonstrates an investigation into the interaction of an object and its surroundings, providing insight into both the object and the environment's composition.

The motion capture applies this study at the scale of a cell, but we would like to develop a tool to investigate the object-environment interaction at the scale

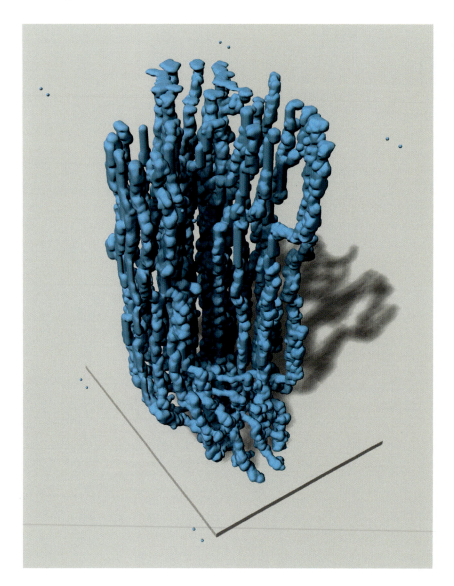

Figure 4.21 Native environment showing history of change with physical artifacts; cell motility 3-D signature over time

of a person. The analogic investigations that were carried out with respect to the motion capture algorithm also provided much insight into how to create a tool that could document the history of the changes between the object and environment three-dimensionally and physically. From the two earlier exercises, our project goal was to create a tool that distinguishes behavioral or physical signatures of external data sets.

We were particularly interested in investigating the human-space interaction in this manner thus creating spatial signatures for people in a particular space or a set of collaborating variables in a space. We stipulate that this could be a definition of personal architecture: the spatial signature of a person or persons in a particular space.

Figure 4.22 Spatial signatures

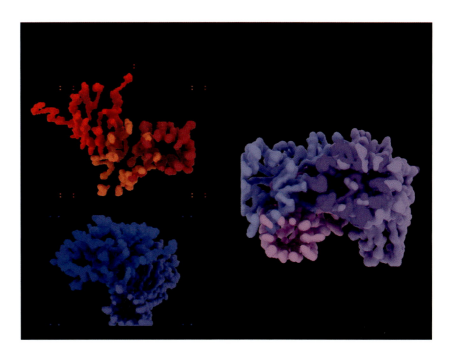

To generate the three-dimensional structure of the signature, we combined the motion-tracking data with an additional agent-based algorithm. This algorithm establishes a series of agents that are designed to evaluate sub-regions of the motion tracking data set over time. Each agent consists of one centroid and several evaluation nodes encircling the centroid. The evaluation nodes examine motion data in their immediate vicinity and are attracted by areas of greater activity. The centroid is calculated after each evaluation node is reported, and the centroid is translated two-dimensionally along the resultant vector of the nodes. In addition to 2-D translation, the agent is advanced along a 3-D vector corresponding to time. Each evaluation node is described three-dimensionally by a triangulated mesh.

The result is a geometric description of motion intensity as agents respond to changes in the data field over time. The resultant geometry was evaluated using ZCorp rapid prototyping technology with color to differentiate the numerous agents active in the system.

Our final application of this toolset was used to look at students moving through a given space during a public event—a record of human movement within a space. The final form has a wealth of information that can be further analyzed. Drastic moments of change in the form can be identified and matched to moments in the input video, thereby establishing a cause and effect of data input to changes in formal output.

Doing this for several environments would be the first step to establish trends in geometries created by the data set. It would also create enough data to

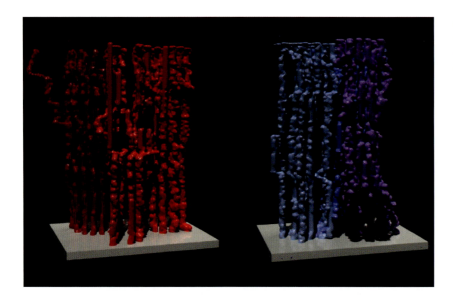

Figure 4.23 Spatial signatures in 3-D; time on vertical axis; native environment (left), denatured environment (right)

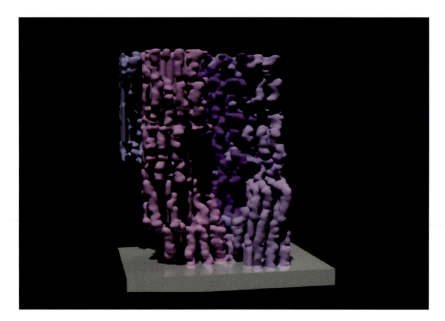

Figure 4.24 Scale-free test of algorithmic process; spatial signature of human movement

cross-analyze the process forms from the different data sets and identify repeating geometries and unique geometries.

At the scale of personal architecture, the spatial signature algorithm has the potential to be a guide for designing spaces—desired movement through space can be sought by processing interactions within the space and drawing correlations between geometries that indicate the desired behaviors. Many of the same principles are applicable at a biological scale. The algorithm abstracts complex (and unknown) data sets into clear signatures of behavior that could potentially be used to identify and pinpoint differences in the extracellular matrix of healthy patients versus those with diseases such as pulmonary hypertension.

Notes
1. Amy Brock, Eric Chang, Chia-Chi Ho *et al.*, "Geometric Determinants of Directional Cell Motility Revealed Using Microcontact Printing," *Langmuir* 19(5) (2003): 1611–1617.
2. Jérôme Solon, Ilya Levental, Kheya Sengupta *et al.*, "Fibroblast Adaptation and Stiffness Matching to Soft Elastic Substrates," *Biophysical Journal* 93(12) (2007): 4453–4461.
3. Ibid.

Case Study: Context-Dependent Cell Deformation
Bo Rin Jung, Yan Cong, and Chia Liao

Nonlinear Systems Biology and Design
Instructors: Jenny E. Sabin, Shawn Sweeney, and Peter Lloyd Jones
Sabin+Jones LabStudio

Based on research of cell motility, the aim of this project was to create an environment that can control and deform the shape of the cell in three dimensions through its own reorganization.[1] Starting from visual observation of recorded cell behavior over time—first through video recordings and then through abstraction—a computer script was developed to simulate cell deformation during movement in two different environments: native substrate and denatured substrate, based on observation and the results from the biological study.

In comparison to the native substrate, we found that in the denatured substrate, cells move faster with less directionality because the adhesion force decreases due to the breakdown of the substrate protein. In the computer simulation, a matrix is set up as an environment, where the cell membrane could move faster and osculate between locations marked as denatured points in the matrix environment. This process was iterated through both two- and three-dimensional geometry, effectively simulating cell motility through geometric and mathematical principles.

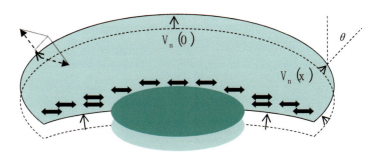

Figure 4.25 Research into the mathematical and geometric descriptions of cellular motility

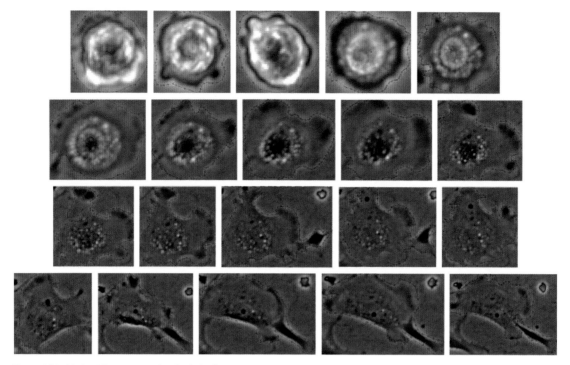

Figure 4.26 Native video capture with individual cell tracing

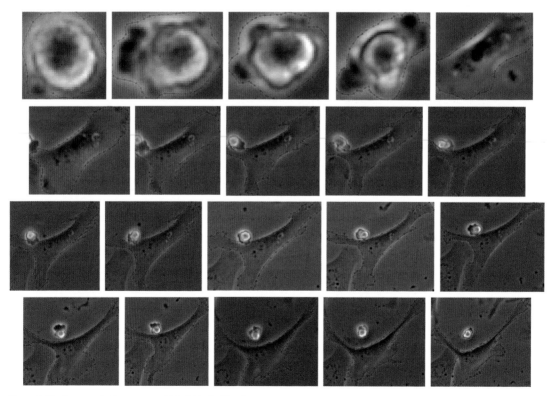

Figure 4.27 Denatured video capture with individual cell tracing

Motility 133

Figure 4.28 Native environment deformation over time

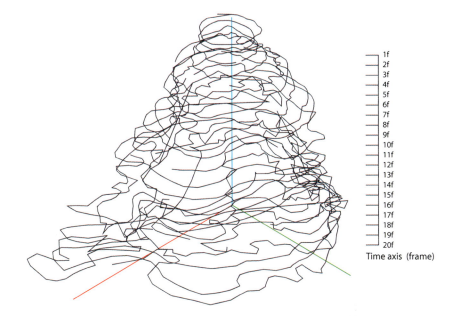

Figure 4.29 Native time frame visualized as a 3-D print

134 Design Computation Tools

Figure 4.30 Denatured time frame visualized as a 3-D print

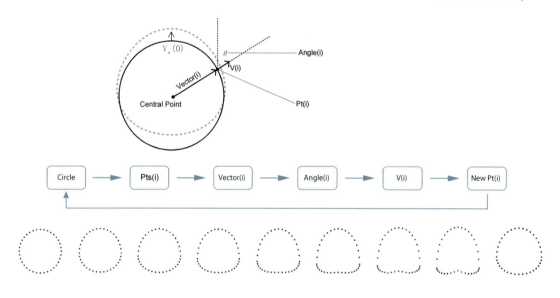

Figure 4.31 Algorithm to simulate cell motility

Figure 4.32 Diagram for cell motility in a native environment

Figure 4.33 Diagram for cell motility in a denatured environment

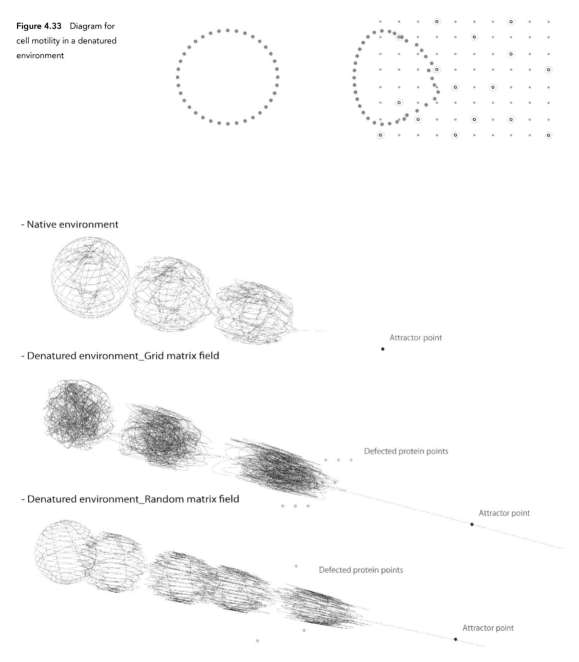

Figure 4.34 3-D experimentation

> **Note**
> 1 Alex Mogilner, "Graded Radial Extension Model and Actin Growth/Myosin Contraction Mechanism." In "Mathematics of Cell Motility: Have We Got Its Number?" *Journal of Mathematical Biology* 58 (2009): 105–134.

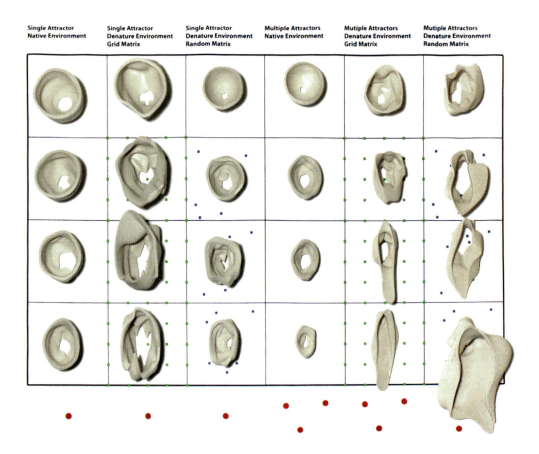

Figure 4.35 Catalog of variations

Topologically Free Cells
Roland Snooks
Invited critic and founding partner of Kokkugia and Director of Studio Roland Snooks; Senior Lecturer, Architecture and Design, RMIT University, 2010

One of the things that is intriguing to me as an architect looking at these cellular structures, is that they are very topologically free. And cells are presumably able to change their relationship to neighboring cells as well as change their own morphologies.

Avoiding Biomimicry
Shawn Sweeney
Guest Instructor and currently Associate Director of translational research at the American Association for Cancer Research (AACR), 2010

One of the things that I think the students are trying to avoid in an outright fashion is biomimicry. The idea is to take the content and then make something that doesn't exactly look like the cellular structure but functions and behaves like it.

Case Study: Cell Motility and the Dynamic Transformation of Structure

Ikje Cheon, Josef Musil, and Dane Zeiler

Nonlinear Systems Biology and Design
Instructors: Jenny E. Sabin, Shawn Sweeney, and Peter Lloyd Jones
Sabin+Jones LabStudio

Although the complete process of cell motility is quite complex, several key features from the provided data sets of human pulmonary artery vascular smooth muscle cells have been observed, predominantly the movement of individual cells and both pulling and pushing motions of the spreading actin filament arms.

This movement is often propelled away from a thin and long projection also known as the "tail". The following process seeks to understand the associated movements of the cell body in response to tail length. Data sets were derived through the comparison of a single smooth muscle cell across 49 frames in both native and denatured environments. This data informed research in several important areas. First, one can conclude a difference in tail behavior in terms of extension length, area, and average angle to extended filopodia. Second, within each cellular environment, there is a continuous pattern of growth and compression of the cell body. Also, the average angle to extended filopodia in the denatured environment is nonlinear across the data set. The trajectory of the research attempts to depict a nonlinear relationship between angle, length, and transformation of structure. This is then applied and translated to a set of networked components where each component is manipulated manually and physically. By allowing the control to be both at an angle and vertical in direction, another layer of complexity and dynamic movement is added.

Figure 4.36   Cell analysis

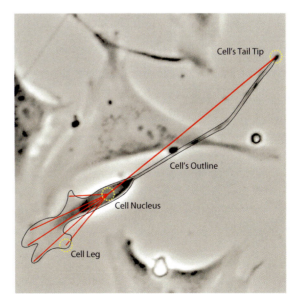

Denatured Cell Analysis

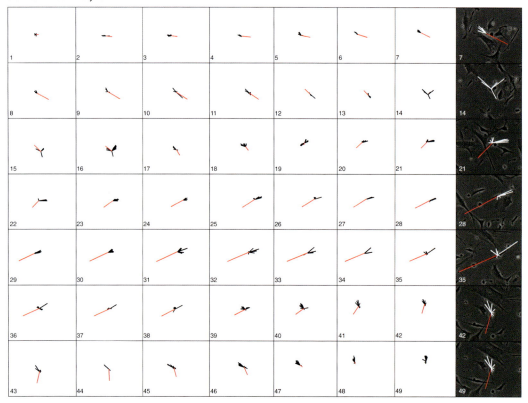

Figure 4.37 Denatured cell analysis

Tail Relationships: Native Conditions

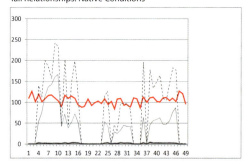

Tail Relationships: Denatured Conditions

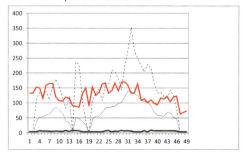

Figure 4.38 Tail relationships over time

Motility   139

Figure 4.39 Overlaid tracings

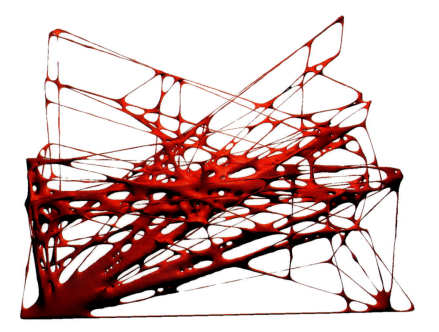

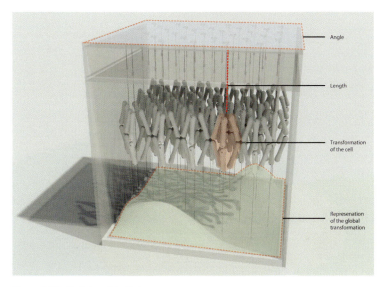

Figure 4.40 Digital model with layers

Figure 4.41 Final physical model

Project Initiatives
(1) Analyze the data extracted from tail qualities looking for overlaps and abnormalities.
(2) Digitally simulate the behavior of a body of components working dynamically as a unit.

(3) Translate the simulation physically with a tensegrity model with controllable parameters. Using the tensegrity modules, the study of the networking between the modules was carried out based on the distance and angle of the attached strings. By manipulating the string, the tension within the module would deform the overall form, and eventually influence the network.

Case Study: Actin Polymerization: Models for Self-Assembly
Philip Tribe, John Wheeler, and Brian Zilis

Nonlinear Systems Biology and Design
Instructors: Jenny E. Sabin, Shawn Sweeney, and Peter Lloyd Jones
Sabin+Jones LabStudio

John Wheeler, Brian Zilis
Independent study with Jenny E. Sabin

Our interest lies in cell motility, specifically actin polymerization and the inherent process of nucleotide hydrolysis and the formation of actin filaments in a three-phase process: lag, growth, and equilibrium.

In the first part of this project, we asked how the essential building blocks of actin polymerization relate to one another, the entire cell, and the extracellular matrix. We focused on the nucleation of actin molecules, the inherent process of nucleotide hydrolysis, and formation of actin filaments over a three-phase

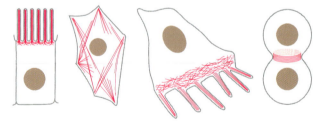

Figure 4.42 Importance of actin polymerization: About 5 percent of all protein in a typical animal cell is actin. Half of this actin is assembled into filaments, while the other half remains as actin monomers. These monomers are essentially held in reserve until they are required. Actin filament assemblies are crucial for most forms of cell motility as well as cell division.

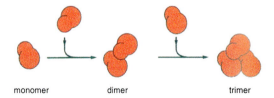

Figure 4.43 Nucleation: Two actin molecules bind relatively weakly to each other. Three actin monomers form a trimer and a much more stable configuration. As more monomers connect to this trimer, the assembly becomes the nucleus for polymerization

Figure 4.44 Phases of polymerization: The lag phase relates to nucleation. The growth phase is the addition of monomers to the polymer. The equilibrium phase is when the addition and subtraction of monomers reaches a steady state.

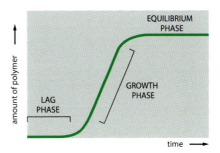

Figure 4.45 Treadmilling: Achieved when a polymer maintains a constant length and occurs during the equilibrium phase when the net flux of monomers to and from the filament is stable.

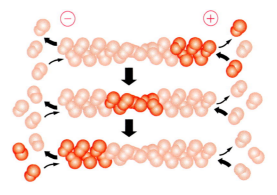

Figure 4.46 Dynamic instability: State of rapid falloff of actin monomers on an actin filament. Polymers fluctuate between a period of slow growth and a period of rapid disassembly.

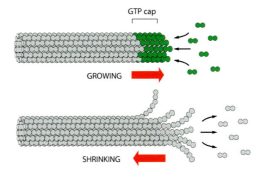

process: lag, growth, and equilibrium. The last phase, more commonly known as "treadmilling," is the result of a balance between actin monomers assembling and disassembling on either end of a filament, resulting in an overall trajectory that is related to the larger cell directionality. We are attempting to understand through various means of modeling and simulation the concept of critical concentration or CC, both rapid polymerization and dynamic instability, and how these concepts may be applied to environmental sciences and systems on a larger scale. Through this research, we developed physical components that were derived from an understanding of this behavior. These components had directionality, polarity, and connectivity.

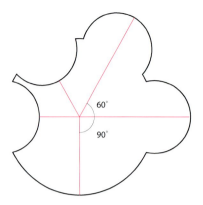

Figure 4.47 2-D Exploration: Allow components to aggregate and accumulate in a single trajectory.

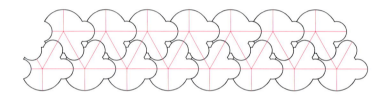

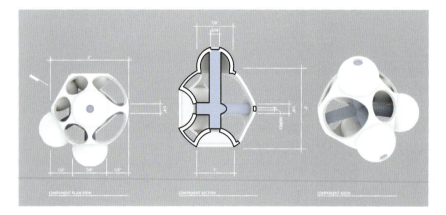

Figure 4.48 Prototype A, dimensions and configuration

Figure 4.49 Prototype A, 3-D print with high strength magnets

Figure 4.50 Prototype parts

Figure 4.51 Prototype A, variegated assembly

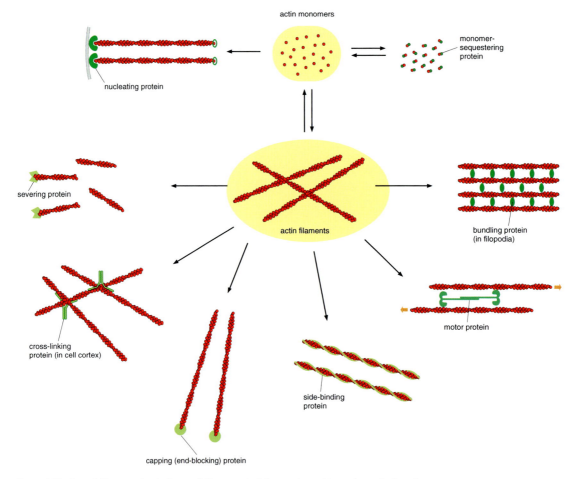

Figure 4.52 Cross-linking proteins: Actin cross-linking proteins influence the packing and organization of actin filaments into secondary structures.

The following independent study sought to exploit the biology further and reiterate developed componentry to become more responsive to contextual elements without losing their passive nature. As an initial prototype, these components are identical in scale and are contained within two static systems that are largely independent from each other: an inner structural system and an outer skin system.

Goals for this exploration include the further development of our initial component, its geometries and connections as well as its inner and outer systems. Also explored here are different methods of fabrication, including those more suitable for mass production and scalar shifts, as well as different material options for different types of performance; elastic versus rigid, translucent versus opaque, etc.

We seek the integration of multiple systems into a cohesive physical manifestation that may have the capacity to self-assemble from a kit of parts. An example would be a structural wall that allows for apertures in conjunction with the circulation for air, water, and/or light. It is our hope that through the

Figure 4.53 Prototype B, rendering of typical assembly

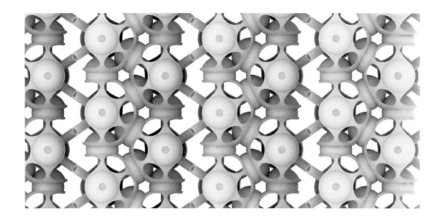

Figure 4.54 Prototype B, top view rendering of larger systems of packing and assembly

Figure 4.55 Prototype B, 3-D print of single component

exploration of 3-D printed components and families of components based upon actin polymerization, we will discover different types of assemblies and methods for self-assembly that take advantage of 3-D printing and nonstandard fabrication while lending themselves useful to their immediate context.

"If you wired the components with electromagnets, you could just flip the switch and have it self-assemble . . ." (Shawn Sweeney, Guest Instructor).

146 Design Computation Tools

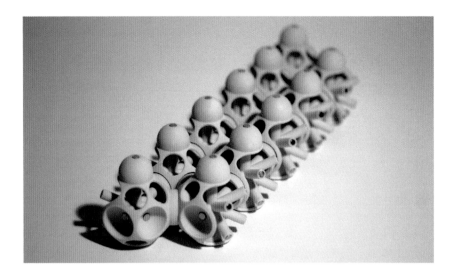

Figure 4.56 Prototype B, regular assembly

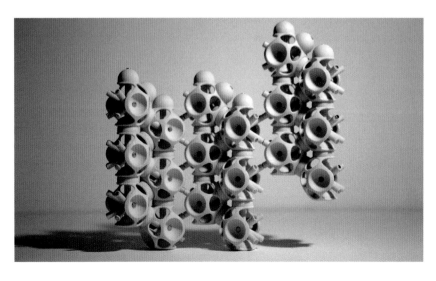

Figures 4.57–4.60 Final prototype assemblies with variegated connection behaviors affecting global form

Visualizing in Another Dimension
Shu Yang
Invited critic, co-advisor, and collaborator; Professor, Materials Science and Engineering and Chemical and Biomolecular Engineering, University of Pennsylvania, 2010

I would completely agree that people in different disciplines will think about things differently and because of these different ways of thinking, it will intrigue our curiosity and expand our research. That's how our collaboration started. And students played key roles to initiate dialogue and complete such interdisciplinary collaborations.

These 3-D printed models and their assemblies intrigued me. It allows us to directly visualize things in another dimension, inspiring us to think of, for example, how polymer molecules could interact with each other beyond computer simulation, and how different shaped particles can be assembled together, how they could respond to environmental changes.

Case Study: Geometries of Change: Mutations and Accumulation in Cell Motility
Yifan Wu, Wenda Xiao, and Ringo Tse

Nonlinear Systems Biology and Design
Instructors: Jenny E. Sabin, Shawn Sweeney, and Peter Lloyd Jones
Sabin+Jones LabStudio

Based upon observation in a controlled experiment of smooth muscle cell motility, we found that cells behave differently in various environments (denatured and native). By tracing the paths of movement, including tracking the centroid of the cell nucleus and the direction of branches of one single cell within a given period of time, we studied the behavior mechanism (morphosis) of native and denatured cells with a time-spatial diagram.

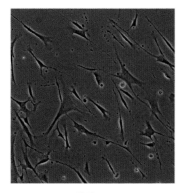

Native

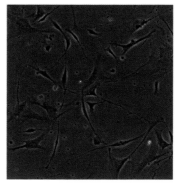

Denatured

Figure 4.61 Human smooth muscle cells in native and denatured environments

Motility   149

Figure 4.62 Movement of a single cell over time in both native and denatured environments

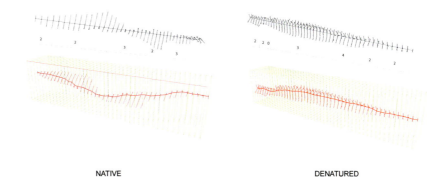

NATIVE   DENATURED

Figure 4.63 Cell signatures; overlay of discrete tracings of cell edge and nucleus displacement

NATIVE   DENATURED

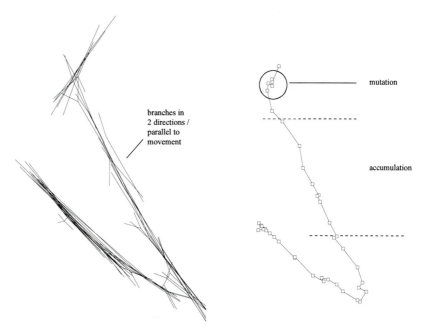

Figure 4.64 Analysis of cell motility over time in two environments. Mutation and accumulation show as distinct features.

150 Design Computation Tools

Internal logic

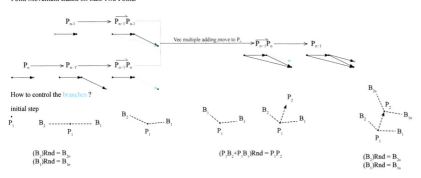

Figure 4.65 Internal logic for cell motility described through vector geometry

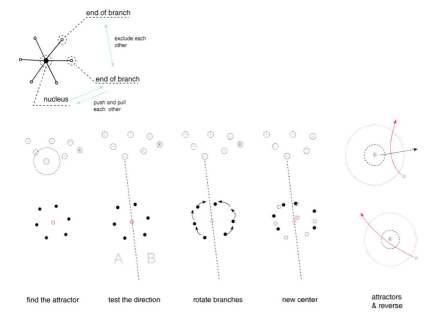

Figure 4.66 Addition of environmental factor to the internal logic.

Based upon these diagrams, we became intrigued by the rhythms associated with cell movement. We considered this rhythmic pattern as a crucial phenomenon manifesting interaction between cells and the extracellular matrix. Initially, we studied the mechanism behind cell motility through the paper entitled "Mechanical Signaling and the Cellular Response to Extracellular Matrix in Angiogenesis and Cardiovascular Physiology,"[1] which explains how the cytoskeleton, cell membrane,

Motility 151

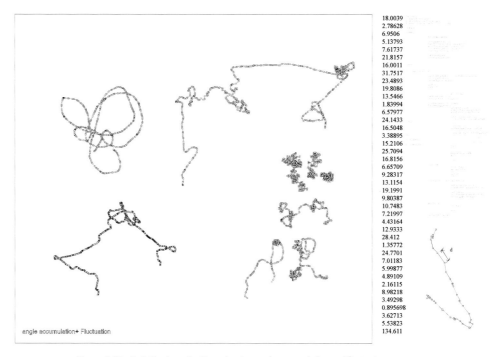

Figure 4.67 Scripting investigations showing angle accumulation and fluctuation

Figure 4.68 Environmental influence

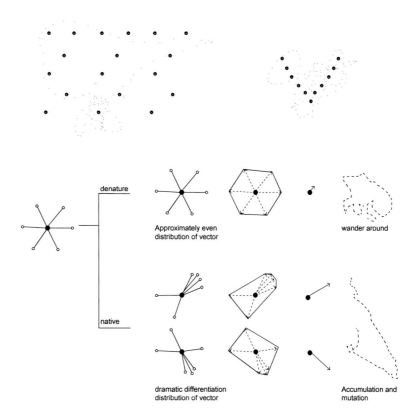

Figure 4.69 Native verses denatured diagram

152 Design Computation Tools

Materialization

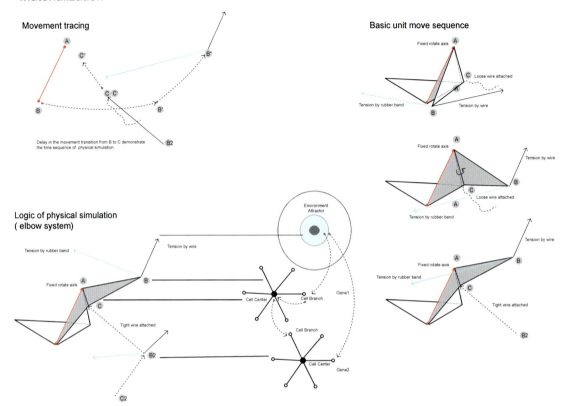

Figure 4.70 Materialization diagram showing folding strategy

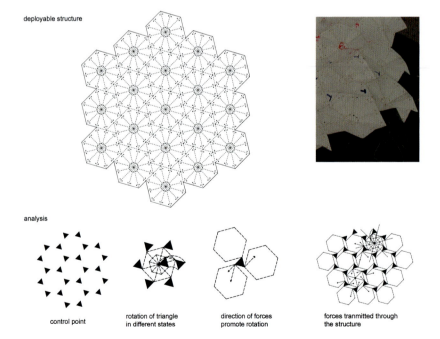

Figure 4.71 Deployable folding Prototype 1

Figure 4.72 System diagram showing folding behaviors and orientation

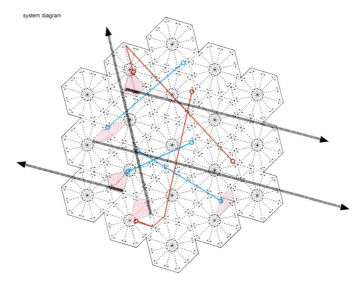

Figure 4.73 Study model

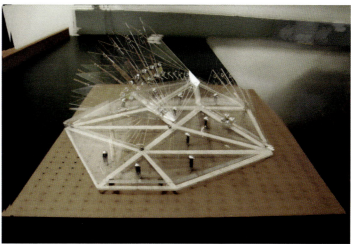

and extracellular matrix (ECM) work together cohesively as a dynamic system. This article applies a tensegrity model to illustrate the local changes in tension that are responsible for the reconstruction of the whole dynamic system and the control of local cell behavior. Next, we investigated the possibility of scripting logics to simulate this mechanism and the motility rhythms that influence and are associated with contextual changes in the environment.

To materialize these variegated motile behaviors and rhythms (a quality of temporality), we transferred the logic to the context of a deployable structure with folding strategies. Components react to and interact with each other based on specific rules; they form a series of specific behaviors within a sequence of timeframes. Here, the rules and behaviors of cellular motility are translated to a new material ground, while taking advantage of path, orientation, and changes in force

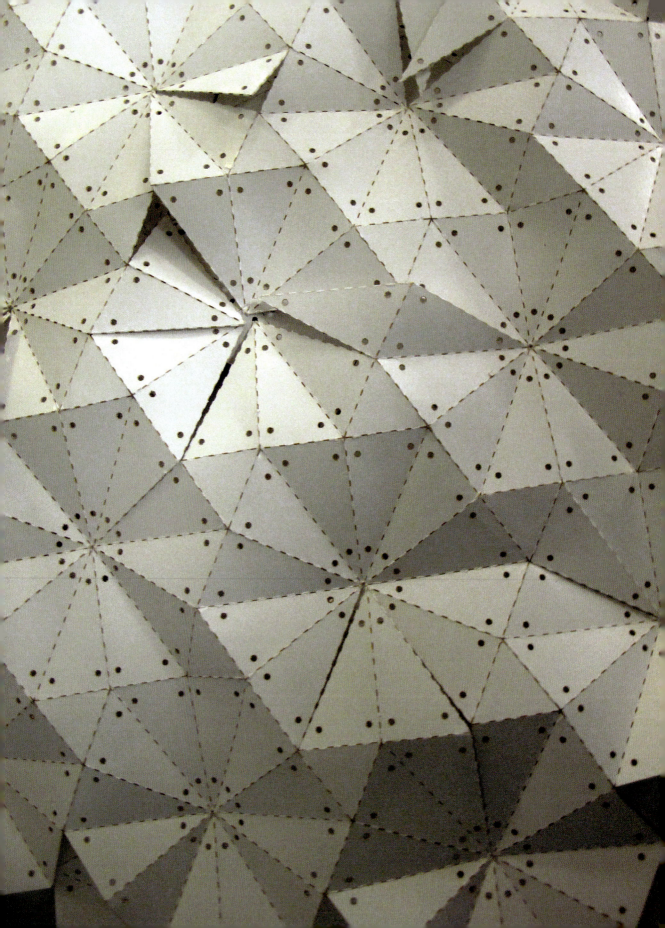

transmission and environment. Some of these principles are taken forward in later research in the context of foldable structures based on cellular logics.

Note
1   Donald E. Ingber, "Mechanical Signaling and the Cellular Response to Extracellular Matrix in Angiogenesis and Cardiovascular Physiology," *Circulation Research* 91(10) (2002): 877–887.

Positioning Mechanism
Peter F. Davies
Invited critic and Director, Institute for Medicine and Engineering, University of Pennsylvania; Robinette Foundation Professor of Cardiovascular Medicine, Professor of Bioengineering, University of Pennsylvania Perelman School of Medicine, 2010

Essentially, this model could be viewed as simulating external loads on a cell, which could alter the position of the center of the structure. This is a reasonable simulation of some of the cytoskeletal properties. The rigidity of the material could be an analog for, say, covalent bonds within and between intermediate proteins, which actually are rigid polymers of protein and required to remain rigid while bound to membrane components. So, it's quite reasonable to have that as a property of the filament as a positioning mechanism which allows the rest of the simulation to be interpreted, or the model to be interpreted alongside the simulation.

Newness
Ronnie Parsons
Invited critic and Partner and Director of Learning Innovation at Mode Lab; Visiting Assistant Professor, Pratt Institute, New York, 2010

One of the issues that I'm picking up on that was present in many of the projects is related to representation. In the early diagrams—or tracings, as you call them—there is a newness present, but when you fall back on the techniques that you are more comfortable with, the terms of this new type of representation are lost. So part of the challenge is not only the simulation and testing of materials, but looking at radical ways to represent the work. You must invent new ways of representing these ideas. That kind of funky wildness that is present in the earlier tracings is where you'd probably find a better connection to the physical models or prototypes.

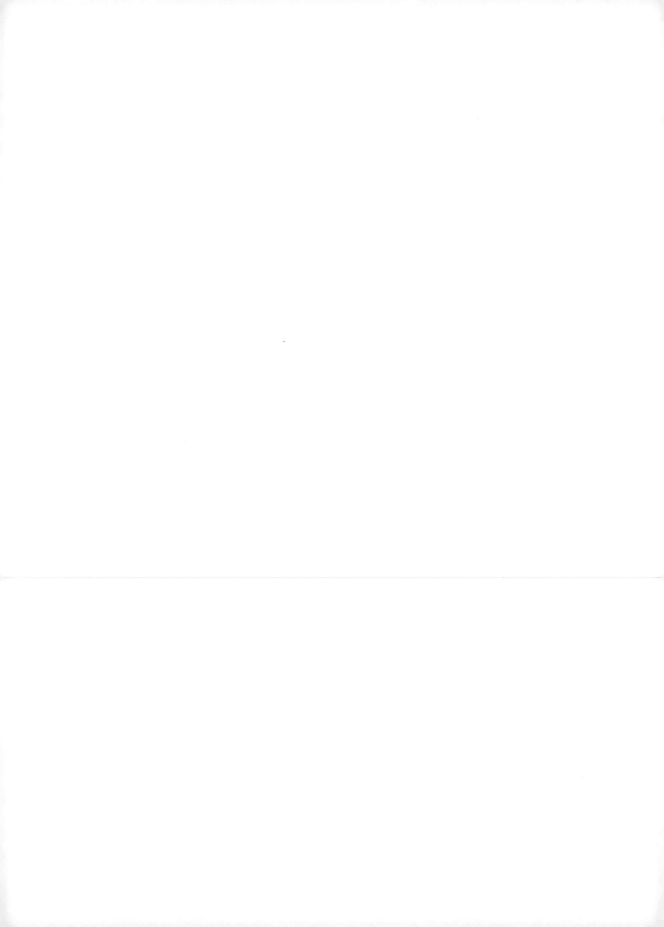

5

Surface Design: The Mammary Gland as a Model of Architectural Connectivity[1]

Jenny E. Sabin and Peter Lloyd Jones

To begin to explore models in complex physical material systems structured as shell and spatial structures and under the topic of surface design, we have used the human mammary gland as a model system.

The physiologic function of the normal adult mammary gland is to produce milk upon demand, and to cease this process following weaning. Accordingly, the mammary gland must expand, differentiate, and then regress in response to its global and local environment. Indeed, at puberty, a rudimentary duct or tube made from epithelial cells transforms into a fractal, tree-like structure, and with pregnancy, the structure of the mammary gland dramatically changes once more. In this instance, a specialized surface membrane structure, made of ECM proteins is produced, and this matrix interacts with adjacent cells, and multiple soluble factors including lactogenic hormones, to promote a massive expansion of the ductal tree. Following birth, milk must be secreted from the ductal cells into a central hollow luminal space; creation of this space occurs via a cellular suicide program within a sub-population of ductal cells that no longer remain in contact with the extracellular matrix. Thus, by controlling cell growth, differentiation and survival, the ECM gives rise to boundaries and space, resulting in extraordinary overall form. This dependence on the ECM environment for normal breast structure and function explains why isolated mammary epithelial cells cultivated on hard, 2-D, chemically-inert, surfaces fail to differentiate, even though they possess the appropriate genes that allow them to do this. When cultivated within a 3-D extracellular matrix, however, ductal cells undergo a normal morphogenetic process. In contrast, with cancer, the integrity and quality of the ECM change, resulting in inappropriate growth responses to this modified matrix environment. Collectively, these and other events lead to the loss of normal tissue architecture, a cardinal feature, and in fact a driving force in breast cancer. Clearly, modeling the behavior of tissues in 3-D represents an important step in understanding their behavior in development and disease.

One of the crucial lessons arising from the above example is that in order for the model to reproduce relevant characteristics of the system being studied, the model has to share similar complexities and constraints of the original system. The design of these constraints, based on logic, intuition and experience,

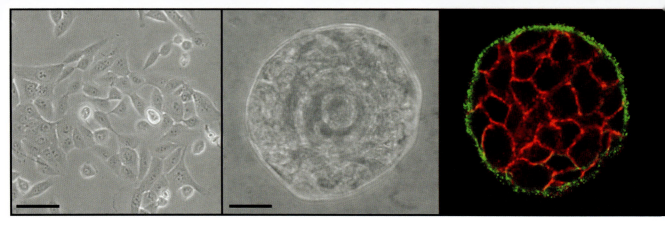

Figure 5.1 This dependence on the extracellular matrix environment for normal breast structure and function explains why isolated mammary epithelial cells cultivated on hard, 2-D, chemically-inert, surfaces fail to achieve a normal form, even though they possess the appropriate genes that should enable them to do this (left panel). When cultivated within a compliant, 3-D extracellular matrix "fabric" within a tissue culture dish, however, ductal cells can be induced to undergo a normal morphogenetic process (middle and right panels).

becomes the nuanced role of the scientist. To approach this, we are studying the interaction of human mammary epithelial cells with the ECM components, laminin and tenascin-C (TN-C). Normal cells rest on a layer of laminin, whereas the cells surrounding breast cells produce tenascin-C (TN-C). Importantly, we have shown that tenascin-C not only alters 3-D tissue architecture, but that it may actively promote the formation of structures and functions associated with breast cancer, including an ability to induce the expression of cancer-associated genes (i.e. oncogenes). To reach this conclusion, our studies made use of a complex 3-D *in vitro* model of breast morphogenesis in which normal cells are cultured within a laminin-enriched gel-like matrix, either with or without the presence of exogenously added tenascin-C protein. Tenascin-C (TN-C) was included in the mix because this protein is produced in ever-increasing quantities around putative and actual breast cancer cells. At the experimental level, whereas control cells formed polarized aggregated structures—designated acini—that to a large extent mimic normal breast tissue structure, complete with a continuous basement membrane and a central lumen (resulting from programmed, site-specific, cell suicide), exposure to TN-C provokes selective loss of the basement membrane and increased epithelial cell growth. To further determine the magnitude and to reveal more detail regarding these changes, we developed an imaging algorithm to generate 3-D renditions of mammary acini, which were then used to assess and quantify acinar topography and volume. Although TN-C increased acinar surface roughness, it had no effect on acinar volume. Based on these results, we hypothesized and thereafter showed that TN-C promotes epithelial cell proliferation within the luminal space. This finding is important, because luminal space filling is a feature of certain forms of breast cancer.

It is likely, however, that additional information regarding the relationships between mammary epithelial cells in acini and their ECM environment probably exists in our model system, yet the current tools and approaches have hampered this trajectory. Thus, finding these new relationships will likely rely upon other techniques. Moreover, abstraction of mammary acini structures is difficult using conventional cell biological computational tools. To resolve this, we have initiated a project to investigate new concepts and techniques that allow further examination

Surface Design 159

Figure 5.2 Our studies make use of a complex 3-D model of breast morphogenesis and cancer in which cells are cultured in a 3-D gel matrix, either with or without the presence of tenascin-C protein.

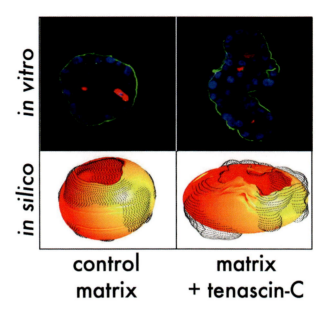

of part-to-whole relationships in the normal mammary epithelium, and in one that is exposed to tenascin-C. It is hoped that these investigations will lead to novel scientific hypotheses and architectural designs.

Overall, the surface design project seeks to quantify and spatialize mammary epithelial tissue contour information through the design of its surface architecture in spherical and elliptical space. Here, the study of relationships found within the closed and open structure of *in vitro* generated tissues gives rise to an abstract understanding of form as it relates to dynamic boundary conditions and biomechanics. The project entails the reconstruction and mapping of the mammary epithelium from a sequence of 2-D images into abstractions that describe the interior composition of individual acini. Via this approach, the architect is afforded new insights into the shifting relationships between surface structure and deeper interior structural concerns. This in turn, may contribute to novel designs of shell and spatial structures that are not only responsive at the level of their surface structure, but also at their deep interior structural cores.

Geometries of Change

In this surface design project, which was investigated by graduate architecture student, Wei Wang, in the first installment of LabStudio during the summer of 2007, we are working with physical and digital algorithms in three different trajectories. The first makes use of digital algorithms: Delaunay Tessellation and the Voronoi diagram. A second trajectory incorporates deployable structures as a testing ground for programmatic information gleaned from the biological model of the study. In the third study, apoptosis, or a programmed cell death, is considered as a modeling system to study the formation and maintenance of luminal space in the normal human mammary gland.

To begin, the Delaunay Tessellation is described in mathematics as the dual tessellation of the Voronoi diagram. In mathematical terms, a Voronoi diagram is a geometric structure or graph that represents proximity information about a set of objects or points. In our case, we are using it as a filter where cellular data are used as input. Given a collection of points or objects and in our case, cells, a given plane is partitioned by assigning to each point its nearest site. Together, these points form the Voronoi diagram. Here, the points on the Voronoi diagram are equidistant to two or more sites. As defined by Boris Nikolaevich Delaunay in 1934, the definition of the Delaunay triangulation of a point set is a collection of edges satisfying what is known as an "empty circle" property: for each edge we can find a circle containing the edge's endpoints but not containing any other points. The Delaunay triangulation is the dual structure of the Voronoi diagram. In mathematics, to dual means to draw a line segment between two Voronoi vertices if their Voronoi polygons share a common edge. This is not all that dissimilar from Le Ricolais' investigations into bimorphism, or the combination of a form with its dual. The planar image of a 3-D structure can be found by graphically representing the forces in its members, once the reactions at boundaries have been determined. This planar image is a structure's dual.

Early modeling investigations for the first trajectory include the analysis of several image Z-stacks derived via confocal microscopy. The first series of 63 images show how cells are distributed in space, based on relationships between an external surface of the acini called the basement membrane, the nuclei which houses the nucleus and the majority of the genome, and the formation of an internal luminal void.

The geometrical centroids of each nucleus are used as reference points to regenerate the structure into a 3-D parametric mesh model. The basement membrane and the boundary of the inner void are traced in a 3-D modeling program. Color channels are used to select pixel-based information from the original image sections. This pixel-based data is refined and sharpened through a series of filters that make adjustments in lightness, contrast, and noise. The processing of pixel-based data enables a more accurate description of the location of the centroid of each nuclei and a clearer definition of the boundary condition.

All 63 sections are mapped and reconstructed, using both the Delaunay and Voronoi meshes. The individual sections of traced points and curves are synthesized by linear stacking with a deviation of 1 micrometer. The first mesh is based on the Delaunay algorithm. This structure models the linear connectivity within acini. Connections are made between acini through shortest path or nearest neighbor information. The second mesh model depicts geometric information filtered by the Voronoi algorithm. This filter visualizes moments of equilibrium found at boundaries under variable pressures. A third algorithm is used which is

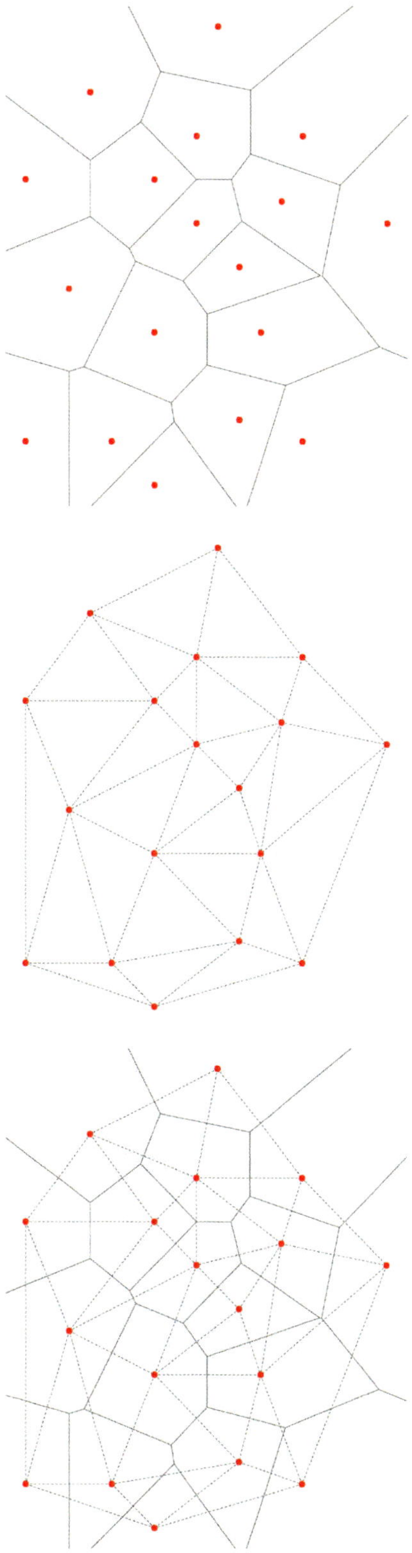

Surface Design **161**

Figure 5.3 Algorithmic tools assist with the filtering and extraction of geometric and site-based information.

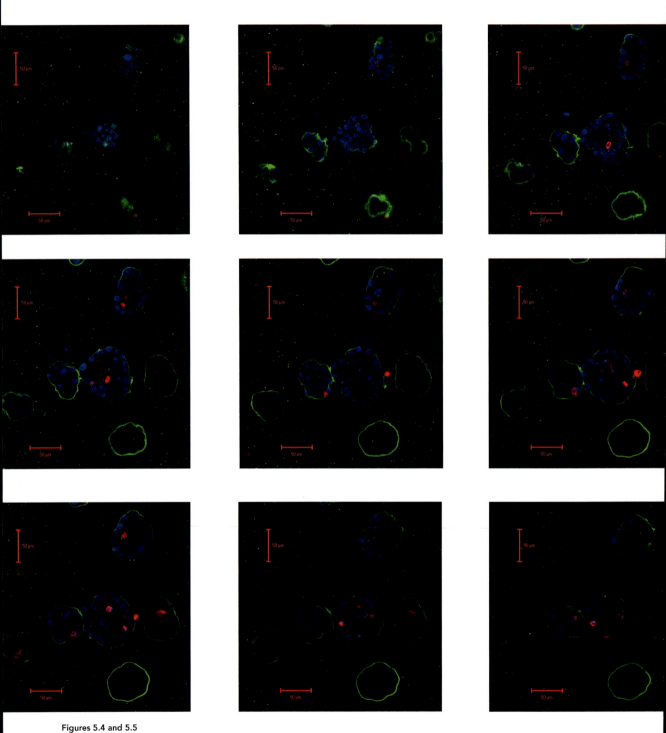

Figures 5.4 and 5.5
The first series of 63 images show how cells are stacked and distributed based on relationships between an external membrane called a basement membrane (shown in green) and the formation of an internal luminal void. Color channels are used to select pixel-based information from the original image sections.

Surface Design 163

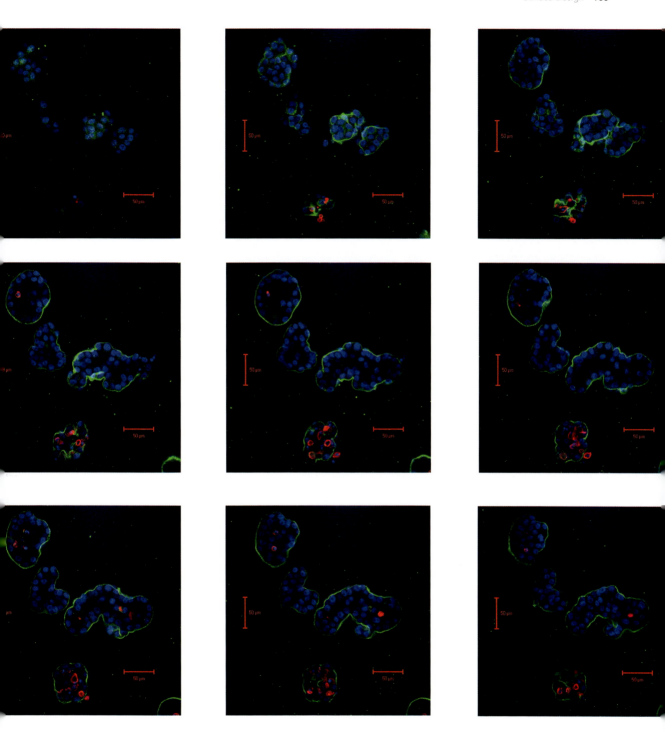

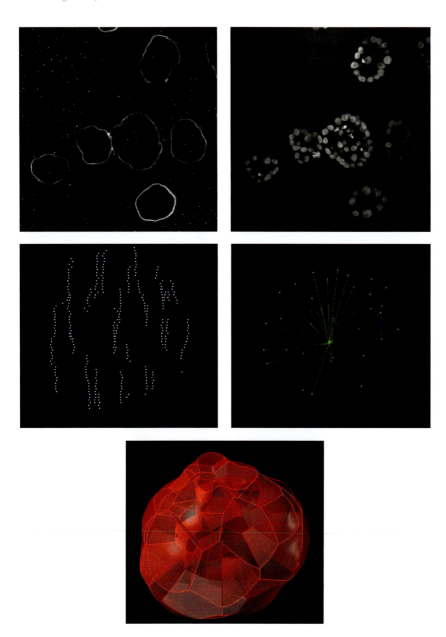

Figures 5.6–5.10 The individual sections of traced points are synthesized by linear stacking with a deviation of 1 micrometer. The blue mesh is based on the Delaunay algorithm. This mesh structure models the linear connectivity within acini. The red mesh model highlight geometric information filtered by the 3-D Voronoi algorithm.

based on the same geometric filter called the Voronoi diagram, but this final step uses a 3-D Voronoi diagram. This allows for the modeling of the 3-D space and structure of equal balanced boundaries between point sets. The algorithm follows the same logic as the 2-D Voronoi, but nearest neighbor paths shift from straight-line connections to equi-distant planes. The red enclosure models the boundaries of cells within the tissue.

In the aforementioned system, we are interested in how the Delaunay and Voronoi filters may aid in the modeling and description of the acinar formation, in terms of minimized energy and dynamics. Here, parametric relationships describe the distribution of forces throughout the structure. For example, a tension deferential exerted upon the external surface tells us about the logic and formation of interior structures. This in turn, provides new insight into the novel design of abstract shell and spatial structures composed of complex surfaces with dynamic interior structures. Subtle adjustments made to interior structures adjust membrane behavior and performance and vice versa.

In the more advanced modeling stages, the exterior basement membrane and the inner luminal void are reconstructed in 3-D digital space from the data abstracted from the confocal image stacks. A single layer of smooth nurbs-based surface is generated from the contours traced from the original images. It should be noted that the basement membrane does not actually look like this, rather, it is amorphous, but these tools allow us to precisely map the position of the basement membrane.

The third step is to use the centroid of the acinus as another reference point or resource, the basement membrane and inner surface as boundaries or limits, and to build a crystal-shaped cellular and spatial structure to abstract and better understand the structure of acini. The logic of locating the boundary for neighboring cells is again based on the 3-D Voronoi diagram, within which a flat surface is generated at an equi-distant position from neighboring points. The crystal structure encloses certain points and is the result of a series of calculations based on distance and point-to-point orientation. A single local crystal module is the result of the synthesis of global behavior across the entire structure. The logic that drives the behavior of the surface boundaries activates, limits, and controls the formation of both individual units and the entire shell structure.

In order to gain more information about the formation of the "crystalline" and spatial structural forms, their surface structures and their parametric relationships, two experimental conditions were modeled and compared—case 1 is a normal control, as described above, whereas case 2 represents acini exposed to both laminin and tenascin-C (i.e., an ECM microenvironment that replicates the tumor microenvironment) as described below.

Figures 5.11 and 5.12 The next step is to reconstruct the surface of the exterior basement membrane and the inner luminal void.

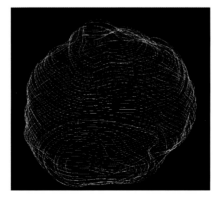

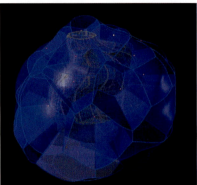

Figures 5.13–5.15 A single layer of smooth nurbs-based surface is generated from the contours traced from the original images. The third step is to utilize the centroid as a resource, the basement membrane and inner surface as boundaries or limits and to build a crystal-shaped cellular structure to abstract formation of acini. The logic of locating and forming the boundary for neighboring cells is again based on the 3-D Voronoi diagram, within which a flat surface is generated in an equal distant position from neighboring points.

The first comparison focused on surface roughness, which increases in acinar cultures exposed to TN-C. Surface roughness is calculated by the slope of the tangent plane on specific points on the surface. The range of roughness is visualized through a color spectrum. Red indicates roughness and blue indicates a smoother surface. In the case of the control models, the surfaces are relatively smooth, whereas the inclusion of TN-C increases the amount and degree of surface roughness.

The second comparison examines distance, packing behavior, and connectivity between neighboring nuclei, features that are routinely assessed by cancer biologists when evaluating the extent of deviation from a normal mode of behavior. A 3-D Delaunay system that is projected to the outer basement membrane surface is used to frame the local connectivity, and the distances between neighboring nuclei can be measured.

The third comparison focuses upon "surface tension." In terms of structure, the tension of a certain position is determined by the combination of its area and radius. Tension increases as area and radii increase. The distance represents the radius from the centroid of the acini to the centroid of the nucleus. The area is determined by how much of the basement membrane is occupied by an individual cell. Based upon observation, the tension in the controlled case is relatively

Surface Design 167

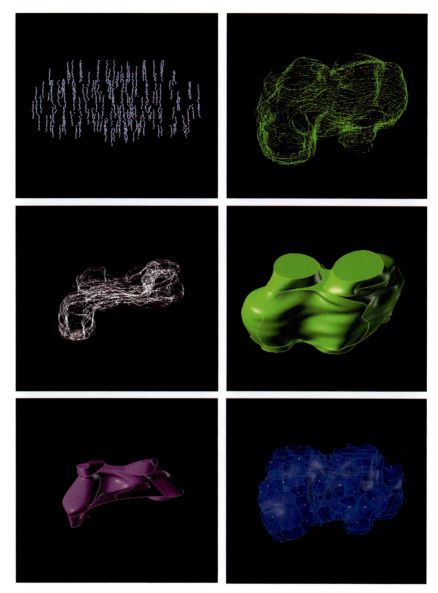

Figures 5.16–5.21 Case 2 with TN-C. Figure 5.16: Stacked nuclei points; Figure 5.17: Stacked contours generated from exterior points; Figure 5.18: Stacked contours generated from interior points; Figure 5.19: Exterior smooth surface; Figure 5.20: Interior smooth surface; Figure 5.21: Voronoi mesh structure, nuclei and interior structure.

evenly distributed, while in the diseased scenario, the tension shifts dramatically. Collectively, these findings may be relevant to breast tumorigenesis in which loss of basement membrane continuity, cell packing, and changes in tensional homeostasis in response to alterations in the ECM and gene expression are known to play a central role. Further, all three studies allow us to envisage potential parallel models where such environmentally impacted crystalline structures may

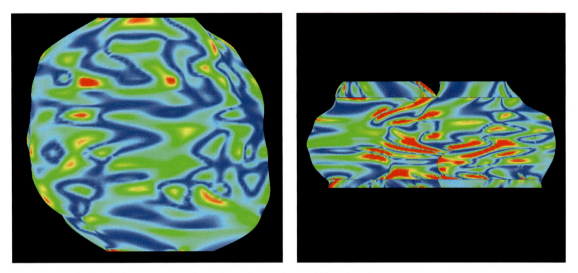

Figures 5.22 and 5.23 The first comparison is focused on surface roughness, which we know increases in breast cells treated with TN-C. Surface roughness is calculated by the slope of the tangent plane on specific points on the surface. The range of roughness is visualized through a color spectrum. Red indicates roughness and blue indicates flatness. Control (Figure 5.22) and TN-C treated (Figure 5.23).

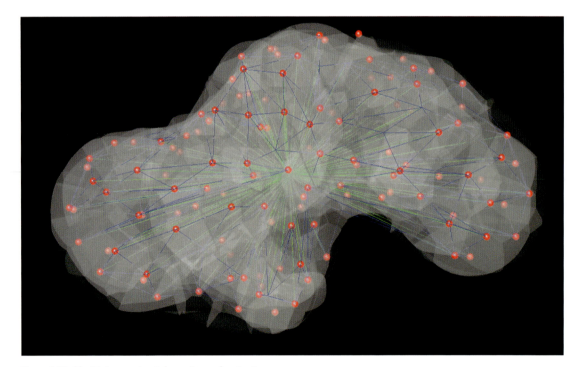

Figure 5.24 The third comparison is focused on surface tension

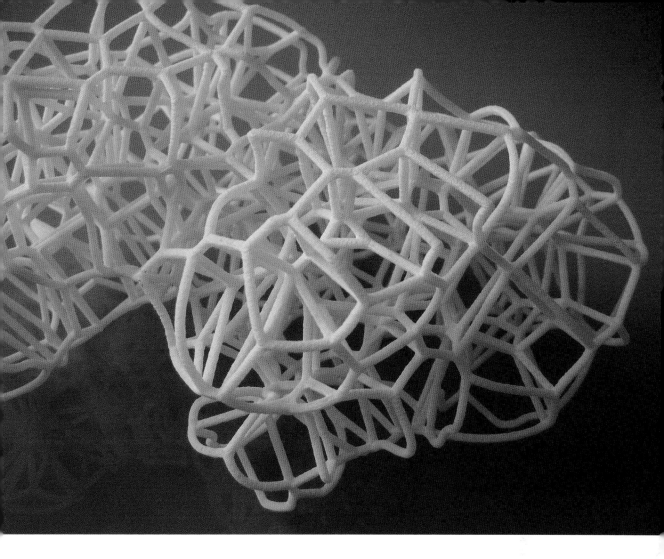

take on new constraints within an architectural context, and at different length scales.

The following research projects continue to build upon this foundation with new questions and scalable constraints. The first explores the behavior of cell-to-cell surface connections as a model for deployable structures.

Note
1   Adapted from Jenny Sabin and Peter Lloyd Jones, "Nonlinear Systems Biology and Design: Surface Design," in *Proceedings of the 28th Annual Conference of Acadia 2008: Silicon Skin, Biological Processes and Computation* (2008), ed. Andrew Kudless *et al.*, pp. 54–65.

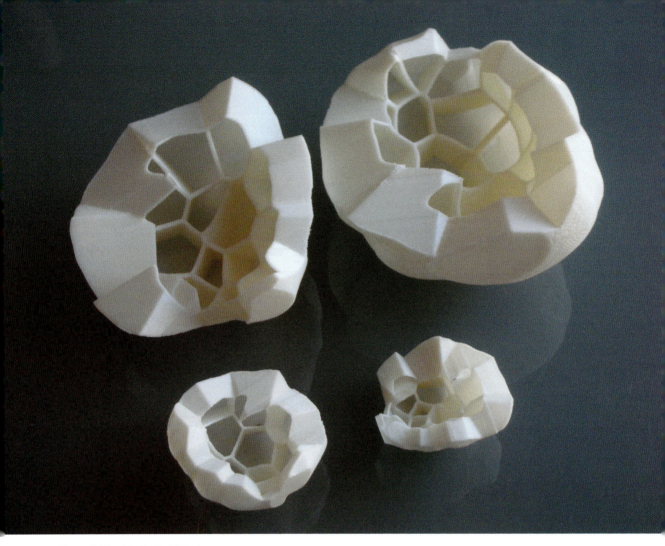

Figures 5.25 and 5.26 Final 3-D prints

Case Study: Adherens Junctions as a Mechanism to Reveal Novel Forms of Structural Deployability
Misako Murata, Austin McIerny, and Wei Wang
With Agne Taraseviciuete

Nonlinear Biosynthesis
Instructors: Jenny E. Sabin and Peter Lloyd Jones
Sabin+Jones LabStudio

Independent Study
Misako Murata

In the second trajectory of the surface design project, deployable structures are incorporated as a testing ground to better study junctions between cell surfaces. Deployable structures are composed of three key elements: structure, mechanisms,

Figure 5.27 Adherens junctions

Cell to Cell adhesion| Basal lateral polarity + understanding direction

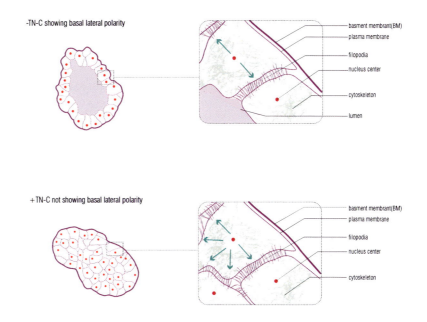

and the programming of such mechanisms. In our case, the information programmed and transmitted through the specified mechanisms comes directly from the biological model being studied.

Adherens junctions are specialized forms of adhesive contacts important for tissue organization in developing and adult organisms, including construction and maintenance of the normal, adult mammary epithelium. Cadherins, a major component of adherens junctions, form protein complexes with cytoplasmic proteins that convert the binding capacity of the extracellular domain into stable cell adhesion between adjacent cells and their surrounding extracellular matrix. The extracellular and intracellular domains of cadherins provide cytoskeletal anchorage between cells, coupling cytoskeletal force generation to strongly adhere to sites on the cell surface and affect the regulation of intracellular signaling events. With breast cancer, however, the stability of these junctions is compromised. In fact, loss of cadherin-based junctions has been shown to contribute to tumor formation. Mechanistically, this relies upon the release of cadherin-associated molecules into the cell interior, where their function is transformed to one that regulates the expression of genes in the nucleus that transform normal cell behavior into an aberrant form. Since tenascin-C affects cell-cell junctions at the level of actin cytoskeleton, we aimed—at the design level—to use this understanding in order to understand how this structural and functional cellular component changes between adjacent cells in control and tenascin-C-treated 3-D organotypic normal, mammary epithelial cell cultures, in the hope that these studies might reveal novel modes of structural deployability.

Filopodia are thin projections from a cell's cyctoplasmic edge containing actin filaments. Central to cell-to-cell adhesion, recent research has found that at coincident membrane sites, filopodia reach and penetrate into adjacent cells, linking them together. Over time, this causes the actin cytoskeleton to remodel. Adherens proteins are expressed on epithelial cells near the apex of the surface showing basal lateral polarity. In our investigation, one hypothesis deals with the degradation of the basal lateral polarity in TN-C-treated cells. Alternations in actin dynamics in response to changes in a surface condition are highly complex and rely upon the superimposition of solid state and soluble scaffolds and signaling hubs, and so for this project, certain interactions are hypothesized based on available data for analysis. Nevertheless, this research has resulted in new ways to define possible variables that could be manipulated, abstracted, and tested in novel deployable structures.

To further reveal the relationships between cell-basement membrane connections and cell-cell interactions that emanate from changes in the former, points of intersection where the cells touch the basement membrane, were extracted and abstracted to visualize the spatial relationships between these points of contact. When applying the deployable structure model, the range, quality, and duration of each deployable performance vary within the normal and diseased contexts. Further, by programming common 2-D and 3-D scissor mechanisms with this biological information, we are able to generate dynamic and differentiated deployable structures. Here, the geometry of the structure transforms from a strictly abstract and predictable state to one that acquires variability and novel response mechanisms to shifts in environmental input. Figures 5.28 and 5.29 are two images of the described deployable structures.

The packing density in a typical 2-D deployable structure may be changed either at the level of the distribution of points within the model being studied, or via manipulating the length of the lines that form connections between neighboring points. Both parameters contribute to the formation and duration of a deployable connection, but they also reference different packing performances. In order to calculate the effective range of both parameters and the length of the basement membrane contour, we built these connection models for every acini section. Here, subtle adjustments to the exterior "basement membrane" change the degree and extent of each local deployment, thus affecting the overall global behavior of the deployable structure.

The final stage of this study involved the scalable reconstruction of the embedded biological behavior in deployable systems (described above) to a series of scaled physical models and finally to a pavilion prototype. The rapid manufacturing of a skin structure composed of water-jet cut aluminum flaps is married with the intricate design and fabrication of steel struts and mechanisms composed of hinges

Surface Design **173**

Figures 5.28 and 5.29
Deployable structures

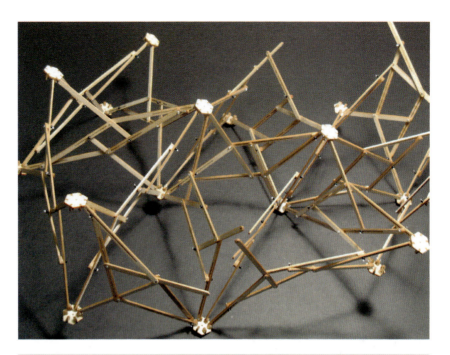

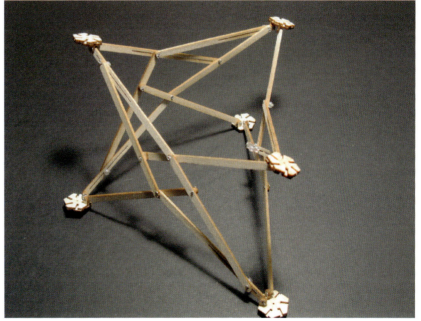

and pins. This deployable system combines two types of connections that allow for a locally responsive deployment. The scissor joint, the first connection type, has one degree of movement. The opposing top and bottom joints are connected by one strut, which forces a simultaneous movement between every top and bottom

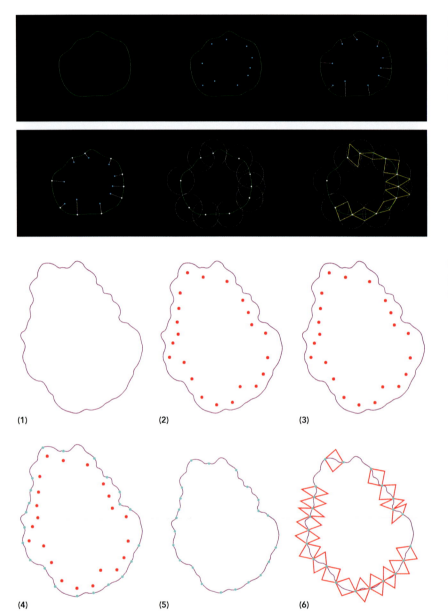

Figures 5.30 and 5.31 Programming of the 2-D and 3-D mechanisms with information from the model of study: the human mammary gland

Figure 5.32 Sequence programming with 2-D scissor mechanisms

connection in the system. A structure built with purely scissor joints is limited to global deployment. The half-scissor joint, the second connection type, has one degree of movement but the top and bottom joints do not need to deploy at the same rate. A structure built with purely half-scissor joints only deploys at the point of stimulus. The overall system has no structure to allow for the top and bottom layers to respond to each other. This type of structure would in fact collapse without any exterior frame or support. Through combining these two types of connections, the resultant deployable system has a clear structure, connecting the top and lower

Figure 5.33 Acini cell organization

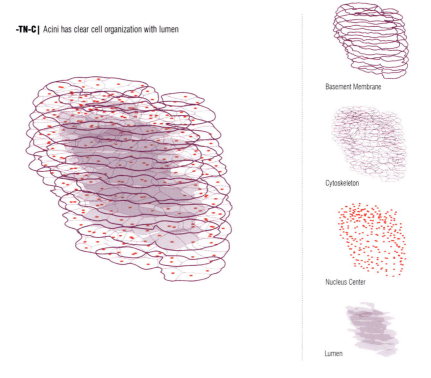

levels, which is able to be used at a specific, local point without causing any general, global deployment.

The surface for this structure is an adaptation of Ron Resch's triangular deployable surface. This particular surface was selected because the deployment pattern is similar to the 2-D hexagonal deployable structure. The hexagonal geometry for the underlying structure features symmetrical, circular deployment in the lateral direction. The Ron Resch surface was redesigned to connect to the underlying structure with these hexagonal joints. A hexagon was placed in the center of the larger triangular pattern within the Ron Resch pattern. The new hexagon was then cut into one large equilateral triangle in the center surrounded by three isosceles triangles. These series of triangles create a point of connection to the base structure while allowing flexibility in the surface for deployment.

In conclusion, a deployable structure is understood through its surface packing and its part-to-part connections. These two readings of deployable structures are applied to two main points of connections within the biological model of study. These are: (1) the cell to stromal extracellular matrix (ECM) connection; and (2) the cell-to-cell adhesion. During the morphogenetic process in breast cell duct formation, the stromal extracellular matrix (ECM) responds and reacts to various stimuli from the external and internal environments, continuously transforming the surface depending on site-specific conditions. The internal cells are connected to the ECM and respond to these changes by creating a continuous

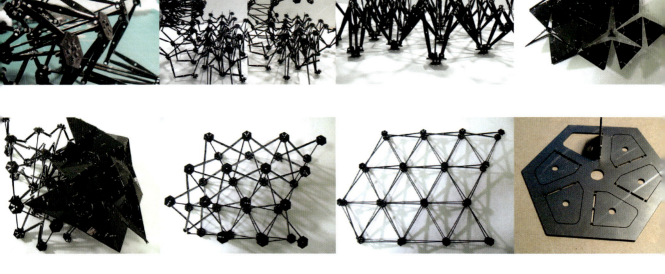

Figure 5.34 The final stage of this study involved the scalable reconstruction of the embedded biological behavior in deployable systems to a series of scaled physical models and finally to a pavilion prototype.

Figures 5.35–5.37 The rapid manufacturing of a skin structure composed of water-jet cut aluminum flaps is married with the intricate design and fabrication of steel struts and mechanisms composed of hinges and pins.

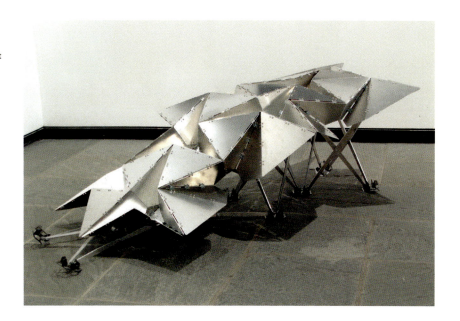

Figure 5.38 Information gained from studying geometry and matter at the cell and tissue level is embedded in the final assembled prototype alongside architectural constraints dealing with issues of scale, material, thickness, and fabrication.

feedback loop. This response system is explored through abstracting the packing behavior and cell-cell relationships of two different cases: (1) tenascin-C-untreated breast cells; and (2) tenascin-C-treated breast cells. Based on data extracted from existing microscopic images, several iterations of responsive systems are tested. The connection of parts to create various components and the organization and distribution of joints are tested through physical models to design how forces are transmitted, which, in turn, affect the surface. Information gained from studying geometry and matter at the cell and tissue level is embedded analogically in the final assembled prototype alongside architectural constraints dealing with issues of scale, material, thickness, and fabrication. The final prototype deploys locally and along nonlinear paths leading to a differential and local experience at a human scale.

Case Study: Self-Organization of Tissue Scaffolds and Formation of Dynamic Surfaces
JaeYoung Lee, Shou Zhang and Jae-Won Shin

Nonlinear Systems Biology and Design
Instructors: Jenny E. Sabin and Peter Lloyd Jones
Sabin+Jones LabStudio

Studies of biological systems show that an organism is developed from a single stem cell characterized by its ability to give rise to different types of cells in a self-regenerative manner ("self-renewal"). During embryonic development, this single cell undergoes a series of divisions to produce progenies, usually containing identical genetic codes. However, it is well established that cells produced from a single parent cell change in terms of their phenotypes and how their genetic codes are read—eventually leading to the hundreds of cell types identified to date.

It is also striking that during development, cells can self-organize to form tissues and organs, observed within a given species and even across different species. This phenomenon can be replicated *in vitro*, where the 3-D structure of mammary acini can be reproduced by seeding mammary epithelial cells into a gel-based 3-D culture system.[1] Since the discovery of DNA as a 'blueprint' of life, it was traditionally believed that encoded in DNA are both the information required for an organism to undergo development and how that genetic material should be read. However, it is now accepted that cells respond to signals from their environment via cell surface receptor proteins. This discovery broadened our understanding of both genetic codes (intrinsic) and environmental factors (extrinsic) that can contribute to how cells are organized. In order to better understand the contribution of extrinsic factors to cell self-organization into tissue scaffolds, we based our study on previous observations made with tenascin-C (TN-C).

As previously described, TN-C is a stromally-derived extracellular matrix (ECM) glycoprotein that contributes to breast cancer by altering epithelial cell interactions with their underlying basement membrane. At a functional level, TN-C binds to specific cell surface receptors on the epithelium to influence their morphology within 3-D acini, resulting in increases in cell proliferation and luminal filling, which are cardinal features of breast cancer. A combined study between biology and architecture by LabStudio investigated the 3-D structure of mammary acini *in vitro* upon treatment of cells with TN-C, and demonstrated that while the surface roughness of the acini increases, the volume remains the same. Given the increased cell proliferation upon TN-C treatment, this led us to the hypothesis that TN-C treatment increases the intercellular pressure of mammary acini, thereby altering acini surface structure while cancer progresses. This alteration is based on previous studies that cell cytoskeletons are coupled to ECM outside the cell via cell surface receptors, and that receptor activation via ECM leads to receptor clustering, thereby affecting cytoskeletal structures and cell surface properties.[2]

More fundamentally, we seek to understand conceptually how a given biological tissue scaffold can be generated based on rules of interactions between cells and ECM. A number of previous studies have identified specific peptide sequences with which a given receptor interacts.[3] Interestingly, previous studies by Rothemund demonstrate that by using the simple rules of complementary DNA interactions (Adenine bonds with Thymine, Guanine with Cytosine), it is possible to fold any arbitrary 2-D shape from a single-stranded DNA with a set of short oligonucleotides that hold the scaffold in place.[4] Here, we demonstrate that any arbitrary 2-D shape of a tissue scaffold can be self-organized from a single or branched strand of ECM binding sites upon their interactions with cells. We also demonstrate that 2-D tissue folding can be extended into 3-D.

To investigate the fundamental relationships between cells and their adjacent ECM in forming a tissue scaffold, we developed a pipeline of investigations to understand and model two phenomena in parallel:

(1) interactions between ECM and cells that can dictate the formation of an overall tissue scaffold;
(2) alterations in scaffold dynamics as governed by cell packing behavior and receptor clustering.

We developed a conceptual system whereby cells and their receptor binding sites in ECM interact to give rise to a self-organized structure in two dimensions. In order to achieve this, a basic unit of interacting microenvironment or niche was designed. Inspired by Delaunay triangulation, owing to its ubiquity in generating a structured mesh, we adopted a triangular niche as a basic tissue-folding unit. As a starting point, a cell was abstracted as a 2-D circle with three distinct receptors separated by 120 degrees; a receptor arm is formed by connecting between a receptor and the center of a circle. An ECM unit was abstracted as a line that contains a distinct binding site at its midpoint. The triangular niche unit is defined as a cell surrounded by three ECM units that are tangential to the cell.

Since each binding site is unique, it is possible to introduce chirality for the arrangement of binding sites: the same set of three binding sites can be arranged differently, either clockwise or counter-clockwise (see Figure 5.39). In addition, we introduced directionality into each binding site, where a cell can bind only in a

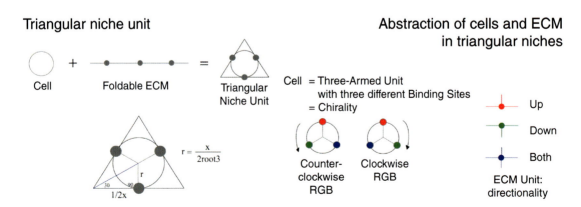

Figure 5.39 Triangular niche unit (left) consisting of a cell with three receptor binding sites separated by 120 degrees and three ECM units that contain distinct binding specificity (right). Chirality can be introduced into a cell and directionality of binding can be introduced into an ECM unit.

specific direction. Therefore, a specific arrangement of binding sites as a continuous line, when mixed with cells with specific chirality, can lead to self-organization into a desired shape as shown in Figure 5.39. The entire 2-D tissue folding system was simulated using Processing 1.0. This system can be extensible by varying the following inputs: length of binding lines, radius of an individual receptor arm in a cell, angle between receptor arms, and number of receptor arms in a cell.

We sought to understand if the 2-D tissue folding system can be extended into 3-D. We discovered that while a regular tetrahedron cannot be folded from a single straight line, the regular tetrahedron can be divided into a di-tetrahedron by adding one extra vertex at the midpoint of any edge; this di-tetrahedron can be folded from a single line (see Figure 5.40). This could be formalized by adapting the geometrical relationship in an Eulerian circuit to create a single bounding strap for a framework. It pinpoints conditions at which a given geometry can be drawn with or without lifting a pen. One necessary condition for a complete Eulerian circuit is that there can only be two vertices with an odd number of converging edges; when this number is greater than two, geometry cannot be drawn with a single stroke.[5] This becomes apparent by counting the number of odd-degree vertices in a tetrahedron versus a di-tetrahedron. In addition, by understanding the geometrical relationship that determines the length of each edge, we could unfold the geometry into a singular flat strap and expand the system. As long as the connection is kept, the length of each bounding wire can be varied, leading the entire system to deform.

In essence, a specific ECM pattern forms the base topology of the model, while each cell in 2-D is revolved into a 3-D sphere. Each tip of the sphere is connected to form a surface deformable by moving the center of a cell cluster. The variables in the model are: closeness of surface to each sphere, degree of attraction

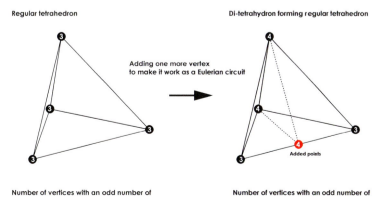

Figure 5.40 Transformation of a tetrahedron into a di-tetrahedron as a unit of a single strand folding in three dimensions

pulling the centroid of the surface closer to each sphere, and density of a point grid on the surface controlling the precision level of the surface. Since this model is based on movable points, the point list is variable and could possibly be expanded to a larger system. The model was implemented with GenerativeComponents (GC).

Lateral deformation of a given cell cluster surface began with a simulation of the ECM controlling the cell membrane. We used a geometry formed by the assembly of either a regular triangle or tetrahedron as an initial input. As we move the points on the tetrahedron, the scale of the cell is changed. After that, we linked a series of spheres through their apogees and nadirs, lastly resolving a three-dimensional continuous surface. Considerations include when the cells move and how the distance between two critical points on the surface will be changed; Delaunay triangles control the movement of the cells. When the distance between two cells is increased, the link between the two apogees lengthens, then we can connect the two points with one triangle by adding the third point in the middle of the two apogees.

Figure 5.41 (a) shows that a single line consisting of a specific sequence of ECM units can be folded into two triangles if they are mixed with two counter-clockwise oriented cells of receptors for ECM binding sites. Although folding a single line may represent an economical implementation of self-organization by obviating the need to synthesize branches of ECM, this can be restrictive, as dictated by the Eulerian circuit. Therefore, we also envision that the 2-D system can be extended to fold a branched ECM as shown in Figure 5.41 (b) for the folding of a hexagonal tissue scaffold.

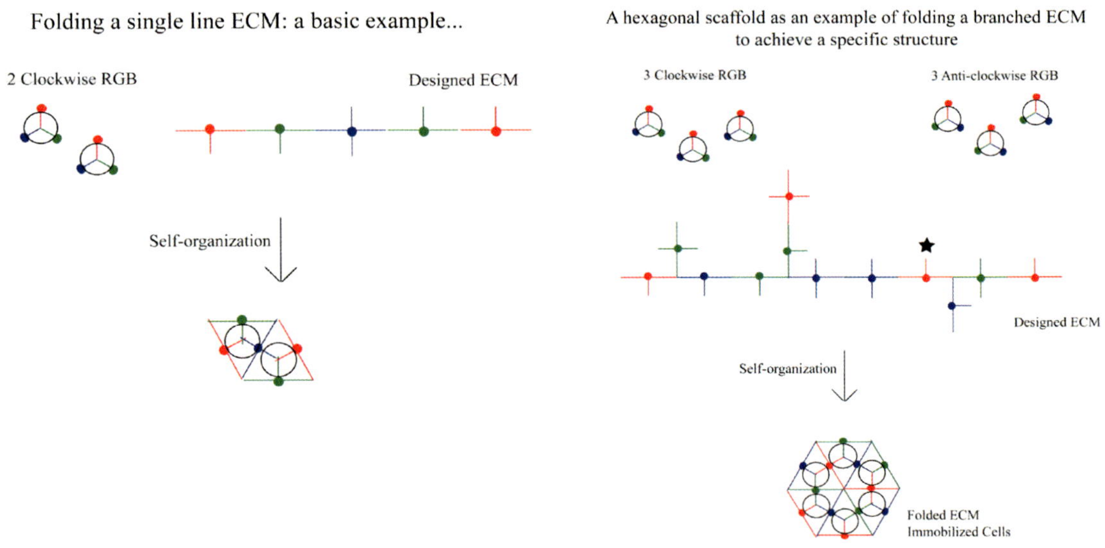

Figure 5.41 2-D tissue folding consisting of triangular niche units for: (a) single-stranded ECM; and (b) branched ECM

After establishing that our system can be used for both unbranched and branched ECM folding, we sought to observe the potential of the system on a mass scale. Both anti-clockwise and clockwise cells were mixed in 1:1 ratio with a single strand made from a repeating ECM sequence. After iterating 1500 frames, the average distance between cells and rotation angles was calculated. Standard deviations of distances and rotation angles were also derived to study the closeness of individual cell values upon self-organization. Individual sites of the repeating ECM sequence were mutated to disrupt their binding ability and the iteration was repeated. Comparison was made between original and mutated sequences using the Student's T-Test. Figure 5.42 shows that the system reliably reflects the effect of mutations on the general folded ECM shape; statistical analysis shows that upon mutation, the average distance of cells increases and rotation angles become more uniform.

To demonstrate how di-tetrahedrons can be used as units to fold a single line into a 3-D shape, the extension of the di-tetrahedron unit was studied. We discovered that as long as the line remains continuous and maintains Eulerian circuit rules, it can be scaled larger, similar to the Sierpinski triangle. Examination of the first degree Sierpinski triangle shows that the number of vertices with an odd number edge is zero, fulfilling the Eulerian circuit rules for a single stroke line. Component deformation can also be achieved by varying the lengths of individual segments to form edges.

We developed a system that allows alterations of the outer-surface based on intercellular interactions. As the average distance between cells becomes larger, the surface topology becomes simpler. This is reminiscent of observations from the 3-D culture of mammary acini; upon treatment of TN-C, the cell count increases, but

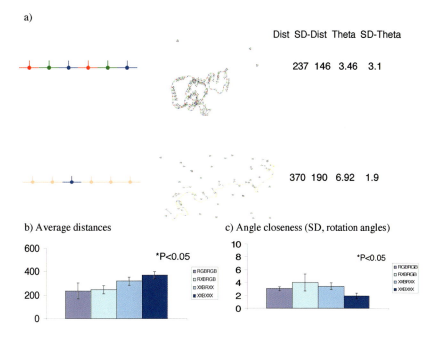

Figure 5.42 Testing the 2-D folding system in mass-scale by mixing 1:1 ratio between clockwise and counter-clockwise cells and the effect of mutation to disrupt binding sites: (a) Mutation of five binding sequences in a repeating ECM (indicated by yellow color) and its effect on the general shape of ECM and quantitative parameters including average and standard deviations of distances of cells and rotation angles; (b) the effect of progressive mutations on average distances; and (c) angle closeness. Each iteration was repeated three times and derived error bars were found. *$p < 0.05$ compared to the un-mutated sequence.

Component Generation
Growth of System
Re-examination
Number of vertices with an odd number of edges : 0 Eulerian circuit exists
Unfolded Bounding Frame

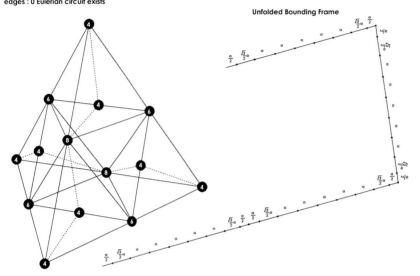

Figure 5.43 Component generation

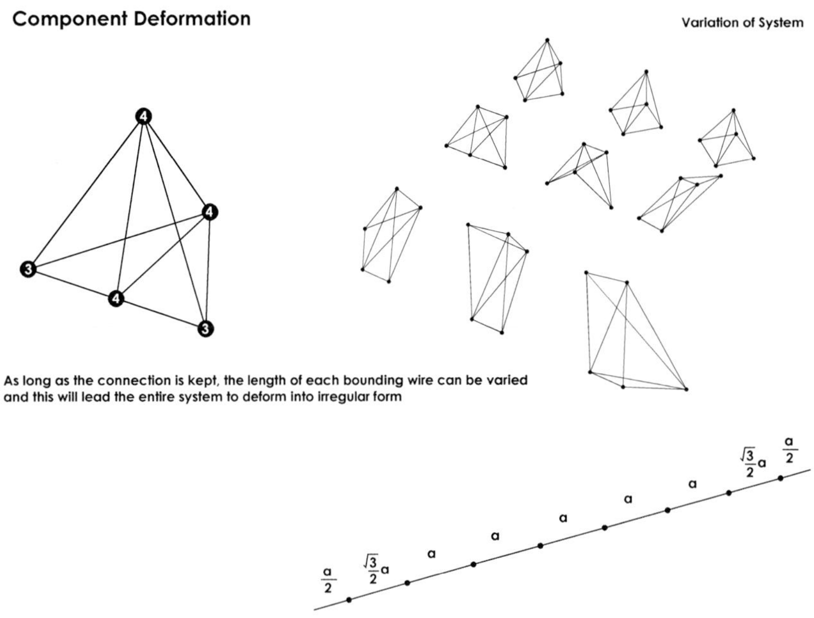

Component Deformation
Variation of System
As long as the connection is kept, the length of each bounding wire can be varied and this will lead the entire system to deform into irregular form

Figure 5.44 Deformation of tetrahedron units to introduce variation

cells are restricted to a confined volume; this increases the complexity of cell surface topology. Figures 5.45 and 5.46 demonstrate lateral deformation with hexagonal surfaces.

Understanding how tissue architecture is formed based on rules of interactions between cells and ECMs will help researchers design an artificial

Surface Design **185**

Figures 5.45 and 5.46 Demonstration using four hexagonal folded units as an initial input to create lateral deformation. As the distance between cells becomes larger, the overall topology becomes simpler.

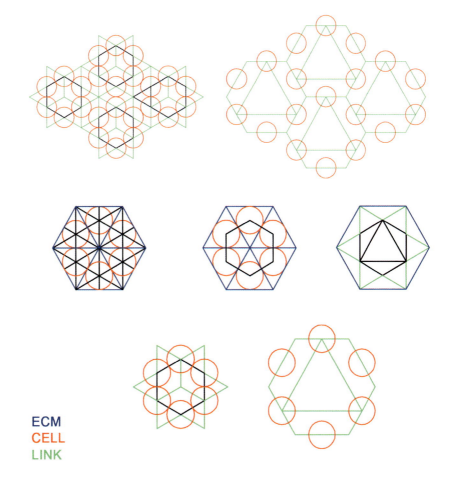

ECM
CELL
LINK

foldable system for the purpose of economical tissue engineering and improved manipulation of cell-cell and cell-ECM interactions. Currently, it is difficult to control interactions among cells in an experimental population on a mass scale. With the system proposed in the present study, this could become more feasible given the current advances in chemical engineering, tissue engineering, and molecular biology for both the generation of artificial ECM fragments with defined binding sites and the expression of specific receptors on cells. This also broadens our understanding of stem cells for their regenerative ability and differentiation potentials in terms of their relationship with their microenvironments.

While the proposed folding system should remain as simple and tractable as possible for the purpose of designing feasible experiments in biology and developing therapeutic strategies, the success of the system in the architecture/design field will depend on whether it can incorporate an evolutionary element to create unexpected variations at a larger scale. Although this remains to be developed further for the proposed system, we demonstrate the possibility that variations of component variables, especially lengths, can be used to introduce larger-scale complexity. Other types of variations, including the introduction of fractals to receptor arms of a cell, could continue to evolve the folding system.

Notes
1. Jenny Sabin and Peter Lloyd Jones, "Nonlinear Systems Biology and Design: Surface Design," in *Proceedings of the 28th Annual Conference of Acadia 2008: Silicon Skin, Biological Processes and Computation*, ed. Andrew Kudless *et al.* (2008), pp. 54–65.
2. Ibid.
3. Bruce Alberts, Alexander Johnson, Julian Lewis, David Morgan, Martin Raff, Keith Roberts, and Peter Walter, *Molecular Biology of the Cell*, 4th ed. (New York: Garland Science, 2002).
4. Paul W.K. Rothemund, "Folding DNA to Create Nanoscale Shapes and Patterns," *Nature* 440(7082) (2006): 297–302.
5. "Eulerian Path," Wikipedia, available at: http://en.wikipedia.org/wiki/Eulerian_path

On Geometry and Cellular Mechanics
Sanford Kwinter
Invited critic and Professor of Theory and Criticism, Pratt Institute, New York, 2008
Peter F. Davies
Invited critic and Director, Institute for Medicine and Engineering, University of Pennsylvania; Robinette Foundation Professor of Cardiovascular Medicine, Professor of Bioengineering, University of Pennsylvania Perelman School of Medicine
Peter Lloyd Jones

> *Kwinter:* I'm going to have to reevaluate my critique because I feel that they took a wrong path really early on and it led them into a bad place. I'll try to articulate it clearly in terms of the question you asked earlier and why it got me scared—did I miss something, did it get past all of our sensors? The suggestion is that both position and orientation, and if you like, the quantitative factors of the morphology of a cellular mass are themselves somehow significant. For example, in your first story, you told us that climatological changes were occurring inside the cellular mass: we had a change in number of cells accompanied by no major change in the bounded space that they were occupying. Normally this suggests changes in pressure and possibly changes in chemistry and possibly changes in all of the catalytic networks, all of which may have been going on with the information matrix. You then ask: "What if we could program, or engineer, the final shape?" and assume that the whole ecology of all of the network biology would somehow be magically reproduced. But it simply wouldn't follow. We've learned that you cannot recreate a forest ecology after a great fire simply by re-introducing, one by one, all of the species that happened previously to be in it because the temporal order is gone and with it the evolutionary-developmental properties. For me, this is the same thing missing in the methods here.

Peter F. Davies: If you open the surface properties of the cell, it's an interface which has a very profound influence upon how the cell behaves.

Kwinter: So by reading the surface you can tell a lot?

Davies: Yes, a lot of it is tied into things we've already discussed, particularly the extracellular matrix that attaches to the outside of the cell at specific sites. The ECM communicates with proteins in the membrane that in turn are connected to the (internal) cytoskeleton. Tension in the cytoskeleton mechanistically drives a lot of the dynamic changes of the surface topography; they're all interlinked. So what intrigues me about this is that you are kind of simplifying it.

Kwinter: They are interlinked. What I am asking is, can you take an actual living cell system and see if those same principles, derived from the triangulation of the points on the surface, actually fit? If you can describe how a stem cell moves from one topography to another by the triangulation, could you then use this geometric simplification as some kind of explanation of a principle or unifying mechanism to further the biology? This argumentation takes one suspiciously back to Egypt or to some sort of Renaissance Quadrivium. It just seems that these shapes belong to Plato's aethereum not to the real world.

Davies: Yes, you can follow the morphometric changes in real time to further the biology. Because they are derived. They're derivatives of derivatives. And that's part of the simplifying process . . .

Kwinter: But you're saying that maybe . . .

Davies: That maybe there are some biological principles there that are applicable to design considerations.

Kwinter: Yeah, just some. Just the parallelogram? Or just the 30-, 60-, 90-, 120-degree angle relationships? You're saying you can actually read . . .

Davies: . . . You only need triangulation to do two-dimensional strength and strain calculations. That's taking a very simplistic non-biological shape, but it's an easier mathematical solution so we can get within the constraints of each triangle. We can calculate the directionality and the magnitude of the strain, for example. Visually, it is well removed from the cell because it's a derived system but it is immensely useful in terms of quantitating, and testing if the quantitation has a biological significance. What I was trying to do is to think about the loop back to the biology as a utility, because I mean that's what the course is about, right? It's a loop from the biology to the modeling and back again.

Kwinter: I would love to have this answered before I leave—it's exactly why I came here today, to listen to the two of you, to hear the story as it currently is best told. Let me use the terms "developmental geometries" or "evolutionary-developmental" geometries or data. Would they not be required in the modeling of a problem like this?

Davies: Spatial cues are absolutely essential for normal development. Peter's an expert on this, not I. I might add it's not just spatial cues, it's chemical cues as well.

Kwinter: That starts adding 4-, 5-, 6-dimensional data into the mix.

Davies: Well, they're there already; however, I don't know that you have to go to those.

Kwinter: You don't?

Peter Lloyd Jones: I mean, just getting back to the neural tube, the fold is created by a series of matrix proteins. But also a fraction of molecules are morphogens that interact with different matrix proteins that form the surface upon which that thing will fold.

Davies: Well, you take away the spatial cues.

Peter Lloyd Jones: Without that, you end up with spina bifida.

Kwinter: I'm not a little brainwashed by my understanding of the history of theoretical biology, to believe that the answers will necessarily lie in a field theory rather than in a simple geometric derivative. I understand that experimentally you've got to isolate the data you're trying to get at and maybe putting it into a classical geometrical framework helps. It's unexpected to me that you guys would consider that as possibly experimentally valid.

Davies: No, no, addressable; validity follows from controlled experiment.

Kwinter: But mostly because you can actually measure and test.

Davies: Yes.

Case Study: Apoptosis: Programmed Cell Death and the Formation of Luminal Space

Pablo Kohan, David Ettinger, and Huishi Li

Nonlinear Systems Biology and Design
Instructors: Jenny E. Sabin and Peter Lloyd Jones
Sabin+Jones LabStudio

This project makes use of parametric and associative design methods to derive a modeling system from elements observed in a cellular apoptotic process.[1] This system works by adjusting distance relationships between points and creating polygons with varying surface areas. Balance exists within the system to control the range in which the overall surface area and volume are able to change.

Normal human mammary epithelial cells (MCF-10A) cultured within a reconstituted extracellular matrix (ECM) form multi-cellular 3-D polarized acini, complete with a central lumen. These cells are enveloped by a continuous endogenous basement membrane. Apoptosis, or programmed cell death, plays a crucial role in the formation and maintenance of luminal space in the mammary gland. The process of apoptosis begins when a cell loses contact with the basement

Figure 5.47 Formation and maintenance of luminal cavity. The third project makes use of parametric and associative design methods to derive a modeling system from elements observed in a cellular apoptotic process.

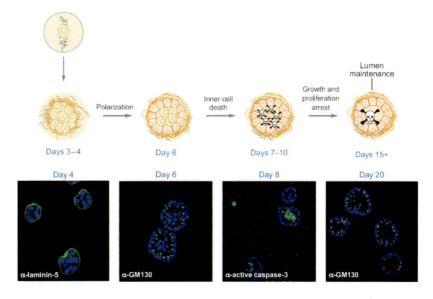

Figure 5.48 Vertical, horizontal, and 2-D deformation of the lattice shown in rendered form

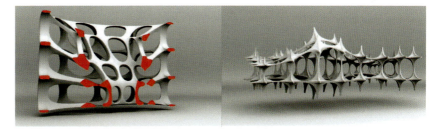

Figure 5.49 Structural lattice component, parametrically modeled

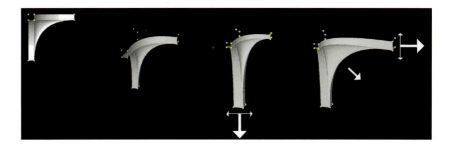

membrane. This disconnect initiates the process of cell suicide. Upon proliferation of new cells within the layer of epithelial cell lining, others are forced off the ECM, beginning the process of cell death. Healthy cells develop during gestation until the layer of epithelial cells reaches a level of stasis in which there is no more cell death or proliferation. In the cancerous condition, proliferation of cells proceeds without the process of apoptosis, eventually filling the luminal cavity. This developmental process was studied to understand the relationships between proliferation and the role of apoptosis in normal human mammary gland health.

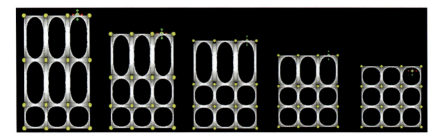

Figure 5.50 Vertical lattice deformations

Figure 5.51 3-D print of deformed lattice. This system works by adjusting distance relationships between points and creating polygons with varying surface areas.

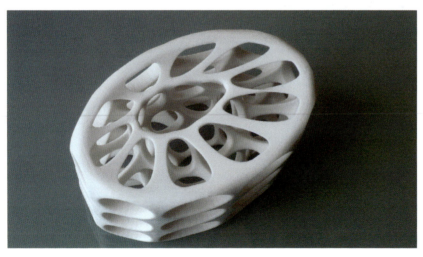

Figure 5.52 3-D print of a cylindrical lattice visualizing the luminal cavity. Apoptosis, or programmed cell death, plays a crucial role in the formation and maintenance of luminal space in the mammary gland.

The first translation of the digital model to physical form resulted in the discovery of structural relevance at the seams created along the lines of collapse. The re-association of the parametric equation from a 2-D surface to a 3-D volume developed what could be a material and structural system that deploys and adapts to specific spatial requirements.

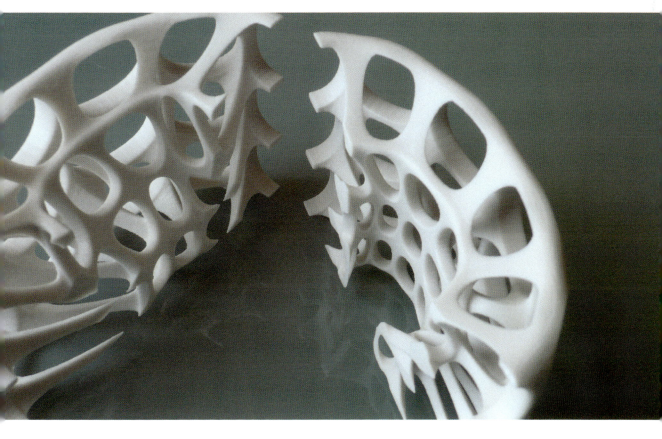

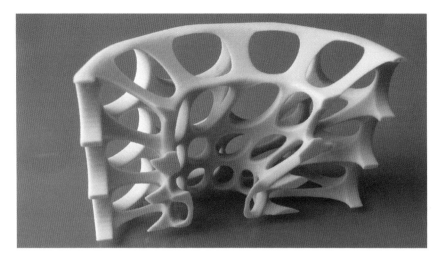

Figures 5.53–5.55 Final 3-D prints. The re-association of the parametric equation from a 2-D surface to a 3-D volume developed what could be a material and structural system that could deploy and adapt to specific spatial requirements.

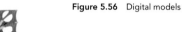

Figure 5.56 Digital models

The relationship between the ECM and the layer of epithelial cells manifests in a double-layered model of gradual difference responding to the luminal remodeling process, which occurs in conjunction with apoptosis.

Note

1   Christy Hebner, Valerie M. Weaver, and Jayanta Debnath, "Modeling Morphogenesis and Oncogenesis in Three-Dimensional Breast Epithelial Cultures," *Annual Review of Pathology: Mechanisms of Disease* (2007): 319, Figure 3.

Case Study: Intercellular Connectivity: Responsive Surface Specificity

Chun Fang, Emaan Farhoud, and Gregory Hurcomb

Nonlinear Systems Biology and Design
Instructors: Jenny E. Sabin and Peter Lloyd Jones
Sabin+Jones LabStudio

In this project, we are interested in exploring the architectural implications of the study of intercellular relationships within the epithelial cells of mammary tissue. Taking inspiration from the structural designer and engineer, Robert Le Ricolais, we

Surface Design 193

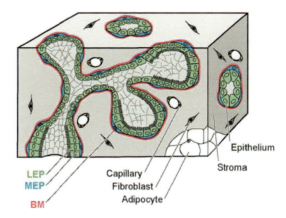

Figure 5.57 Epithelial cells of mammary tissue

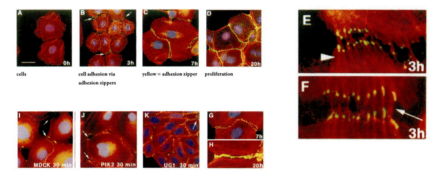

Figure 5.58 Examining cell-cell interactions—through the careful study of their connective processes, along with their relationships to signals from their micro-environment—has implications for architecturally relevant surface topologies

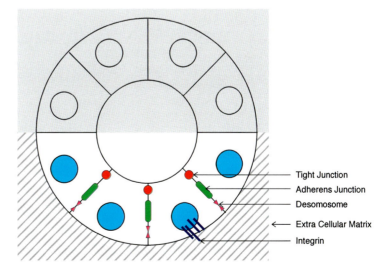

Figure 5.59 A stable cell-cell adhesive process results in normally functioning cells and cell lumen. Attachments between cells and cells to the ECM are mediated through adherens junctions and integrin connections.

investigated the possibility of developing synergetic structural models in which the sum of each assembly of components yields much greater results than that of each individual part. We believe that examining cell-cell interactions through the careful study of their connective processes, along with their relationship to signals from their microenvironment, has the potential for implications in architecturally relevant surface topologies. First, cell adhesion processes and the formation of the lumen in normal polarized cells are studied, in conjunction with the breakdown of the lumen and the ECM (extracellular matrix) through tenascin-C (TN-C)-dependent changes in the basement membrane. Here, we explore pathways between an ordered healthy cell state, with basement membrane, polarity, and lumen intact. This is followed through an intermediate state and finally to a disordered condition where cells move inward. This last condition disrupts the lumen due to the introduction of TN-C and basement membrane distortion. The architectural implications of a switch (TN-C) that alters the exterior surface condition will then affect the number of connections and the quality of connections (thickness or thinness) for the overall model. We also consider in our analysis the relationship between biological process, environments, and their potential manipulations in producing a variety of iterations.

Second, epithelial cells of mammary tissues were examined, specifically cells in normal breast tissue. Some of these cells exist in a polarized state and form a lumen internal to themselves due to a healthy basement membrane and functioning cell-to-cell interactions in the form of adherens junctions, desmosomes, and tight junctions (connecting cell-to-cell), as well as integrins connecting the cell to basement membrane. We then compare these cells to lumen breakdown and luminal filling due to the introduction of TN-C into the system. Of interest here is the process by which cells move from an ordered state to a disordered one. Through

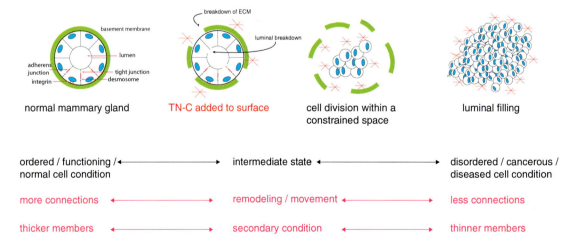

Figure 5.60 Biological condition: we hypothesize that TN-C-dependent change in the basement membrane (basal side) leads to distortion of the lumen. There is a reduced surface area as well as fragmentation of that surface through cell multiplication, which leads to luminal filling.

the addition of TN-C to the healthy cultured cells, there is a breakdown of both the cell to basement membrane interactions and connectivity. Additionally, there is a breakdown of the lumen and the cell-to-cell interactions that help to form the normal polarized condition where a healthy tissue could be considered in an ordered state. As the ECM further breaks down and polarity ceases to exist, there is an increase in cell division and further filling of the lumen with an increased number of cells.

Architecturally, we were interested in how TN-C operates as a switch mechanism in the disassembly of the system and how changes to this disassembly could affect the number and quality of connections (thickness, thinness, and length). Similarly, to switch from an assembled and organized cell state to a more disordered one, a structured architectural system could be disassembled and structurally altered. By working with an original grid of points and connections, we constructed three different models based on density, connectivity, and location.

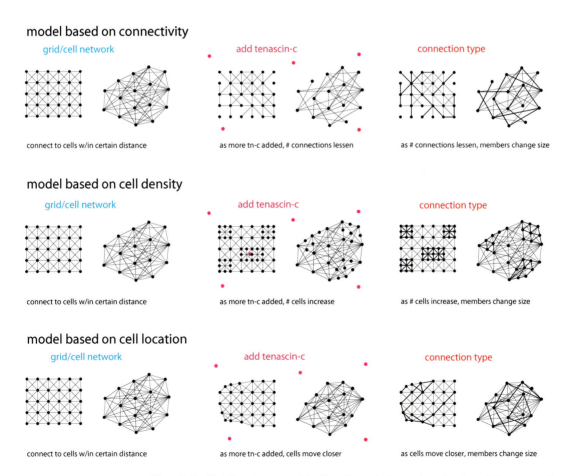

Figure 5.61 Model based on connectivity. Here, changes in the exterior surface through the presence of a switch (TN-C) either increase or decrease the number of connections, and the quality of connections.

These three models then underwent the disassembly process with the addition of TN-C. In the model based on connectivity, as the disassembler (TN-C) was increased, the number of connections decreased, while in the section order operation, as the connections decreased, the members changed size/thickness.

In the model based on cell density, the number of points/cells was changed within the same bounding box. As the disassembler moved through the system, the number of connections and types of conditions altered accordingly with an increase in number of cells, thus increasing connections. The second-order operation was not achieved in this model, but with the increased number of cells, there would have been a similar operation of variable thickness—increasing thickness of members due to increased distance. In the third model, the number of cells remained the same and the disassembler caused a change in the bounding box where the proximity of cells to one another increased and correspondingly, the number of connections increased in this constrained space.

In the model based on connectivity, we worked with two conditions of points, one an organized collection of points in a grid and the other a random collection of points—both organized in a square matrix, or a bounding box within which the points could aggregate. The strength of the disassembler was kept constant and moved through both the organized grid condition and the

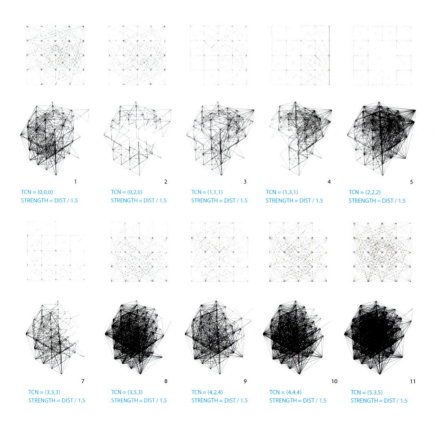

Figure 5.62 Matrix drawing of iterations. By working with an original grid of points and connections, three different models are constructed based on density, connectivity, and location. The three models are subjected to the disassembly process with the addition of TN-C.

Figure 5.63   Rendered detail

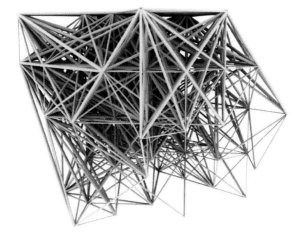

random point connection condition, creating moments of rupture and disassembly at various junctures. The aim was to create holes and conditions of variability. This model also reached the second-order operation where the thickness of the connections was altered based upon the distance between the cells. This was carried out through a change in the radius of a pipe connection between these points.

For the model based on cell density, as the disassembler increased, the number of cells increased and as cells increased, the members changed in size and thickness. In the model based on cell location, as the disassembler increased, the cells moved closer and in turn, connection members changed size and thickness.

These models were constructed in order to begin to examine the potentials of inhabiting and manipulating a surface as a sectional apparatus. In connection with the ideas of Robert Le Ricolais, we were interested in the possibility of designing and building with holes. In each model, a grid of points simulates a collection of cells that undergo a process of disassembly to create changes in structure, member thickness, presence or absence of voids, and density. This disassembler is uniquely organized through the logic of biological disorder caused by the presence or absence of the molecule TN-C. While not fully resolved, these models and simulations begin to address the possibilities of biologically-driven design methods organized and rationalized through the algorithmic and coded logic of discrete processes enacted from a simple set of rules for disassembly.

Conclusion and Next Steps

How might the aforementioned modeling and simulation investigations enable new understandings in how a surface structure may respond dynamically to its environment and in turn be tuned by its deep interior structure through feedback mechanisms? Could studying cellular motility give rise to new models and methods for negotiating dynamics, programmable matter, and change in architecture? Further, how can the development of visualizations and simulations of complex biological networking systems give rise to new modes of thinking in design that engage complexity, energy systems, material plasticity, the structure of information, and even issues of perception and the personalization of space? Associative and parametric software packages are enabling architects and now cell biologists to model architectural and structural relationships inclusive of environmental, material, and physical parameters. Never before have architects so readily been able to shorten the path between design drawing and fabricated output. In fact, the act of drawing is arguably a modeling exercise in and of itself where fabrication and construction instructions become output. Furthermore, current generative modeling tools enable designers to model and visualize natural forms quite readily. The extracellular matrix as model affords a deep understanding of the dynamics of context in the generation of form. As the structural designer and father of the space frame Robert Le Ricolais asked, "Why would we convert radiolarian structures into buildings?" Le Ricolais claimed that nature is our greatest teacher in dealing with the problem of form. By immersing oneself in biological design problems, such as those described in the previous sections, and abstracting these biological relationships into code-driven parametric and associative methods and tools that are also enmeshed with feedback, it is possible to gain new insights into how nature deals with dynamics and environment within cell and tissue structures. Certainly, we do not aim to generate the shape of a building after a cellular structure, but perhaps architects might learn from these biological models such that architecture acquires "tissueness" or "cellness" and is not merely "cell- or tissue-like." We believe the tools and methods produced and designed throughout this process will potentiate alternative applications in architecture and most importantly, foster new habits of thought in design and science. Current topical applications include: New and emerging technologies in programmable matter; adaptive architecture; simulation of nonlinear nano-to-micro material properties at the architectural scale; design, performance testing and bioinspired digital fabrication in architecture; biomimicry in architecture; bioinspired sustainable and ecological design; multi-functional nano- and microstructured soft materials; 3-D printing and programmable matter; materially-directed generative fabrication; nonstandard and experimental structures; and material scalability (nano to architectural scale).

The abstract models described in this chapter offer up novel approaches and methods for the design and fabrication of macro-scale shells, spatial and deployable structures capable of shape-shifting in alternative and scalable contexts. The projects in

Part III begin to scale and translate these relationships through a series of materialized and fabricated architectural prototypes. How do the constraints of making and constructing productively contaminate and inform these biological processes with architectural thought and projection? Before delving into these human-scale datascapes and prototypes, the next three case studies feature three advanced research projects produced in LabStudio and the Jones Lab under the leadership and advising of Sabin and Jones. Each project explores and generates advanced applications and speculations in science and architecture including novel visualization and simulation tools for both diagnostic and architectural design purposes and micron-scale fabrication. Each project is influenced, steered and specified by a unique hybrid design science research space as so eloquently addressed in Sanford Kwinter's statement below. Finally, this section will conclude with several projects based upon bioinspired materials and design, setting up a transition into speculative and scalable applications in architectural prototyping and materials for design and synthesis, which will be discussed at length in Parts III and IV.

Biological Data and Intuition
Sanford Kwinter
Invited critic and Professor of Theory and Criticism, Pratt Institute, New York, 2008

What's interesting is that boundaries persist between the domains of architecture and biology as well as the fact that they have been deliberately removed in this studio and their interpenetrations protected at least for the duration of an experiment. I see the principal question to be the following: how much can geometry be used to function as an analytical device to make *behaviors* available to our intuitions? Architects are increasingly becoming interested in this problem, sometimes blindly, and often without knowing at all where it's going to end up. Where the students were successful, and where they may also have failed, is in their employment of traditional means of mapping as practiced in their own field (architecture). They may have missed something, something hard for us in architecture to achieve, to figure out how to use geometry to render information about movement. I refer here to a technique in common use for over a hundred years now in math and physics that deploys a different type of coordinate system, one that maps and creates a "phase space." This technique allows one to move from one coordinate system seamlessly to another of a different type. One transposes values to a coordinate system of a different order where the relations between X, Y, and Z can be different and defined freely by an alternate "parametrization." So one reprograms principally the correlations . . . I suspect we need to practice this for decades in architecture before we will have completely changed intuition. We gain too little advantage in applying the criteria of advanced science to the familiar experiments of 3-D form. What we are looking at here today could come to be a foundation of a different way of entering data in histological milieus in such a way as to make it available to the senses in a totally different type of format. There is potential revelation here of great import, but it's more than likely going to happen by accident, it's going to appear in the space of a moment, through one of those flashes where suddenly something that was done in a studio here, one or two years previously, or a particular model suddenly returns to mind and illuminates . . . We know the story of the double helix structure, assembled by several teams piecemeal in parallel until—through theft!—the global solution appears.

It's an interesting game, particularly to develop the ability to move fluidly from microscopic to macroscopic scales . . .

Figure 5.64 Final review for Nonlinear Systems Biology and Design. Critics seated left to right: Peter Lloyd Jones, Sanford Kwinter, Ferda Kolatan; Graduate Department of Architecture, School of Design, University of Pennsylvania, 2008.

Case Study: Understanding Behavioral Rule Sets Through Cell Motility

Erica S. Savig, Mathieu C. Tamby, Jenny E. Sabin, and Peter Lloyd Jones

This research study was motivated by a shared curiosity about the intricate relationship between the physical behavior of human cells and the material environment that envelopes them. The overarching aim was to define the "space and structure" of the movement, or motility, of smooth muscle cells (SMCs), and to develop novel tools that would allow us to ascertain, at an unprecedented level, whether cell motility is dependent upon the organization of the surroundings.

This approach was inspired by architectural research previously conducted at PennDesign, that looked at the space and structure of ice dancing.[1, 2] In that work, the choreographies of ice dances were mapped using points on dancers' bodies. Using a custom algorithm within three-dimensional architectural modeling software, visual signatures were generated, displaying the unique organizations of each dance. Starting with unknown cell choreographies, we translated this process through a bit of reverse engineering. We first captured and visualized cell movements, which we did through the development of new techniques and algorithmic tools. Then we devised fine-detailed ways of breaking down and measuring cell motility. We also sought to visualize cell motility in less traditional ways, including with 3-D printing. Throughout the process, we discovered and defined an array of cellular dance choreographies. With the newly derived tools, even subtle moves made by the same cell when faced with different local environmental conditions could be revealed.

The study results appropriately convey data at a disciplinary boundary—in the form of 3-D digital models, diagrams, statistics and 3-D physical models—invoking an understanding of cell motility through a variety of modes. Working closely together on the project, a graduate architecture student (Savig) and cell biology post-doctoral scholar (Tamby) integrated a process of architectural computational design with a scientific method of experimental research. Different languages, conceptualizations, lines of questioning, and analytical methods merged into a new scientific model of design research.

Biological Context and Research Aims

As introduced in the beginning of Chapter 4 on Motility, this expanded LabStudio project examined the mimicked effects of alterations in the pulmonary arterial vasculature ECM upon SMC motility, in the context of pulmonary hypertension.[3–6]

The research study was shaped using three guiding aims: The first was to *develop new tools and methods* for the analysis of videos of live cells moving across native or degraded type I collagen material substrates. The analytical process would be highly visual and intuitive, would fit into the scientists' existing investigative process, and would meet the quantitative standards for assessing biological data, requiring the derivation of new metrics. For the second aim, we sought to *create signatures of cell motility* that would visually and quantitatively indicate unique characteristics of a cell and the nature of its microenvironment. Since the underlying molecular mechanisms driving specific forms of local and global cell motility behavior are fairly well understood, our visualizations would also support a data exploration and hypothesis-generation process. Additionally, they would introduce new ways of seeing and conceptualizing cell motility systems. The final aim was to *identify specific patterns of cell motility, or cell choreographies*. That is, to describe coordinated movements of live cells. This would also require new ways of communicating cell motility behaviors, visually and verbally.

Experimental Strategy

We questioned the effect of an altered tissue microenvironment on SMC motility behaviors.[7, 8] Various experimental methods are used to assess cell migration, including the transwell and scratch wound healing assays. These methods assess the speed and/or directionality at which cells move, but do not account for the more discrete and highly local changes to cell morphology that occur on the membrane at the edge of a cell, which we suspected would be important when studying cell-ECM interactions. As such, we developed a strategy to simulate normal and PH diseased microenvironments. As cells freely traverse the substrates, their changing morphologies and global movements would be captured via video-microscopy and analyzed for differences through image analysis techniques.

Thin films of intact or native fibrillar type I collagen had previously been engineered for use in vascular smooth muscle cell experimentation. Although cells cannot fully embed within thin films, their adherence to the collagen molecules triggers many of the internal molecular changes that cells undergo while interacting within a three-dimensional ECM. The inability of cells to move around in three dimensions is an ideal characteristic for video-microscopy. The capture of only a single optical plane could represent the cells' important interactions with the ECM, substantially reducing data. We thus modeled normal and PH microenvironments as either native fibrillar or degraded nonfribrillar type I collagen thin films, respectively, and normal human pulmonary artery SMCs were obtained and cultured on these films as previously described by our group.[9]

Following plating of cells on thin films, they were immediately placed in a temperature- and humidity-controlled biochamber attached to a Nikon inverted microscope. The microscope was equipped with a camera and software-controlled motorized stage that enabled the capture of videos of motile live cells, from multiple locations on each thin film. We captured 14 hours of footage from each of the two thin film conditions. Images were acquired through the phase contrast microscope, at 5-minute intervals, and 16-bit tiff images were extracted from the video-microscopy software.

Filtering and Extracting Motile Behavior

Using the videos, we observed organizational differences in cell morphologies as they first attached and then moved across the two different substrates. In particular, the finer-scale movements of filopodia (i.e., membrane extensions from the body of the cell that probe and feel the surrounding ECM) appeared to be more coordinated on native collagen. Thus, we decided to further examine the outlines for each cell, in order to further scrutinize filopodial movements (Figure 5.65). To accomplish this, we randomly selected then outlined five cells on each substrate, using images taken at 15-minute intervals over 14 hours. This resulted in 57 sequential images, which were then imported into Adobe Illustrator. Using a Wacom graph tablet, the outlines of SMCs were traced and represented in vector format. This procedure allowed us to accurately capture a level of morphological detail missed by automatic tracing algorithms.

Geometric Tooling

Using Familiar Frameworks

The 57 vector-based outlines per cell were then imported into Rhinoceros 3-D modeling CAD software. After considering various ways of representing morphological change (such as the classic depiction of change through time in a series from left to right), we decided to build on the biologist's common understanding of cell motility as a mapped trajectory of the entire cell body on a 2-D plane. As such, we kept the cell outlines in their original x–y positions, so that when viewed in a Plan view within Rhino (in 2-D, from top to bottom), they would show exactly where the cells were moving within the image plane. We imported a microscopy image taken at time = 0 to further orient the viewer. Next, we spaced the outlines in a third

Figure 5.65 Reduction of cell motility to cell outlines. The outlines of five cells were traced from each of the selected movies on both fibrillar and nonfibrillar substrates. Shown are three images from different points in a nonfibrillar substrate video sequence.

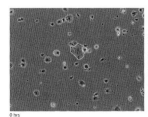
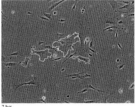
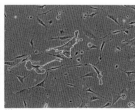

Figure 5.66 A spatio-temporal framework for cell signatures: Perspective view of all outlines obtained from the fibrillar experimental condition, where the z-axis represents time. The microscopy image at time = 0 was used to ground the user spatially onto a 2-D plane, and contextually with a more familiar view of live cells.

dimension, along a z-axis, ordering the outlines chronologically from bottom to top (Figure 5.66). This 3-D representation immediately revealed the hierarchical nature of the cells' morphological changes, showing the smaller scale movements of filopodia relative to the larger-scale movements of the entire cell. This spatial orientation for the data became the foundation for a number of cell signatures.

(De)constructing Cell Morphology

In an attempt to analyze the cell morphologies, we began by deconstructing them through the creation of an analytical drawing, mapping new geometries (e.g., points, lines, circles) onto the cell outlines. The intention was to "geometrize" each cell outline, to be able to quantitatively compare them. In one particular example, each outline was reduced to a series of points (Figure 5.67). A polar grid with intervals of 7.5 degrees was drawn at the centroid of each cell outline. The intersections of each outline on the grid lines were marked with a point, resulting in 47 points per cell ("outer reach points"). Local changes to morphology could now be quantifiably tracked, as differences in polar coordinates of each of the 47 points over time.

Bridging of Geometry from Image to Numbers

Leveraging easily accessible object attributes within Rhino (such as the length of a curve), and making use of basic Rhino commands (such as Rhino.Distance, which calculates the distance between two points), we were able to extract quantitative data

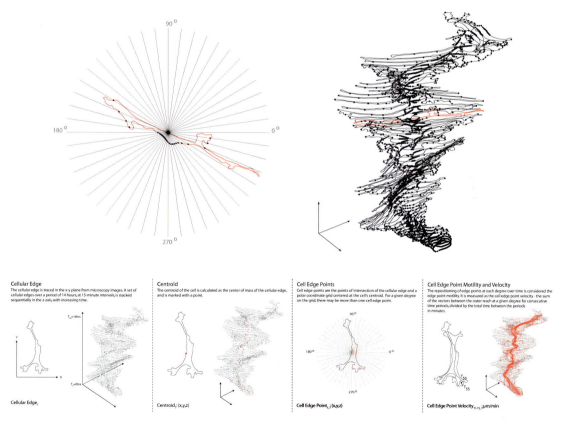

Figure 5.67 Deconstructing cell morphologies with geometric analysis: Cell morphologies were reduced to 47 points each (a). Starting with each outline (b), a centroid was calculated and represented with a point (c). A polar grid with 7.5 degree intervals was placed at the centroid of each outline. Points of intersection of the outline with the grid were marked with points (d). The consistent number of points enabled the tracking of changes to the cell outline over time (e).

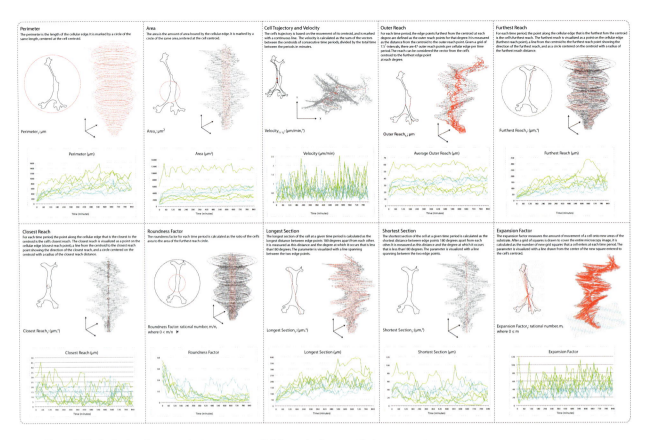

Figure 5.68 Derivation of new quantitative parameters: Building on basic geometric shape parameters, such as centroid, area, and perimeter, an array of new metrics were defined.

from the new geometries. This led to the derivation of a series of new parameters for measuring cell morphology and motility (Figure 5.68). Throughout our analytical process, we continued to create new parameters that would quantify particular characteristics of motility. For example, the "Expansion Factor" was derived to capture a cell's ability to expand into new space as it moved. Larger reaches into new space and total cell movements would have higher weights compared to more minor reconfigurations of morphology in a case where a cell appeared to stay put.

Working within RhinoScript, we automated the analytical drawing and extraction of this numerical data *en masse*, exporting it into an Excel spreadsheet, where we performed statistical analysis. Nine of 10 new parameters showed statistically significant differences (*p* value < .05) in mean values between cells on fibrillar vs. nonfibrillar substrates (in a two-sample t-test considering separate variances, with Bonferonni adjustment, as calculated within Systat 13 statistical software, comparing data from all 57 time points for the five cells on each substrate) (Figure 5.69).

Cellular Choreographies Revealed
In our search for variegated choreographies, we wanted a visualization that would home in on the relationship between local filopodial movements and the global

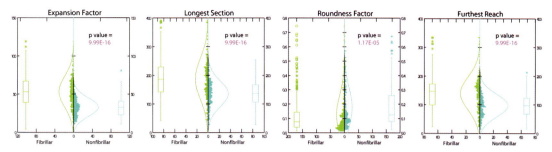

Figure 5.69 Newly derived quantitative parameters show statistically significant differences between substrates: Two-way student t-tests were performed to compare data for each of the five cells on each substrate, over all 57 time points. Nine of the 10 parameters showed statistically significant differences (p < .05) between the mean values. Seven of the 10 parameters had p values < .001. Shown are four parameters that were derived to quantitatively represent particular characteristics of newly observed cell behaviors.

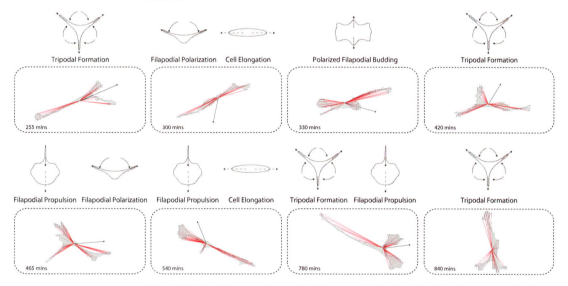

Figure 5.70 A cell choreography revealed: Changes to the location of a cell's outline were visualized as new areas entered on a Cartesian coordinate grid (marked by red lines drawn from the cell's centroid to the new area entered). These sequences revealed patterns based on number and direction of filopodia, additionally relative to the direction of movement of the entire cell (marked by an arrow). The selected sequence shown demonstrates repeated changes (i.e. "dance steps", diagrammed and named above each time point), that make up a cellular dance choreography.

trajectory of the entire cell body. To accomplish this, we used a technique of merging two different frames of reference: a polar coordinate system around the centroid of each cell, and a Cartesian coordinate system bound by the microscopy image frame. The Cartesian grid partitioned the image area into small squares through which we could track x–y movements, both globally and locally. Using scripting techniques as previously described, new grid squares into which cell outlines entered at each time period were marked with a line drawn from the cell centroid to the grid square.

On observing the results, we recognized a sequence of cell behaviors, with expected variations displayed by multiple cells on both substrates (Figure 5.70).

The sequence begins with three filopodia stretching away from the main body of the cell; we termed this "tripodal formation." Next, one of the filopodia retracts, and the cell moves in the direction of the release, which we designated "filopodial propulsion." The remaining filopodia then rotate around the cell and polarize into a common spindle SMC morphology, deemed "filopodial polarization." At times, the cell would then elongate ("cell elongation"). The pattern would repeat as a rhythmic, hierarchical, and coordinated series of movements—a cellular choreography.

Abstraction of Behaviors
Another visualization approach focused on more radical abstraction, taking away any cell-like context (Figure 5.71). Our intention was to offer representations of cell motility that would force viewers to more attentively orient themselves through pattern finding, as they try to make sense of the never-before-seen structures. Expanding the exploration to our sense of touch and other spatial sensitivities, we also 3-D printed digital models into physical form (Figure 5.72).

In the examples shown (Figure 5.71), we "unraveled" the cellular outline, so to speak, stretching it across an x–y plane, while maintaining information about its shape. The visualizations could be read as 3-D graphs, with the cell's edge mapped according to time along the y-axis, to location about the cell (i.e., degree around centroid) along the x-axis, and to distance from the cell's centroid along the z-axis. Continuous lines were first drawn to connect the corresponding 47 outer reach points of a given cell for each time period (called "motility lines"). These motility lines represent three pieces of information: (1) each corresponds to a single degree around a cell's centroid; (2) time is represented along the length of each line; and (3) the jagged nature of the lines results from the distance of the cell from its centroid. The motility lines were then aligned side-by-side, ordered on the y-axis by the degree around the centroid that they represented. They were then rotated so that movements were oriented towards the z-axis, the flat 2-D plane below becoming an expansive representation of the cell centroid. Short lines were drawn to connect the motility lines for each given time period, essentially very loosely "filling in" the rest of the cell's edge that was missing information.

What resulted was a planar visualization that can be interpreted as an unraveled cell. The map highlights areas of intensity and change, as well as how these changes propagate over time and around the cell's edge. Evident to most viewers is the amount of change and intensity of the cell on fibrillar collagen compared to that on nonfibrillar collagen.

Each new visualization prompts data exploration. An interactive 3-D modeling CAD space provides convenient functionality for doing so. One can direct the viewing angles and simulate a "walk-through," mitigating much of the difficulty in analyzing 3-D visualizations that are presented statically as a 2-D image. Layers of information can be turned on and off, offering data filtering

Case Study in Behavioral Rule Sets **209**

Cellular Edge Unraveled

This signature is derived from the cell edge point motility curves, as seen in blue. The curves were aligned sequentially in the x-y plane according to the degree around the centroid (from 0 to 360), thus the x-direction represents the degree around the centroid, the y-direction represents time, and the z-direction represents the distance of each cell edge point from the centroid. The curves were each broken into 100 segments, and black lines connected the curves at each corresponding segment. Although the lines are equally spaced across each cell edge point motility curve, the spacing inherently varies from curve to curve, thus creating densities and sharp directional changes in areas of relatively major change in cell shape, both over time and around the cell.

Cellular Edge Points Unraveled

This visualization repositions points along the cell's outline into a 3-D graph, with the degree around the centroid in the x-direction, time in the y-direction, and the distance from the centroid in the z-direction. These points were taken at equal distances along the cell's edge, resulting in accumulations of points in areas around the centroid where filapodia are located, and empty spots where the edge is relatively smooth.

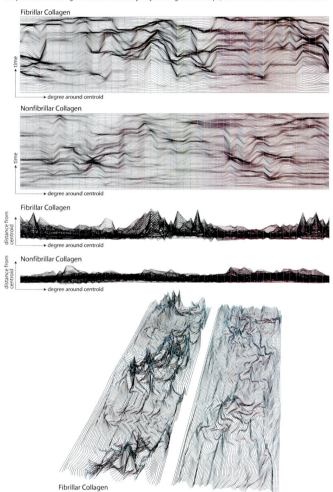

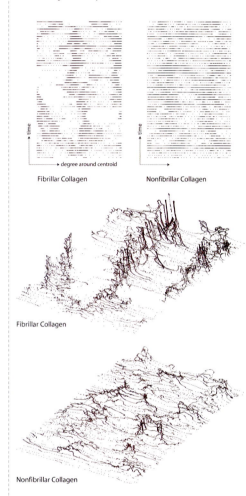

Figure 5.71 Radically abstracted cell signatures: In a more radical approach to visualizing cell motility, abstract representations were drawn, out of context of the single cell shape and its surroundings.

Figure 5.72 Additional models were 3-D printed, for tactile and alternative spatial exploration.

functionality. Pieces of data can be temporarily isolated or moved closer to other data, for easier visual comparison. Even the sketch capabilities, of quickly annotating the visualizations with shapes and text, offer a way to keep track of observations, supporting the persistence of an exploration process.

Biological Interpretation

From Geometry to Cell Behaviors

Immersed within such abstraction, it was an essential part of our analytical process to continuously reference the original microscopy videos. In the videos we could relate observations to events and environmental conditions, leading to the discovery of multiple distinct cell behaviors within the context of the cells' surroundings. For example, a dramatic cell surface change was found to be the retraction of a filopodia in response to touching another cell. A cell's persistence in maintaining a long, polarized shape was found to mimic the directionality of a large underlying collagen fibril. The density of cell edge points close to the centroid (Figure 5.73) was a condition where most of the cell body would stay close to the cell's center of mass ("cell cohesion"), despite multiple filopodia reaching and retracting over relatively long distances.

We may then ask whether these behaviors can be explained within the context of cellular adhesion, internal mechanics and ECM biology, formulating new scientific hypotheses. How do particular tensile forces resulting from strong or persistent ECM adhesions or from a large central cell mass, affect cellular movement? Do such forces allow for particular explorative filopodial behaviors? Does an altered collagen molecular structure expose additional ECM binding sites, offering the potential to form stronger ECM adhesions? Does cell morphology with a relatively large and dynamically changing perimeter, and a generally static area, indicate a cellular edge complexity driven by actin reorganization? How is such actin reorganization affected at the molecular level by surrounding collagen organizations? A series of biological experiments can be formulated to answer each of these questions, further contributing to better knowledge of the environmental-dependence of cell motility, with implications for our understanding and potential treatment of diseases such as PH.

Communication of Cell Behaviors

On starting the project, we were inspired by the ice dancing reference. Setting up the relationship between a cell and a human body was effective for our ability to make sense of and recall some of the observed complex patterns of cell motility. In conversation, we personified cells, describing them as swimmers doing some rendition of a breaststroke, or as floundering around. We also noticed that the prevalence of the more coordinated swimmers seemed higher on fibrillar collagen, and questioned if the disorganization in the nonfibrillar substrates contributed to the lack of coordination of their movements.

Case Study in Behavioral Rule Sets 211

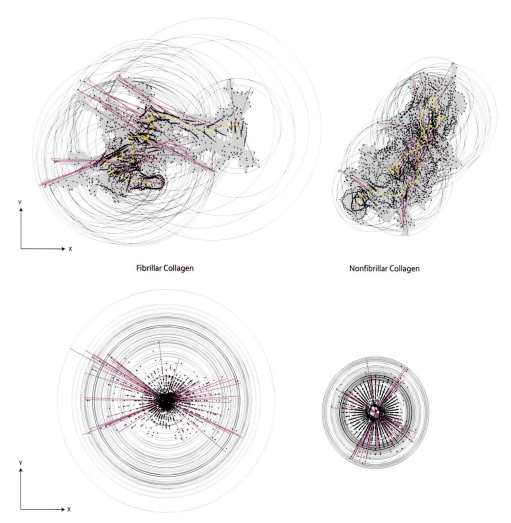

Figure 5.73 Interpretation of cell signatures: A 3-D signature of the "furthest reach" and "cell edge points," viewed from the top-down, in 2-D (above). Normalizing the signature by putting all of the cell centroids onto the same point shows another view (below). The "furthest reach," is represented with a circle that emphasizes the distance from the centroid and a magenta line that indicates the direction of the filopodia. A signature for a particular cell on fibrillar collagen (left) indicates that its filopodia reach away from its centroid much further than a particular cell on nonfibrillar collagen (right). Despite the larger furthest reach over time, the cell on fibrillar collagen shows a much higher density of cell edge points close to its centroid, while those of the cell on nonfibrillar collagen are more dispersed. This behavior was observed in multiple cells in the videos, and is called "cell cohesion," where the cell maintains a large core center of mass, while its filopodia stretch great distances.

We defined these found choreographies hierarchically, as local movements at the scale of a hand, for instance, as more global changes at the scale of the body, or as the combination of both. As no standard method existed within the field to keep track of such spatio-temporal motility behaviors other than written descriptions, we created diagrammatic sketches for each, named them, and wrote a detailed description (Figure 5.74).

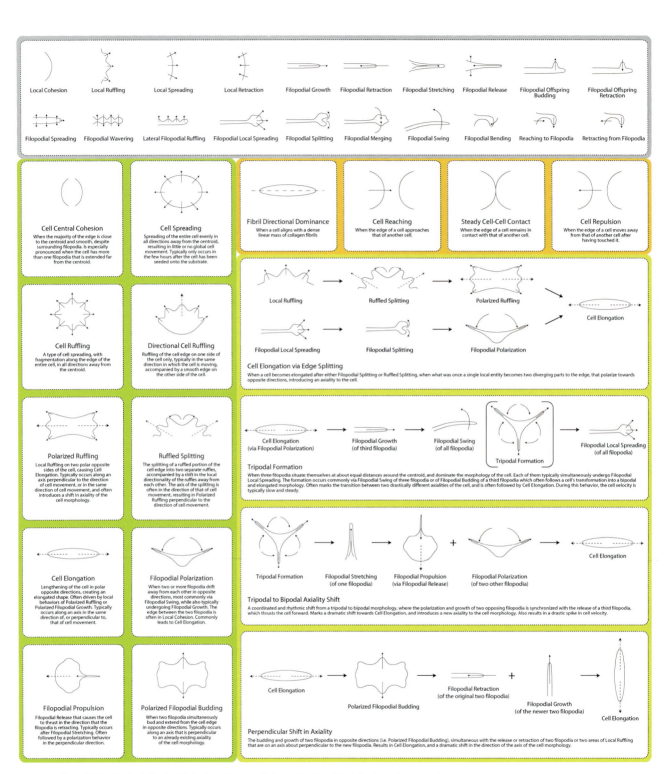

Figure 5.74 Newly defined cell behaviors: An array of observed cell motility behaviors, diagrammed, described and named, at various levels of hierarchy of the cell body: locally, akin to hand or arm movements; globally, as changes to the entire cell body; movements relative to cell surroundings; and sequences of behaviors making up choreographies.

Conclusion

Findings and Implications

Centered on a computational tooling process, this study demonstrates an effective approach for both designers and scientists to explore, make discoveries within, and more deeply understand biological systems. The complexity of changing cell morphologies (Figure 5.66) was reduced to manageable geometries (Figure 5.67), translated into abstract cell signatures (Figures 5.70–5.72), quantified into an array of newly derived parameters (Figure 5.68), and statistically shown to have significant differences on fibrillar vs. nonfibrillar thin film collagen substrates (Figure 5.69). It led to the discovery of new cell motility behaviors—or choreographies—and a novel means of describing them with names and representative diagrams (Figure 5.74). These preliminary results suggest that the architecture and organization of a cell's extracellular matrix do seem to impact its own motility behaviors.

The techniques of customizing analyses through tooling in 3-D CAD software with geometric analytical drawing can be replicated in similar studies. The battery of template cell motility signatures and the diagrammatic representations of cell behaviors can be reused. The concept of the cell motility signature can ultimately be leveraged in medical applications, as another means of obtaining cell-specific understandings of an individual patient. Finally, this project has shown that designers and scientists can effectively merge their research processes into an integrated approach. Visual experiential modes of representation combine with quantitative analytical explanations, expanding the researchers' familiar ways of understanding.

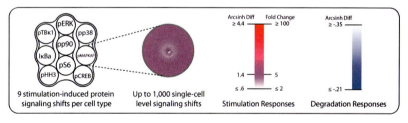

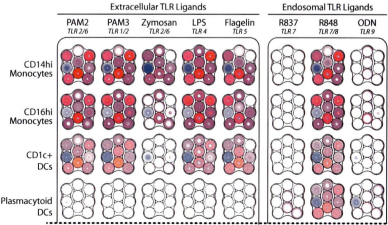

Figure 5.75 Data visualization developments for mass cytometry: Continuations of the visual data exploration process introduced by the cell motility project led to a visualization tool that produced this signaling network heatmap, showing for the first time signaling shifts at the single-cell level. Represented are stimulation-induced cell signals measured from single cells in human whole blood, via mass cytometry. Shown is a selection of four cell type populations (of 11; in rows), under 8 stimulation conditions (columns). The visualization led to discoveries in human toll-like receptor signaling.

Further Developments

Additional developments in both architecture and biology followed this research. Within LabStudio, the work led to investigations into material architectures that would respond at the micro-scale to human tactile interactions, envisioning the potential for more "life-like" architecture.[10] As described in Chapter 6, this spawned a trajectory of research into material-engineered responsive architectural surfaces, through the experimentation, simulation and manipulation of material behaviors.[11, 12]

Developments in multidimensional data visualization continued at the Stanford University lab of Garry Nolan,[13, 14] leading to discoveries of new immune cell signaling behaviors, including responses to cancer therapeutics.[15] Millions of numerical, non-spatial data points on the intracellular molecular behaviors of single cells, produced by mass cytometry technology, can now also be visually explored *en masse* at the single cell level (Figure 5.75).[16]

Notes
1. J. Wong, "Dance and Space," advanced research seminar, *Form and Algorithm*, C. Balmond and J.E. Sabin, PennDesign, Philadelphia: University Pennsylvania, 2006.
2. Jenny Sabin and Peter Lloyd Jones, in conversation on J. Wong, "Dance and Space," advanced research seminar, *Form and Algorithm*, C. Balmond and J.E. Sabin, PennDesign, Philadelphia: University Pennsylvania, 2006.
3. Peter Lloyd Jones, Julie Crack, and Marlene Rabinovitch, "Regulation of Tenascin-C, a Vascular Smooth Muscle Cell Survival Factor That Interacts with the α v β 3 Integrin to Promote Epidermal Growth Factor Receptor Phosphorylation and Growth," *The Journal of Cell Biology* 139(1) (1997): 279–293.
4. R. Chapados, K. Abe, K. Ihida-Stansbury *et al.*, "ROCK Controls Matrix Synthesis in Vascular Smooth Muscle Cells: Coupling Vasoconstriction to Vascular Remodeling," *Circulation Research* 99 (2006): 837–844.
5. Kyle Northcote Cowan, Peter Lloyd Jones, and Marlene Rabinovitch, "Elastase and Matrix Metalloproteinase Inhibitors Induce Regression, and Tenascin-C Antisense Prevents Progression, of Vascular Disease," *Journal of Clinical Investigation* 105(1) (2000): 21–34.
6. K.N. Cowan, P.L. Jones, and M. Rabinovitch, "Regression of Hypertrophied Rat Pulmonary Arteries in Organ Culture Is Associated with Suppression of Proteolytic Activity, Inhibition of Tenascin-C, and Smooth Muscle Cell Apoptosis," *Circulation Research* 84(10) (1999): 1223–1233.
7. Nancy J. Boudreau and Peter Lloyd Jones, "Extracellular Matrix and Integrin Signalling: The Shape of Things to Come," *Biochemical Journal* 339(3) (1999): 481.
8. Nancy J. Bordreau and Mina J. Bissell, "Extracellular Matrix Signaling: Integration of Form and Function in Normal and Malignant Cells," *Current Opinion in Cell Biology* 10(5) (1998): 640–646.
9. John T. Elliott, John T. Woodward, Kurt J. Langenbach *et al.*, "Vascular Smooth Muscle Cell Response on Thin Films of Collagen," *Matrix Biology* 24(7) (2005): 489–502.
10. Erica Savig, Master of Architecture thesis, University of Pennsylvania, PennDesign, advised by Jenny Sabin, Peter Lloyd Jones and Shu Yang.
11. Jenny Sabin, Andrew Lucia, Giffen Ott, and Simin Wang, "Prototyping Interactive Nonlinear Nano-To-Micro Scaled Material Properties and Effects at the Human Scale," *SimAUD '14 Proceedings of the Symposium on Simulation for Architecture and Urban Design*, Article 22 (2014).
12. Ibid.
13. Martin Krzywinski and Erica Savig, "Points of View: Multidimensional Data," *Nature Methods* 10(7) (2013): 595.
14. Supported in part by a NSF Graduate Research Fellowship to Erica Savig.
15. Bernd Bodenmiller, Eli R. Zunder, Rachel Finck, *et al.*, "Multiplexed Mass Cytometry Profiling of Cellular States Perturbed by Small-molecule Regulators," *Nature Biotechnology* 30(9) (2012): 858–867.
16. William E. O'Gorman, Elena W.Y. Hsieh, *et al.*, "Single-cell Systems-level Analysis of Human Toll-like Receptor Activation Defines a Chemokine Signature in Patients with Systemic Lupus Erythematosus," *Journal of Allergy and Clinical Immunology* 136(5) (2015): 1326–1336.

Case Study: Motility and the Observation of Change[1]

Andrew Lucia with Jenny E. Sabin and Peter Lloyd Jones

"There is no objective experience." In so saying, Gregory Bateson reminds us what Every Schoolboy Knows . . .[2]

Building upon research between architecture, the biological sciences and the foundations described in the previous sections, this chapter explores observations of dynamic organizations of matter in a multi-dimensional, microscopic-scale human cellular system and a human-scaled perceptual environment from an information theoretical framework. Exploring an alternative to descriptive geometric languages in the pursuit of pattern recognition, this work argues for mechanisms of *difference* in relation to the roles of history and framing that ultimately affect the *information* content inherent to data arrays generated within dynamic material systems; here, information is defined as a function of observed temporal difference within compared arrays of data. This work ultimately underscores a fundamental shift away from an explanation of environments in descriptive and projective geometric terms to one based on spatiotemporal order and disorder, and ultimately *structural information* content—as a function of the encounter of variant or invariant elements within a given data set that give rise to the perception of material plasticity.

As part of research undertaken between the arts and sciences, the aims of the tooling presented broadly fall under a category of visualization, particularly data visualization. The distillation of patterns within data is sought—the meaningful organization within these data that allows one to discern salient features embedded within its form. To compare and contrast, two different case studies are presented. The first is "objective and scientific" whereby the behavior of moving cells under

Figure 5.76 *Draughtsman Drawing a Recumbent Woman*, Albrecht Dürer, 1525.

a microscope is observed. The second concerns the "subjective" experience of a space traversed. Purposefully, these extremes are used to elaborate on larger preconceptions that come to bear through the very mechanism by which the visualizations presented are produced—*difference*. The issues broached in these studies are at the very origin of how one frames their presence within any context. In doing so, the Motility Project within LabStudio provides fertile ground for issues of concern to both designers and scientists alike. What follows is a way of thinking, a way of understanding observation leading to a particular tooling methodology rooted in the very essence of change. Why is this relevant to architecture and science?

In design, notions of change and context are often forced into languages that are not properly suited to their study, particularly due to disciplinary traditions and their accompanying tools—namely descriptive and projective geometries and the analytic Cartesian spaces in which they reside. And while in the arts and sciences these geometries and abstract models of space have been and will continue to be incredibly useful as a basis for describing shape and boundaries within a comparative language, so too they inherently produce a gross reduction of systems of far greater complexity. What is at issue here is an overreliance on these tools and modes of thinking that have hampered an ability to pose critical underlying questions about the dynamic systems we observe and with which we interact as active agents—as participants in an ecology within which we are embedded. On a more fundamental level, conventional tooling languages and their environments continue to operate within a terrain dominated by a geometric underpinning at a time when many non-geometric issues are equally pressing, namely those of light and aesthetics—arenas of energy and experience.

As observers in the world, both designers and scientists alike are particularly good at seeing, intuiting, and synthesizing form in terms of the shape and morphology of things, specifically things within space. Historically, these morphologies and their containers have tended to be represented geometrically and likely for good evolutionary reasons; pertaining to the mental processes that distill shape from a scene, Gregory Bateson notes, "William Blake tells us firmly that wise men see outlines and therefore they draw them."[3] And while this may hold credence from a cognitive standpoint, it also presents for us problems of undue oversimplification.

The Usefulness of Information

Implicit to visualizing any system is a notion of observation, as such, the current discourse works from a basis of light-borne phenomena. From a scientific standpoint in the case studies presented here, that observation occurs via the microscopic observation of cellular behavior in tissue culture through time (see Figure 5.77); the biological challenge at hand was to discern an observable difference in smooth muscle cellular behavior given the same cell culture in varying contexts, native and

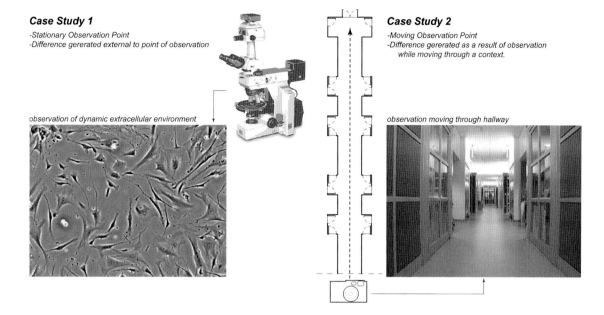

Figure 5.77 Case Study 1, (left) and Case Study 2 (right) for data acquisition; hallway inside the Institute for Medicine and Engineering by Venturi, Scott Brown at University of Pennsylvania.

non-native environments. In doing so, the Motility Project confronts issues that are inherently beyond its immediate scope while having far broader implications to both the arts and sciences—those of context and change. A second case study is presented, one in which difference arises as a result of the movement of a camera through a static space, here a hallway (see Figure 5.77, case studies 1 and 2). Importantly, the tooling and methodology deployed for both of these case studies are identical (light-borne artifacts captured through a digital camera and analyzed as pixels), save the origins of *difference* generated within the systems.

The novel digital tools within this project, while dealing with particular criteria of the biological material in question, were developed within a broader scale-free context dealing more generally with the notion of change within environments. These tools do not privilege the designer or scientist, but rather act as a probing mechanism to better understand how one detects patterns of change within a given context—patterns that are generated at the loci of difference within any data array. Here, the tethered mode of pattern generation between the two case studies portrays a scenario that puts into question any distinction between notions of an objective or subjective endeavor. To this, classically held notions of autonomy and reductionism in design and the sciences are confronted (notions that are continually embedded in the very tools and models of space on which we rely) via relativistic principles of information rooted in observation and the encounter of unique events.

Information as Spatial and Temporal Structure
Borrowing concepts from perceptual psychology and information theory, this research synthesizes two uses of the term information, notably those of the

Figure 5.78 Gradient extraction from a natural scene. The vector gradient (as arrows) contained within the cropped and enlarged regions correspond to differences of adjacent light intensities within this natural scene. Gibson's definition of the ambient information array pre-dates this particular technique, and as such he did not depict the array in these terms. However, of this representation it should be noted that the vector gradient emphasizes all salient features of the scene while placing emphasis on adjacent regions of greatest intensities of difference.

psychologist James J. Gibson and the engineer Claude Shannon, respectively.[4] In this borrowed usage, the former distills regions of spatial variation corresponding to the salient features of the visual world at any unique point of observation (see Figure 5.78), while the latter computes moments of uncertainty in the temporal difference of those visual arrays.[5]

Essential to either of these uses of the term information is a notion of difference. It is required of Gibson's information array to specify spatial structure (see Figure 5.85), while uncertainty specifies temporal structure from Shannon's information theoretic standpoint (see, for example, Figure 5.79).[6] It is precisely this difference that is entangled within the structure and quality of information. Though fundamentally devoid of meaning from a linguistic standpoint, difference yields both the order and organization of available information spatially and temporally. Importantly, that difference is inherently linked to uncertainty and therefore is probabilistic in nature. If no difference is detectable in a given context, an observer must be in a state of change (encountering difference and ultimately generating information) in order to be made aware of that context's material. As such, Gibson's and Shannon's ideas are synthesized here through a sympathetic model of information (see Figure 5.80). Order, organization, redundancy, and the structure of that redundancy underlie the form of dynamic systems and their outward appearances.

This suggests an implicit, non-trivial, sympathetic union between an observer and the data about material (here in the form of structured ambient light) being observed; a communion through which information is generated. The information arising for a scientist peering through a microscope at a given context (here moving

Figure 5.79 In this example of visual sensory adaptation known as Troxler Fading, the observer is asked to fixate on the red dot in the center of the image. After a brief period the outer hazy ring will appear to vanish. The effects of the microsaccadic movements of the eye are overcome in this "weak" signal example (i.e., this effect will not be produced if the outer ring is of a higher contrast within its context or depicted as a hard line). From an information theoretic standpoint this adaptation may be thought of as analogous to the weak signal's low probability of change within the visual array, here exhibiting a high likelihood of stability in the visual field.

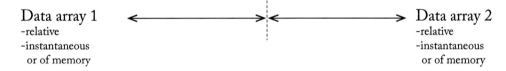

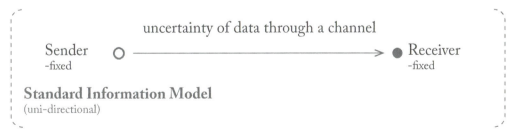

Figure 5.80 Standard vs. sympathetic informational models. Note the distinction from the classic standard model of information with that of the sympathetic model proposed here. The latter relies on the comparison of two relative data arrays while the former describes a unidirectional flow of information through a discrete channel. Corresponding to the studies presented here, the sympathetic model presupposes an engagement between a relative observer and the surrounding dynamic informational world rather than implying informational flow through a communication channel, for example, telecommunication.

cells) is of the same fundamental type as that of the information a person would receive while walking through a space—as a function of difference. Taken at a primary level, information in both of these contexts concerns the predictability of the organization of light being observed. The distinction between these two scenarios is the stationary vantage of the observer in the former (case study 1, the science lab) and the changing vantage of the observer in the latter (case study 2, moving through space). These are extreme poles of the same relativistic issue (that of a fixed or dynamic observer within a context) while generally observations exist somewhere between—within dynamic environments and under non-static points of observation.

At issue here is precisely how we "see" the world about ourselves, or how that "subjective vision" of the world arises from our interaction with it. Correspondingly in the sciences we are met with issues as to how we observe and visualize the data from an "objective scientific" standpoint. At their loci there is no distinction between these two modes of operation. If there is a distinction made (and there most certainly is), it arises due to an engrained ideology stating that the data produced in a scientific lab is somehow more objective then the data we interact with as a person in the world. Data produced in a lab may be more controlled, or the parameters that are allowed to vary in the process of experiment are kept to a minimum, but this does not make the information generated any more or less objective.

The Language of Change

Drawing is not the same as form, it is a way of seeing form.

(Edgar Degas)[7]

Figure 5.81 A study for *Dancers at the Barre*. Edgar Degas, 1876–1877.

Taking this insight by Degas further, it may be contended that drawing is a way of seeing form's appearance at a given instance (its shape), and that *form in its totality* is beyond precise location in space or time rooted in fundamental aspects of difference arising from change.[8] What then might be an appropriate informational manner in which to deal with this change if indeed traditional geometric means are not well suited to do so?

In the biological sciences a commonly held practice in determining cellular characteristics, be they motile or static, is the determination of shape known as *segmentation* (see Figure 5.82). The technique relies on the determination of boundary conditions separating the edge of the cell from its environment, typically through the use of pixel-based data originating as light. Oftentimes a person completes these efforts manually, or more often are made to automate this in an effort to inject less human bias into the operations. Not limited to biology, in broader contexts such as astronomy or any discipline in which the detection of discrete bodies (even human surveillance) are the subject of interest, this is known as *object detection*.[9] In doing so within cellular studies, a critical error of assumption is made—that there is a strict distinction between a thing and its environment or between the communicating cells in that environment. It is not clear, particularly in the case of cell edges, where the boundary between the body and the environment exists, if indeed there is one at all (see Figure 5.83). Add to this that most experiments relying on segmentation will deplete a cellular population to a point at which cells can be examined in isolation, thereby also negating the more native behavior of cells communicating in such quantity whereby boundary divisions of the individuals are impossible to make.

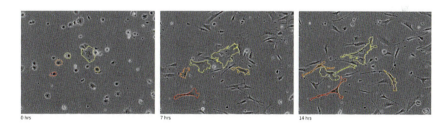

Figure 5.82 Example of cell segmentation, whereby the boundaries of cells are outlined

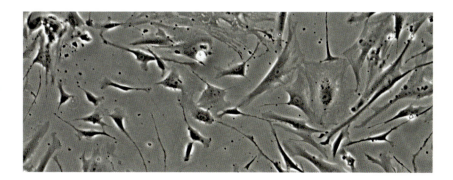

Figure 5.83 Smooth muscle cells under microscopy in vitro. Note overlapping cells and hazy character of cell edge conditions demonstrating the difficulty (if not impossibility) of segmentation techniques.

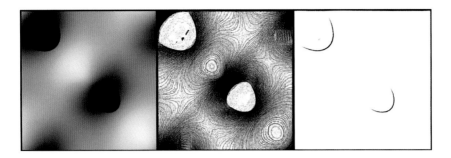

Figure 5.84 Reduction from gradient to line

Geometry inherently relegates boundaries to strict edges when deployed as a description of something's shape. In doing so, potentials within that material world are severely limited. If we accept that the fundamental questions needing to be asked of any dynamic system are ones of energy and communication, then we must come to the messy terms of that world and operate within. This is a noisy and statistical world, one not easily defined in terms of strict limit, rather one marked by hazy threshold (see Figure 5.84). Take, for instance, the above example of Troxler Fading in which a visual array is not demarcated along a rigid boundary and only perceptible through probabilistic mechanisms. Indeed, most of the data available to our senses is of this type, even those extending beyond the visual domain through to the tactile or auditory such as heat and pressure differentials. Similarly, on a cellular level, either in natural contexts or in laboratory settings, transference of energy and the production of information do not occur within and along clean limiting bounds between bodies, and moreover are an equal byproduct of the context in which the organic materials reside. A reduction to a boundary, or even number of boundaries, is a common reaction most persons have when thinking of the world as objects within space. These objects could be within the built environment or cells within a microenvironment. Taken to an extreme, if one is to accept that the emergence of our sensations of the world arise through a mutual engagement between the observer and their environment, then the fundamental language of this type is one of field-based difference, not of discrete object detection. It is precisely this sort of problem that renders conventional means of geometric tooling, a reduction to geometric languages, unfit for dealing with complexity and dynamic systems.

Distillation of Patterns within a Field of Change
Part I: The Tool and a Biological Impetus

The prior discussion should be met with an open recognition that a statistical or informational way of dealing with energy, matter, and observation is by no means novel. Indeed, such methods are commonly deployed in the sciences and design disciplines, but rarely, if ever, in a truly ecological sense, particularly at the aesthetic level.[10] What is argued for is a reframing of the contexts in which we conceptually place ourselves as observers and active agents, and therefore a reframing of the tools' functions that mediate these contexts.

Thus far, this discussion has posited the issues at more-or-less two opposite extremes on a spectrum. In both there is an observing mechanism and material being observed. In Case Study 1 the observing mechanism (the microscope) is stationary while the context is in flux (the cells in their environment). In the latter scenario of case study 2, change is generated as the byproduct of a rigid environment and moving observer. In both cases, the data captured is of a light-based digital nature, pixels within a frame.[11]

Taken as a data field, events may therefore occur only within each discrete pixel across a temporal array with spatial structures and positions remaining constant. Only discrete pixels are taken into consideration temporally (the ordinal spatial structure of the pixels is considered to be absolute and unalterable), thus taking into consideration the likelihood of events within discrete signals (individual pixel states through time). The spatial structure of the overall pixel arrays on each discrete picture plane is maintained. Given this pixel-based approach, the fidelity of the spatial measurement of any array in question would therefore be governed by the resolution of the associated pixel array, while temporal fidelity is taken to be the duration between each captured frame (see Figure 5.85). Thus, each time state (frame) may be compared against its temporal neighbors and with what has changed historically. This common technique is termed *image differencing*. What is significant about this simple operation is the production of information prior to a simplification of the data; information is generated as a product of the change within the data in its raw state and in doing so does not privilege field or object. Rather, the light-borne artifacts of the system are taken in total; there is no reduction of the field to discrete objects.

The information mappings presented in case studies 1 and 2 represent the cumulative information produced over all time states. In these mappings, pixel intensities depict moments which are considered as the "common, moderate, and rare-events" within the respective systems for all time states (see Figure 5.86). The probabilities of an event occurring within a given mapping are evaluated

Figure 5.85 Schematic diagram of Pixel Event Space, Difference, History, and Information

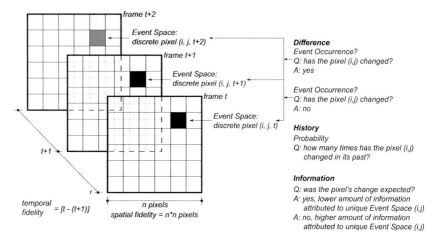

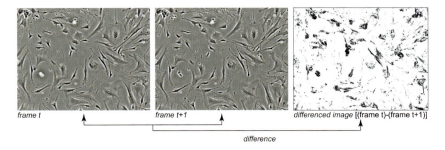

Figure 5.86 Difference in pixel contrast (right) between consecutive cell images (left and middle) under microscope (captured with video microscopy)

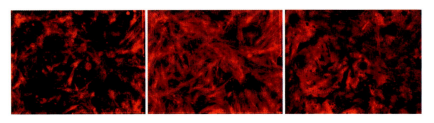

Figure 5.87 Cumulative summation values for information maps derived from the same video that demonstrate weighted "standard" [(p)log$_2$(p)] (left), "moderate" [log$_2$(p)] (middle), and "rare" [(1/p)log$_2$(p)] (right) event occurrence information for smooth muscle cells in the same non-native environment.

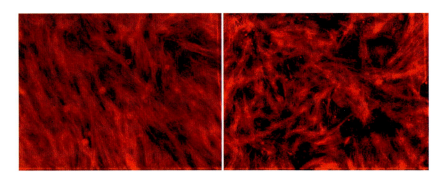

Figure 5.88 "Moderate" information maps for smooth muscle cells in two unique environments, native (left) and non-native (right)

internally to each system; the uncertainty of a pixel changing is a factor of its own history. Taken further, information that is generated in these systems may be comprehended as patterns of the probability of difference through time, whereby information is not simply the agglomeration of difference over time. Rather, information is the likelihood that difference will be encountered, and therefore is of a temporally structured nature (see Figure 5.87). Moments of surprise, or events within the field, can be tailored to highlight this likelihood (here common, moderate, or rare) within the data, resulting in weighted emergent patterns (see Figure 5.88).

These information maps visually demonstrate the predictability of change through time in a given environment, not just the layering of changing object morphologies through time. Here, the ability to discern changes in the behavior of cells in a given context is immediately distinguishable to both a lay and trained eye, becoming immediately evident via a visual understanding of the data presented. In the sciences, however, mere visual diagnostic results are only part of a broad set of indicators, both qualitative and quantitative, and thus a common question arises: "Can this be quantified?" Indeed from a statistical standpoint, the "p-values" latent in the field-based data of difference are of much higher significance than those typically produced via segmentation techniques, ushering in an alternate method by which these motile scenes may be distilled. But one must not overlook the inherent qualities of the visual maps produced, which in and of themselves offer an overwhelming account of behavioral signatures. Taken over multiple trials, macro-patterns are preserved though different; the intuitive eye is able to recognize the patterns present within these two cultured environments even though exact pattern is not preserved (see Figures 5.89 and 5.90). Here, number gives way to quality and ultimately structural pattern associated with fluent material in communion with its environment.

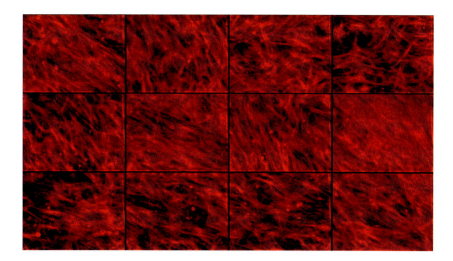

Figure 5.89 Twelve unique instances of moderate-event information maps for smooth muscle cells in a native environment

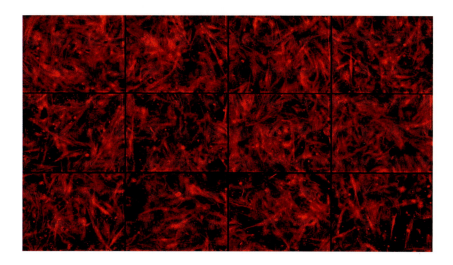

Figure 5.90 Twelve unique instances of moderate-event information maps for smooth muscle cells in a non-native environment

Part II: The Table Turned

Case study 2, demonstrating an observing mechanism through a hallway, has undergone the exact process outlined above (see Figure 5.91 hall difference diagram). While this type of experience would commonly be deemed "subjective," here there is no distinction made between this process and one in which difference is recorded from a stationary standpoint; the advent of difference within a field of data is the only factor of concern, be it "objective" or "subjective"—terms that bear no value here. Using the same tool, but having the generation of change propagate via the movement of the observer through space, a datascape begins to emerge (see Figure 5.92)—one that is much more closely linked to the emergent perceptual processes we undergo on an everyday basis.[12] Furthermore, one could begin to speculate upon the underlying structure of these mappings. To what do they correspond?

In *The Ecological Approach to Visual Perception*, James Gibson articulated (though diagrammatically) both the structure and flow of his *ambient information*

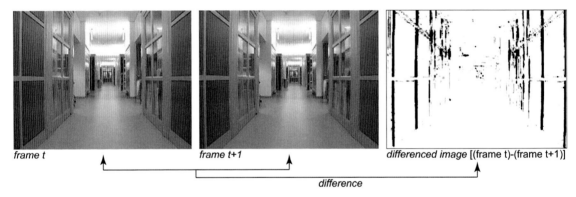

Figure 5.91 Difference in pixel contrast (right) between consecutive images (left and middle) while moving through hallway (captured with video camera)

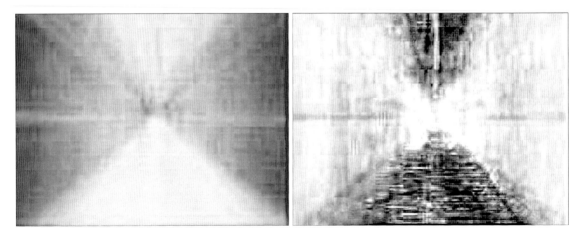

Figure 5.92 Cumulative summation values for information maps derived from a video as a camera is tracked down a corridor that demonstrate weighted "moderate" [$\log_2(p)$] (left) and "rare" [$(1/p)\log_2(p)$] (right) event information for an observer traversing a hallway

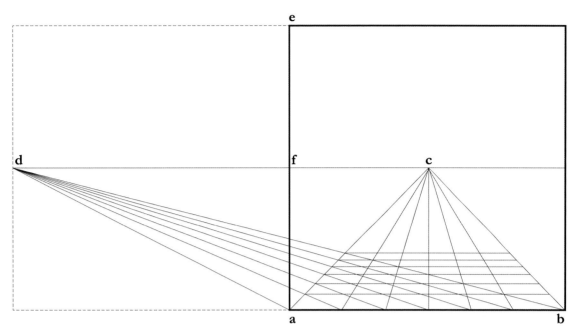

Figure 5.93 Diagram of Perspective Construction after Leon Battista Alberti: De Pictura (conceived 1435–1436): (ab) ground plane; (c) vanishing point (centric); (d) viewer's eye; (ae) picture plane; (df) viewing distance; (dfc) horizon line

array (see Figure 5.94).[13] Strikingly, one can draw immediate correlations within the information maps of movement through space (see Figure 5.92) with those resembling Gibson's ambient optic array diagram. Taken further, one can also discern distinct parallels between these maps and constructed perspectival space (see Figure 5.93), a point Gibson was certain to make but also distinguish as merely similar. For Gibson, rather, this phenomenon is a result of the variant and invariant structures of information-based experience within the environment. In Figures 5.92, the mappings produced are a result of the information generated while traversing space, specifically space with material reflecting structured ambient light, and not constructed through ideal means. While this pattern has arisen as a byproduct of a space created under Cartesian logics (the hallway), the reader should be reminded that this mapping was not fabricated by abstract means of geometry but rather generated as a consequence of engagement between the data within an environment and the observation of that data through time. This not only underscores the prescient nature of Gibson's theory of structured optical flow, but may also begin to suggest that the underlying form of vanishing point geometry (projected to a picture plane) is perhaps more fundamentally explained as an invariant aspect of the structure of information as one traverses space and therefore does not necessarily need to be relegated to the confines of formal geometric languages.

The studies and ideas presented in this research extend beyond disciplinary boundaries by way of scrutinizing the underlying field-based dynamic data present

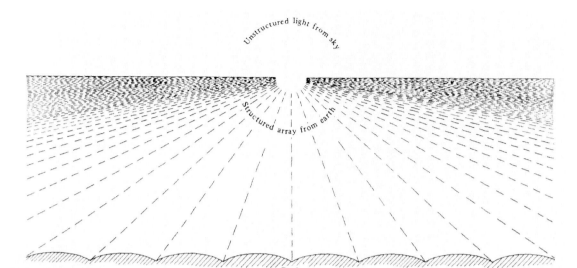

Figure 5.94 The ambient optic array from a wrinkled earth outdoors under the sky
Source: James J. Gibson from *The Ecological Approach to Visual Perception*.

in material systems, be they biological or architectural, through a trans-disciplinary approach to design and research that has allowed for the projective application of these speculative methods to find relevance beyond purely theoretical grounds and within seemingly disparate disciplines. Suggested above is a fundamental shift in how we conceive of and approach designing objects, environments, and affects. The studies presented, rooted in difference and information, serve to demonstrate a shift in thinking from the construction and production of objects and environments in geometric terms to one which places emphasis upon thinking organizationally through time about the aggregation of material systems and their phenomenal affects. While descriptive and projective geometric principles (and the analytic container in which they reside) continue to serve the architecture community, they are doubly burdened by the remnants of an idealized and reductive geometric world. Furthermore, these geometric principles and abstractions also provoke an *ex post facto* description and simplification of the world about us that also require a qualitative description to be ascribed to the entities (empty geometric husks) within that world. As an alternative, an approach rooted in information production inherently suggests an instantaneous perceptual construction of the world through which we traverse. This latter direction also takes into account the affectual attributes of the material organizations and their qualities with which we interact as participating agents while providing a framework and rigorous methodology in support of a discourse by which we may approach a set of design problems surrounding material formation and phenomena removed from their symbolic inheritance. This approach necessitates considering the actual parameters at play in the production of material organizations and affects by examining the much larger implications of difference, history, and information upon our perception of systemic environments as we shift away from an object-biased approach to the analysis, comprehension, and construction of the world in which we are actively a part.

Notes

1. Portions of this chapter were adapted from Andrew P. Lucia, Jenny E. Sabin, and Peter Lloyd Jones, "Memory, Difference, and Information: Generative Architectures Latent to Material and Perceptual Plasticity," *15th International Conference on Information Visualisation (IV)* (2011): 379–388. © 2015 IEEE. Reprinted, with permission, from Proceedings of the 15th International Conference on Information Visualisation (IV), 2011.
2. Gregory Bateson, "Every Schoolboy Knows. . .," in *Mind and Nature: A Necessary Unity* (Cresskill, NJ: Hampton Press, 2002), p. 28.
3. Ibid., p. 189. For Bateson, this would be more or less an admission that number is arising from the quantity. Here, the line is the number that is the distillation of vast quantities of light thresholds within a scene. And while for Bateson this provides evidence of his cybernetic model of the mind, for us it presents a problem in the form of tooling when one does not wish to be limited to that line.
4. James J. Gibson, *The Ecological Approach to Visual Perception* (Hillsdale, NJ: Lawrence Erlbaum Associates, 1986). Claude E. Shannon, *The Mathematical Theory of Communication*, (Urbana, IL: University of Illinois Press, 1949).
5. For a complete introduction, rationale, and background to the theories adapted and synthesized for this research, see Lucia *et al.*, "Memory, Difference, and Information . . .".
6. See George Mather, *Foundations of Sensation and Perception* (Hove: Psychology Press, 2009), p. 108. "Microsaccades seem to be essential for refreshing visual responses, perhaps preventing adaptation to unchanging stimulation. When the retinal image is perfectly stabilized visual sensations disappear entirely." Paraphrasing E.G. Heckenmueller, "Stabilization of the Retinal Image: A Review of Method, Effects, and Theory," *Psychological Bulletin* 63(3) (1965): 157–169.
7. "*Le dessin n'est pas la forme, il est la manière de voir la forme.*" From Paul Valéry, *Degas Danse Dessin*, trans. David Paul (New York: Princeton University Press, 1989).
8. Adapted from Gregory Bateson with the assumption that difference underlies form. Bateson does not make this specific claim of form, rather, the attribution to Bateson in this reference asserts that "difference is a nonsubstantial phenomenon not located in space or time; difference is related to negentropy and entropy rather than to energy" (Bateson, *Mind and Nature*, p. 85).
9. "Object detection is a computer technology related to computer vision and image processing that deals with detecting instances of semantic objects of a certain class (such as humans, buildings, or cars) in digital images and videos." from "Object Detection," Wikipedia, available at: https://en.wikipedia.org/wiki/Object_detection (accessed September, 8, 2015).
10. Notable examples would include the use of image differencing in astronomy, statistical methods in particle physics and fluid dynamics, and energy simulation to name a few. An exception to this discussion from a design standpoint would be Michelle Addington's study in which a compelling argument is made for CFD software and a reframing of the observer within space. However, Addington's argument does not directly broach issues of aesthetics, rather leaving the discussion firmly situated in issues of energy transfer and efficiency. Michelle Addington, "The Phenomena of the Non-visual," in *Softspace: From Representation of Form to a Simulation of Space*, ed. Sean Lally and Jessica Young (London: Routledge, 2007), pp. 39–50.
11. As this is being written, the advent of workable virtual reality technology may allow for truly N-dimensional experiential immersions, those that move beyond projection to a picture plane. While moving beyond this planar registration, an important step in overcoming geometric boundaries, this technology is not an exception to the argument made here, that of the role of information within the imminent formation and quality of material bodies.
12. The studies presented here do not aim to mimic the exact physiological circumstances under which the body's perceptual systems operate and therefore are not a simulation of information as it specifically acts within sensory apparatuses. Rather, this is a way of thinking and working through types of informational operations that parallel perceptual processes. In doing so we may begin to probe the types of patterns that arise under such statistical circumstances. The advanced reader is invited to read Kenneth Norwich's

Information, Sensation, and Perception, a theory of information theoretic perception from a neurological standpoint, the basis of which could be utilized to effectively create such a simulation. Kenneth H. Norwich, *Information, Sensation, and Perception* (San Diego, CA: Academic Press, 1993).

13   According to Gibson:

> This is an optic array at a single fixed point of observation. It illustrates the main invariants of natural perspective; the separation of the two hemispheres of the ambient array at the horizon, and the increasing density of the optical texture toward its maximum at the horizon. These are invariant even when the array flows, as it does when the point of observation moves.

Case Study: Microfabrication: Spatializing Cell Signaling and Sensing Mechanisms

Keith Neeves

The following case study explores fabrication, architecture, feedback, and cell signaling within biological systems at multiple length scales. Keith Neeves was a post-doctoral fellow in the Scott Diamond lab at the Institute for Medicine and Engineering, University of Pennsylvania. He now directs the Neeves Lab in Chemical and Biological Engineering, Colorado School of Mines.

The behavior, or phenotype, of cells is dictated by their response to external chemical and mechanical signals. Cells have a variety of receptors on their surface that can sense chemicals and feel forces. Cells have a characteristic length of 1–10 µm and their receptors have dimensions on the order of 10–100 nm. Consequently, to understand why one set of external signals yields a unique phenotype requires manipulation of matter at these length scales. To achieve this precise spatial control, researchers have increasingly adapted micro- and nanofabrication methods, which were initially developed for the microelectronics industry, to the study of biological systems. These fabrication methods are a means of mimicking the architectural, chemical, and mechanical environment of normal and diseased tissues.

The external signals that cells sense can be separated into three categories: (1) surface-bound molecules; (2) soluble molecules; and (3) mechanical forces. Surface-bound molecules are physically or chemically bound to the extracellular matrix or surface of other cells. Soluble molecules are free to move about in the extracellular space or be transported in the blood and lymphatic system. Mechanical forces include phenomena like shear forces imposed by flowing blood on vessel walls and the compression of cartilage in the knee. We are usually unable to independently manipulate these signals in the human body or animal models, so *in vitro* models are necessary to determine mechanisms.

The conventional *in vitro* method for measuring cellular behavior is to culture a single cell type in a dish and measure its behavior in response to one or more signals. For example, if we are interested in the role of a certain soluble molecule, then we add it to the growth media. If we are interested in a certain surface-bound molecule, then we grow cells on a surface coated with it. What these studies neglect is the spatial presentation of these signals. In the body, soluble molecules are typically presented as a finite burst from adjacent cells or as a sustained gradient to guide motility. Similarly, the geometrical presentation of surface-bound molecules can tell a cell whether to differentiate into another cell type, divide, or die. Microfabrication methods provide a bridge between the physiological and complex *in vivo* environment and the controlled, but oversimplified cell culture methods. In this case study we will present one example where these methods have been useful in elucidating mechanisms and diagnosis: the formation of blood clots.

Measuring Platelet Adhesion in a Microfluidic Vascular Injury Model

The formation of a blood clot is a complex process that involves all three of the aforementioned external signals: soluble molecules, surface-bound molecules, and mechanical forces. In a normal vessel, endothelial cells line the wall and secrete several soluble platelet inhibitors (Figure 5.95 A).

These inhibitors keep platelets quiescent as they move through the vascular system. At the moment of a vascular injury, platelets transform from passive particles to dynamic building blocks for a clot. A vascular injury is typically characterized by the exposure of the subendothelial matrix that lies underneath the endothelial cells. The subendothelial matrix is composed of a number of surface-bound proteins that platelets adhere to, in particular, collagens and von Willebrand factor (VWF) (Figure 5.95 B). The mechanism by which platelets adhere to the subendothelial matrix is shear stress-dependent.[1] Shear stress, the force that platelets feel due to blood flow, can vary over two orders of magnitude

A. Patent vessel

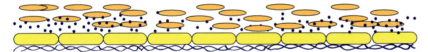

B. Platelet adhesion to subendothelial matrix of an injured vessel

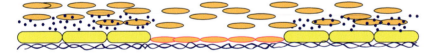

C. Generation and release of soluble platelet agonists from adhered platelets

D. Growth and stabilization of clot

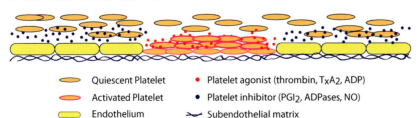

Figure 5.95 The external signals that lead to the formation of a blood clot. A. In a patent vessel the endothelial cells secrete inhibitors that keep platelets quiescent. B. Upon injury of the vessel, a protein rich bed called the subendothelial matrix is exposed. The surface bound proteins in this matrix promote platelet adhesion. C. Following initial adhesion, platelets secrete soluble agonists that recruit new platelets to the clot. D. The clot is stabilized against shear forces by the interactions between platelets within the interior of the clot.

between the venous (low shear stress) and arterial (high shear stress) vessels. At low shear stresses, platelets can directly adhere to the different types of collagens in the subendothelial matrix. However, at high shear stresses, platelet-collagen bonds cannot form fast enough before platelets are washed away. For platelets to adhere at these high shear stresses, they need to bond to VWF. Platelets can then roll along VWF to slow themselves down before firmly adhering to collagen.

To study platelet adhesion *ex vivo*, we need to recreate the shear stresses of flowing blood and a finite presentation of subendothelial matrix proteins. The first step is to define the injury by micropatterning the matrix proteins. Micropatterning refers to a group of methods used to pattern molecules on the micrometer scale, which includes microstamping,[2] lift-off,[3] and microfluidics.[4] Here, we'll focus on microfluidics because it is best suited for growing thin collagen films in a defined area. Microfluidics are simply small fluid channels with characteristic dimensions of 1–100 μm. There are many ways to make microfluidic channels, in this example, we use soft lithography. Soft lithography is a method in which microfabricated features on a wafer are molded into an elastomeric polymer.[5]

To pattern the collagen fibers, we introduce a solution of soluble collagen into the channel and allow it to polymerize into fibers on a glass surface (Figure 5.96 A).[6]

Following polymerization, we remove the microfluidic channels and rinse away any unattached collagen. Next, a second set of channels that runs perpendicular to the collagen strips is reversibly bonded to the glass slide. Whole blood is then infused through the channels at a desired shear stress (Figures 5.96 B and 5.96 C). We can label the platelets with a fluorescent tag to track their accumulation on the collagen strips as a function of time and shear stress (Figure 5.96 D). Following the experiment we measure total platelet accumulation using light microscopy (Figure 5.97).

We then remove the channels and conduct high resolution light or electron microscopy on the resultant blood clot to study platelet morphology (Figure 5.98).

This procedure of flowing whole blood over micropatterned collagen mimics two external signals that act on platelets. First, mechanical shear forces that are similar to physiological shear forces are felt by the platelet due to the fluid motion

Figure 5.96 Schematic of collagen patterning and blood flow experiment. A. A strip of type I collagen (100 μm) is deposited and immobilized by microfluidic patterning along the length of a glass slide. B. A second microfluidic device with a set of channels is oriented perpendicular to the patterned collagen. Blood is then infused at physiological shear rates. C. Microfluidic device with thirteen microfluidic channels is vacuum-bonded to a glass slide and mounted on a fluorescence microscope. D. Deposition of fluorescently labeled platelets on micropatterned collagen.

Source: Modified from Neeves *et al.* (2008) and reproduced by permission of Blackwell Publishing.

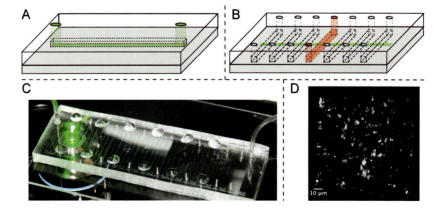

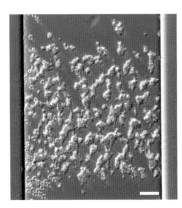

Figure 5.97 Platelet accumulation on patterned collagen within a microfluidic channel. Islands of platelet aggregates cover about 50% of the surface at a wall shear rate of 100 s^{-1}. Scale bar = 50 μm.

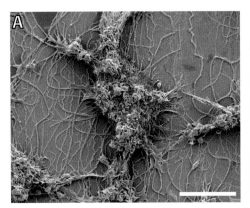
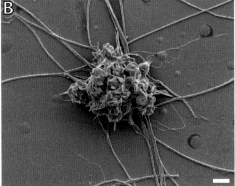
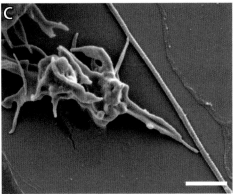

Figure 5.98 Scanning electron micrographs of platelet aggregates formed under flow on type I collagen at 2500X, scale bar = 10 μm (A); 5000X, scale bar = 1 μm (B); 20,000X, scale bar = 1 μm (C).

through the microfluidic channel. Shear forces are those forces caused by adjacent fluid sheets slipping past each other in the direction of flow. In a channel, the shear force is highest at the walls and zero in the middle of the channel. Therefore, platelets feel the highest forces near the wall, which is where they must adhere to collagen, the second external signal. The beauty of the clotting system is that platelets only stick to surfaces that are signals for injury, in this case, collagen and VWF. Individuals with defects in the VWF gene have a disease called von Willebrand disease, which is the most common bleeding disorder. In most cases of VWD, platelets do not adhere and aggregate as well at high shear stresses. Therefore, the microfluidic vascular injury model has a possible use as a diagnostic for this disease. We have found that this model is capable of discriminating between blood samples from healthy individuals and individuals with bleeding disorders.

Simulating the Release of Platelet Agonists

The recruitment of new platelets to a growing clot is, in part, a function of soluble molecules released from or generated on adhered platelets. Platelets have receptors for these soluble molecules, called agonists, on their surface. The most potent agonists are thrombin, ADP, and thromboxane A2. Many common drugs inhibit the action of platelet agonists. Anticoagulants like heparin inhibit thrombin. Common nonsteroidal anti-inflammatory drugs (NSAIDs) such as aspirin and ibuprofen prevent thromboxane production. Clopidogrel (Plavix®), which is used to inhibit clotting in individuals with vascular diseases, binds irreversibly to the ADP receptor on platelets. In a flow experiment where platelets adhere to a prothrombotic surface like collagen, it is difficult to measure how much of these agonists are produced endogenously. Therefore, it is unknown how platelet agonists act synergistically under flow to promote clot growth. In the microfluidic vascular injury model, we showed how micropatterning allows for control over the spatial presentation of surface bound molecules. In a second model, the membrane microfluidic model, we have developed a method for controlling the spatial presentation of soluble molecules (Figure 5.99).[7]

In this model, a membrane is sandwiched between two microfluidic channels. In the bottom channel, we flow blood at a controlled wall shear stress, similar to the microfluidic vascular injury model. Instead of flowing blood over collagen to promote platelet adhesion, we introduce a platelet agonist through the membrane into the flowing blood. The amount of agonist that comes through the membrane per unit time per unit area is called the flux. We can adjust the flux by varying the pressure gradient across the membrane and/or the concentration of flux in the top channel.

Figure 5.100 shows an example of how the agonist flux affects platelet aggregation. In the experiment, we infused human whole blood at a wall shear rate of 250 s^{-1} and introduced ADP at fluxes of 1.5, 2.4, and 4.4×10^{-18} mol/µm^2 s. The area and height of the platelet aggregates depend on ADP flux. There are no platelet aggregates formed at the lowest flux. At the middle flux, we observed aggregates of 20–250 platelets, which tended to be single layers of cells. Aggregates of up to 1000 platelets were observed at the highest flux. These large aggregates contained multiple layers of platelets.

The relationship between solute flux and stable clot formation is unknown because no previous methods have allowed for controlled fluxes of agonists into

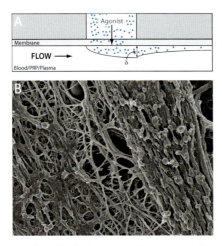

Figure 5.99 A membrane-based microfluidic device allows for the controlled presentation of soluble agonists to flowing streams of blood. A. The device consists of two microfluidic channels running perpendicular to each other and separated by a membrane. The properties of the membrane and the relative pressures in two channels dictate the rate, or flux, that the agonists are introduced into the flowing stream of blood. B. Scanning electron micrograph of a clot formed when thrombin is introduced into whole blood flowing at a wall shear rate of 250 s^{-1}.

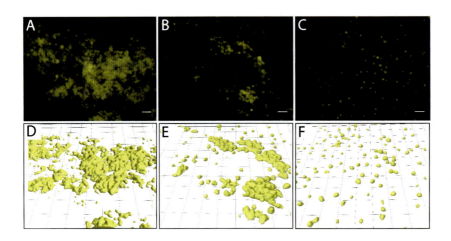

Figure 5.100 Spinning disk confocal images (A–C) and volume reconstruction (D–F) of platelet aggregates formed by introducing ADP into flowing whole blood at a defined flux; 4.4×10^{-18} mol µm^{-2} s^{-1} (A,D), 2.4×10^{-18} mol µm^{-2} s^{-1} (B,E), 1.5×10^{-18} mol µm^{-2} s^{-1} (C,F). Scale bar = 10 µm. Modified from Neeves et al. (2008) and reproduced by permission of the Royal Society of Chemistry.

flowing blood. The membrane microfluidic method fills this technology gap. There are a few diseases where dysfunction in agonist flux leads to bleeding or thrombotic episodes. One example is hemophilia, where deficiencies in clotting factors attenuate thrombin flux, which ultimately leads to unstable clots. The membrane microfluidic method provides a means of manipulating solute flux that could be useful for studying hemophilia and other bleeding disorders.

Microfabricated Models of the Interstitial Space of a Blood Clot

Unlike the interactions between platelets and adhesive surfaces outlined above, little is known about the interactions between adjacent platelets during the growth and

Figure 5.101 To investigate how drugs move through a clot, we developed microfabricated blood clot mimics in silicon wafers. A. The polygons of silicon represent the volume occupied by platelets in a clot. Here we have modeled an occlusive clot in a microfluidic channel. Scale bar = 500 µm. B. The spaces between the polygons represent the interstitial space through which drugs can penetrate. Scale bar = 100 µm.

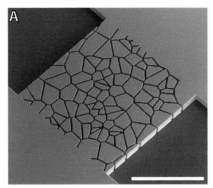 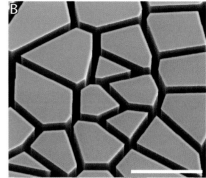

stabilization of a clot. One pertinent question is how molecules move between the very small gaps between cells. This question is important because drugs that are designed to break up clots must penetrate into the interior of a thrombus through these gaps. Ideally, we want to formulate drugs to enhance their penetration into the interior of clots. This is difficult to do in tissue or even *in vitro* because we cannot independently manipulate or measure the geometry of the interior of a clot.

An alternative technique is to take what we know about the architecture of a clot and then recreate it in a completely synthetic system. Since we are concerned primarily about geometry, rather than biological function, we will replace cells with microfabricated features that mimic the size and shape of cells (Figure 5.101). The orientation and proximity of these features to each other mimic some of the features observed in actual clots. For instance, a post-mortem analysis of an occlusive thrombus in the coronary artery reveals that platelets have a gap width of 10–100 nm near the vessel wall. Platelets are ellipsoids with a major axis of ~2.5 µm, a minor axis of ~1 µm, and thickness of 500 nm.

While we can certainly recreate the exact dimensions of the platelet and the gaps between platelets, that small scale does not lend itself to observation by light microscopy. Since the wavelength of light is ~100s nm, it is difficult to resolve features smaller than 200 nm. To recreate this geometry in a microfabricated device, we are going to use a common engineering concept of scaling analysis. Scaling analysis is used to either scale a process to a larger or smaller size, while maintaining the same operating conditions. For example, to design an oil pipeline, we would begin with characterizing the flow in a pilot scale pipeline 1/100th of the size of the final pipeline. In this pilot scale pipeline, we would measure the pressure drop as a function of pump rate. We would characterize these parameters in terms of non-dimensional numbers. Non-dimensional numbers describe the underlying physics independent of size. In the case of pipe flow, the Reynolds number describes the flow in the pipe independent of its size. The Reynolds number is a ratio of the inertial forces (how fast the fluid is moving) to the viscous forces (how much the fluid resists flow). When we scale-up from our pilot scale pipeline that is 1 inch in diameter to pipeline that is 10 feet in diameter, we hold the Reynolds number constant.

We can apply a similar scaling argument to our synthetic blood clot. If we scale up our blood clot by a factor of 100, then the platelets have a characteristic length scale of 100 μm and the gap width becomes 10 μm. Our work with synthetic blood clots is in its nascent stages. From early results we have found that drug penetration is dictated by the size, geometry, and connectivity of the interstitial spaces between cells. When we embed these synthetic clots in a microfluidic channel, we observed that only small molecular weight drugs can penetrate deep into the interior of a clot. This result suggests that some of the cutting-edge therapeutic strategies that involve large proteins or nanoparticles may not be relevant for treating large thrombi such as those that form following the rupture of atherosclerotic plaques.

Conclusion

Microfabrication methods allow us to perform experiments on biological systems that were not previously possible. One of the primary features of these methods is control over the presentation of external signals at physiologically relevant spatial scales. In this case study, we presented a few examples of the power of this approach. Each example included one or more external signals known to play a crucial role in the formation of a blood clot. The initial event of clotting is the exposure of a prothrombotic surface-bound molecule to which platelets will adhere. The mechanism of adhesion is a function of the mechanical stress that platelets experience at the vessel wall. We have modeled platelet adhesion using a combination of micropatterning and microfluidics. Micropatterning provides a means of presenting a finite stimulus to flowing platelets. Microfluidics channels were used to recreate the shear stresses and geometry of the microvasculature. Following this initial event, adhered platelets recruit new platelets to an injury by releasing soluble agonists. We have modeled this process using a membrane-based microfluidic device in which fluid from one channel was introduced into another channel at a defined rate. Finally, we have effectively mimicked the interior of a blood clot through the design and fabrication of a silicon chip. We can use this mimic to design better drug delivery systems and protocols for clot dissolution.

Notes
1. Z. M. Ruggeri and G. L. Mendolicchio, "Adhesion Mechanisms in Platelet Function," *Circulation Research* 100(12) (2007): 1673–1685.
2. John C. Chang, Gregory J. Brewer, and Bruce C. Wheeler, "A Modified Microstamping Technique Enhances Polylysine Transfer and Neuronal Cell Patterning," *Biomaterials* 24(17) (2003): 2863–2870.
3. Rebecca J. Jackman, David C. Duffy, Oksana Cherniavskaya, and George M. Whitesides, "Using Elastomeric Membranes as Dry Resists and for Dry Lift-Off," *Langmuir* 15(8) (1999): 2973–2984.
4. Isabelle Caelen, André Bernard, David Juncker et al., "Formation of Gradients of Proteins on Surfaces with Microfluidic Networks," *Langmuir* 16(24) (2000): 9125–9130.
5. J. Cooper McDonald, David C. Duffy, Janelle R. Anderson et al., "Fabrication of Microfluidic Systems in Poly(dimethylsiloxane)," *Electrophoresis* 21(1) (2000): 27–40.
6. K. B. Neeves, S. F. Maloney, K. P. Fong et al., "Microfluidic Focal Thrombosis Model for Measuring Murine Platelet Deposition and Stability: PAR4 Signaling Enhances Shear-resistance of Platelet Aggregates," *Journal of Thrombosis and Haemostasis* 6(12) (2008): 2193–2201.
7. Keith B. Neeves and Scott L. Diamond, "A Membrane-based Microfluidic Device for Controlling the Flux of Platelet Agonists into Flowing Blood," *Lab on a Chip Lab Chip* 8(5), (2008): 701–709.

6
BioInspired Materials and Design
Jenny E. Sabin

Biology and materials science present useful conceptual models for architects to consider, where form is in constant adaptation with environmental events. Here, geometry and matter operate together as active elastic ground—a datascape—that steers and specifies form, function and structure in context. Through direct references to the flexibility and sensitivity of the human body, we are interested in developing adaptive materials and architecture where code, pattern, environmental cues, geometry and matter operate together as a conceptual design space.

It is the embedding of material systems with biological relationships and behavior to generate analogic models that are enmeshed with feedback.

(Jenny E. Sabin)

Introduction

Building upon the foundation laid over four years of collaborative teaching in the course entitled "Nonlinear Systems Biology and Design," a fifth iteration of the course was introduced in the Department of Architecture at Cornell University in 2012 entitled "BioInspired Materials and Design." Similar to the previous course, this installment investigates the intersections of architecture and science, but with a special focus upon the potential application of insights and theories from biology and computation to the design of material structures. The course draws heavily upon our collaborative research project entitled eSkin (see Part IV) and the development of tools and methods for speculative bioinspired materials and adaptive building skins. Overall, the course investigates biologically informed design through the visualization of complex datasets, digital fabrication, and the production of experimental material systems for prototype speculations of adaptive building skins. Part seminar and part workshop, the elective serves to deepen knowledge of biological complexity from a radically different viewpoint, and to develop design thinking and tooling through the study of complex systems of matter, alongside an introduction to advanced scripting logics in parametric and associative software and the production of hybrid material systems, including 3-D printing fabrication.

As with the eSkin project, our emphasis rests heavily upon the study of natural and artificial ecology and design, especially in observing how cells interacting with pre-designed geometric patterns alter these patterns to generate new surface effects through passive means. Central to this is the analysis and modeling of environmentally specified changes in geometry, pattern, and material at multiple length scales. Patterning options are highly flexible, ranging from geometric meshes, gradients, aperiodic and periodic patterns, and infinite color variations. The following student projects present design explorations into bioinspired adaptive and programmable materials. In some of the featured projects, such as Structural Color at the Architectural Scale, material features and effects at the micron scale are studied, translated and applied at the architectural scale, while in other projects, such as Dynamic Degeneration and Reconstruction of the ECM, the work remains speculative and potentially multi-scalar. In all cases, the intent was to instill new habits of thought in design, enriched and rigorously informed by material agency, the dynamics of cellular biology, and alternative models for addressing the topic of adaptive architecture.

Case Study: Structural Color at the Architectural Scale
Ross John Amato, Gabriel Wilson Salvatierra, and Elena Sophia Toumayan

BioInspired Materials and Design
Instructor: Jenny E. Sabin
Visiting lecturers and critics: Kaori Ihida-Stansbury, Jae-Won Shin and Shu Yang

In basic terms, photonic crystals are nanostructures that refract light at differing indices. When light waves hit a photonic material or structure, they create optical interference, which causes the surface to appear colorful and luminous. The surfaces of certain organisms use these structural patterns that interfere with wavelengths to produce a more vibrant and unfading appearance than a pigmented surface. This project looked at datasets where the effectiveness of these structures is governed by the specific orientation of these sub particles and the way they refract light.

The two datasets provided by Shu Yang and her group at the University of Pennsylvania showed photonic crystals being deformed by two different external forces: physical stretching and heat pressing (in the project referred to as "Physical

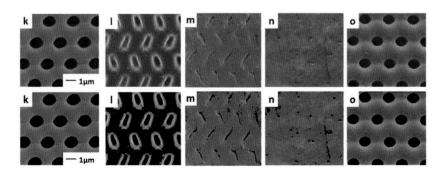

Figure 6.1 Research background

Source: Drawings and diagrams based on data generated by Shu Yang Group at University of Pennsylvania.

Exertion" and "Subject to Heat" respectively). The structural and photonic phenomena present in the datasets are limited to their specific length scales because they exist exclusively at a scale that interferes with wavelengths of visible light. This constraint factor restricts models to this super-molecular scale in order for them to remain functional.

After analysis of the datasets and a translational rule-set, we produced a model informed by relationships found within the data. A series of different analytical methods and translation techniques were employed to aim for a reactive built prototype that could be realized at multiple scales and for multiple uses. Though the translational model is printed at the macro scale and does not function within the wavelength of light, the prototype begins to have spatial and visual effects arguably similar to the outputs of the biological precedents. Embedded within the final model are notions of light filtration, gradient, operability, and visual control.

The first steps of the analysis involved the extraction of curves from the two datasets and modeling the movement of photonic crystals from open to closed. These two datasets were compared to an Idealized Deformation of the geometry: a perfect circle deforming into an elongated ellipse at 45°. In Idealized Deformation, the circumference remains constant while the area decreases at a constant rate. The Physical-Exertion dataset showed similar trends in the deformation of the geometry, with some expected deviations due to data imperfections. The Subject to Heat dataset exhibited a significant jump in circumference and area, going from open to closed but remained fairly regular, going from closed to open. Furthering these investigations, the curves of the three datasets were extruded against time and lofted to generate 3-D visualizations of the behavior.

Since photonic crystal systems exist in the form of arrays, we generated arrays from the analysis of a single aperture to represent a whole system. This is done, assuming that the apertures remain constant throughout the array. The next step considered the varying of apertures throughout the system. Though not yet proven to exist in reality, we thought the potential for a gradient photonic array to be compelling. As a result, gradients were generated using the varying apertures from each dataset. Gradients were also visualized as another documentation

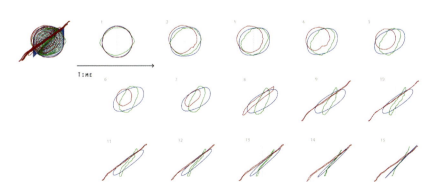

Figure 6.2 Distortion of geometry overlay; these drawings directly compare the relationships of the three datasets. It shows deviation from the origin and changes in circumference and area in the last two time frames.

Note: Red = Subjected to heat; Blue = Idealized deformation; Green = Physical exertion.

of the rate of aperture change. For example, gradient transition for Idealized Deformation is constant while Subject-to-Heat exhibits noticeable change when the gradient transition is abrupt. The same method for generating arrays was implemented for the Subject-to-Heat dataset that maps the stress area found in the system for each degree of aperture. This stress-area dataset was broken down into its component parts, generating a chart showing the percentage of stress mapped against video frames. Similarly, embedded in the translation model are relationships found within the data itself. Constructed along the lines of a Cartesian grid, the formal qualities of the translation contain numerical values generated from analysis of the data.

Figure 6.3 Generated array and simulation of aperture gradient

Figure 6.4 Generated aperture array. Subject-to-heat stress area data; the same method for generating arrays and an aperture gradient was used with the subject-to-heat video dataset that maps the behavior of the systems in terms of stress.

BioInspired Materials and Design   243

Figure 6.5 Generated aperture gradient

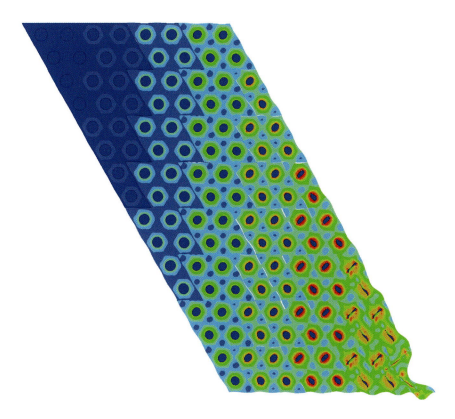

The logic behind the translation breaks down the Subject-to-Heat dataset along the x–y grid. Each curve was divided into 100 points mapped along the x-axis, 15 frames going from opened to closed mapped along the y-axis. At each of the points on this grid, a deformation occurred based on three methods of analysis, thickening each individual point in the x, y, and z axes:

- x-axis deformation compares the ideal to the real. This method of analysis calculates the distance between idealized and actual geometries to generate the x-axis deformation of the grid point, demonstrating differences in rate of change of the openings.
- y-axis deformation shows point displacement from one frame to the next, calculating the distance between each individual point (0–100) as it moves through the 15 frames to generate the y-axis deformation of the grid points. This shows comparative variations in elongation of individual frames.
- z-axis deformation analyzes roughness and imperfections in the dataset by calculating the distance between the imperfections of the real curves compared to their smoothed-out counterparts, generating z-axis deformation of the grid point. Impurities and oscillations of the real curve can be compared to an idealized version that maintains the same rate of change, unlike the idealized geometry.

This translation step produced a scale-less physical model. Though a representation of trends found in the data, the transformed model demonstrates properties that exhibit potential uses for solar radiation reactivity of a building façade.

Limiting the model to the x–y dimensions, this prototype was altered to react to a solar radiation video, with instances of higher radiation producing a more closed condition on the façade and instances of lower radiation producing a more open condition. Two methods of reactivity were used. The first is "Long Duration Extremes," which remaps the values of aperture based on the extreme radiation values over the course of one day (entire video). This results in lower variation of aperture from one side to the other. The second method is "Momentary Duration Extremes," which remaps the values of aperture based on the extreme radiation values present in one instance (movie frame). This produces a gradient range of completely open to completely closed apertures in each frame. It maps the highest and lowest radiation values present in that individual frame rather than relative to the entire video. The architectural intent is to generate an adaptive façade condition that is reactive to solar radiation conditions.

Finally, we speculated how a reactive façade might function. Essentially, the distribution of solid to void is shifting from place to place on the façade according to external conditions. In order for this redistribution of aperture to function, the volume of the entity may not be altered. Mass may not be added or subtracted to the unit. The existing mass may only be redistributed to produce the reactive qualities. As a result, the two-dimensional x–y prototype in the

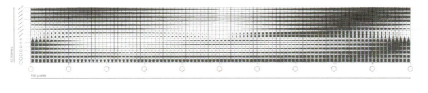

Figure 6.6 Logic of translation: Constructed along the lines of a Cartesian grid, the formal qualities of the translation contain numerical values generated from analysis of the data.

Figure 6.7 Solar radiation stills: the transformed model demonstrates properties that exhibit potential uses in solar radiation reactivity for a building façade.

BioInspired Materials and Design 245

Figure 6.8 Volumetric application in all axes

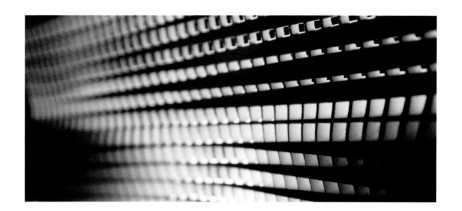

Figure 6.9 Final model

previous slide gains a third axis in the z direction to compensate for redistribution of aperture, maintaining a constant volume. The z-axis variable also considers structural integrity as the wall thickens where the members start to become too thin in the x–y directions and vice versa. This system could be realized at various scales, though specific material or medium is not specified. However, the rule-set governing the model is based on general knowledge of volumetrics in order to propose a system that could be reactive to certain conditions through redistribution of a medium.

Case Study: Dynamic Degeneration and Reconstruction of the ECM
Sean Kim, Ryan Peterson, and Maria Stanciu

Bioinspired Materials and Design
Instructor: Jenny E. Sabin
Visiting lecturers and critics: Kaori Ihida-Stansbury, Jae-Won Shin and Shu Yang

The central focus of this study was to investigate the process of degeneration in the cellular structure of diseased cells. The basis of our exploration is derived from research by Kaori Ihida-Stansbury in Peter Lloyd Jones's lab on the use of scaffolding to repair the extracellular matrix (ECM) of lung cells afflicted with pulmonary hypertension (PAH).[1] Ihida-Stansbury provided two images, one

Figure 6.10 Normal artery matrix (left) versus PAH matrix (right)

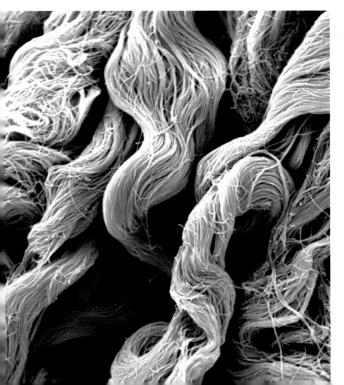
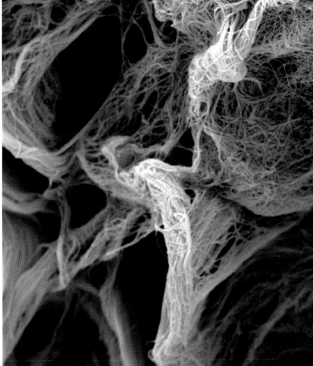

BioInspired Materials and Design 247

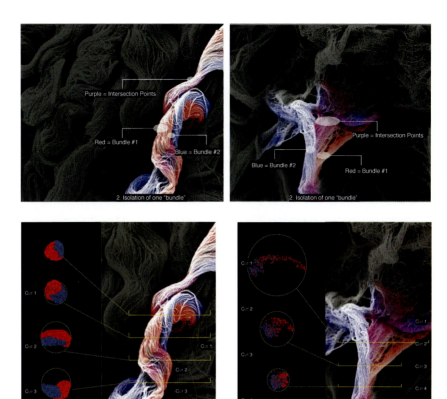

Figure 6.11 Normal artery matrix (left) versus PAH matrix (right). Dividing the isolated, larger bundle into two bundles; blue and red represent the two different groups, while purple indicates where the two are speculated to intersect.

Figure 6.12 Sections along the matrix demonstrate the various movements (weaving, twisting, bundling, expanding) of the two different groups of smaller bundles, in relation to each other.

showing the ordered bundles of structure in a normal artery matrix, and the other portraying the chaotic, frayed structure of a PAH matrix. What we found intriguing was that there was no evidence showing the stages that the ECM undergoes as it deteriorates from a healthy to a diseased cell. Our aim was to identify relationships between structural members in the images of the ECM and establish parameters that would illustrate the degeneration of the ECM bundles.

Our analysis of the images concluded that the ECM is composed of a series of bundled fibers twisted around one another. Within each of these bundles, fibers are also grouped into smaller bundles, which in turn rotate around each other. The structure of a normal artery matrix consists of varying scales of relationships between fibers. Although there is no uniform shape and rotational property within the bundle structure and sub-structure, there is an order present between each element that allows for a unified composition of non-uniform bundles. In the PAH cell there appears to be an absence of order. Some of the individual fibers appear to cluster together but the relationships between them are not uniform like those in the normal matrix. It also appears as if the ends of each of the bundles are no longer confined in a tight bundle but instead have sprawled or frayed outward. This creates drastically larger spaces between each bundled element and appears to create the expansion of the cell that is prevalent in cells experiencing hypertension.

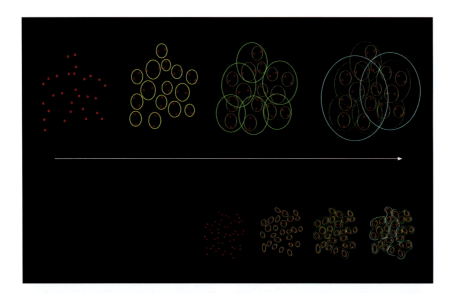

Figure 6.13 Grouping: Diagram showing how the individual threads group themselves on various different levels.

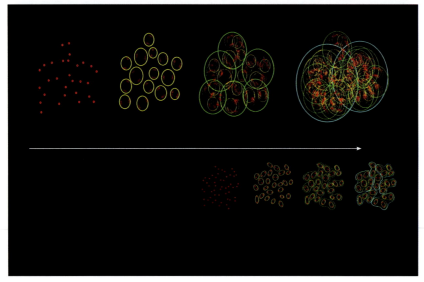

Figure 6.14 Rotation: Diagram showing how the individual threads rotate independently, leading to further rotations within the different levels of groups.

In mapping the ECM in a three-dimensional environment, we subdivided the bundle into a series of control points along horizontal sections through the bundle. We created an ideal bundle that was composed of five circular sections of equal radius to represent the form of the overall bundle. Each one of these circles contained multiple circles to simulate the sub-structure bundles, each populated by a number of points. A line could then be interpolated between each corresponding point on these sections to create the individual bundle fibers. To demonstrate the twist of the bundle, the initial circles were rotated along with interior circles at higher rates so that both sets of rotational patterns could be observed.

Our goal was to show the degeneration of structure that was absent in our two initial photos, so instead of creating an entirely new script for the diseased

BioInspired Materials and Design   249

bundle, we reused our computer script simulating normal cells, but added factors to the parameters to produce effects present in the diseased ECM. In this adapted script, the radius of the circles changed at the two polar ends of the bundle; an increase in the bundle rotation would therefore coincide with an increase in the expansion and displacement of control points along the ends of the bundle. This

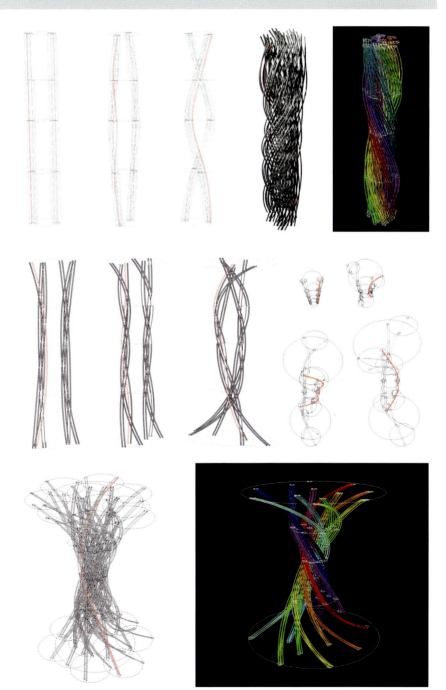

Figure 6.15 Sequence normal artery matrix: Isolation of one thread. One large bundle: Different colors indicate the smaller groups of bundles within the larger bundle. Within the different colored bundles, each tube represents the threads.

Figure 6.16 Sequence PAH matrix: Isolation of one thread. One large bundle: Different colors indicate the smaller groups of bundles within the larger bundle. Within the different colored bundles, each tube represents the threads.

displacement was at first uniform, but a randomizing vector and distance were added later to the movement of these outer control joints in order to accurately simulate the behaviors of diseased bundles.

We also explored physical modeling of the bundles to account for materiality and other characteristics that were absent in the scripted Rhino Grasshopper model. In the 3-D model, when two fibers collide, they pass through one another, whereas in the real world these fibers would actually collide, and alter each other's movement. Our initial physical model was composed of a line of strings bound in a box. These strings were connected to a secondary system of strings at specific points, which were in turn threaded along a large bolt. As the bolt was twisted, it would cause the secondary system of strings to pull the main system of tensioned strings so that they would rotate about one another. Instead of controlling rotation at the poles as in the 3-D model, rotation was orchestrated by the manipulation of the interior control points of the fibers. Our final physical study was an analog representation of the Grasshopper model, where three sets of bundles rotate around a central axis and rotation is controlled only at the ends of the strings.

Our research did not conclude with the discovery of how cells actually degenerate, but provided an educated speculation as to how this degeneration might occur. In exploring the script in a physical environment we encountered problems in creating a system that could be both easily controlled and complex enough to portray the different scales of rotational relationships. We were able to portray the inner rotational relationships but never the uniformed overall rotation of the bundle. This was because we lacked control of interior points, whose location distance and proximity value could not be controlled, and therefore, collisions increased at the center of the bundle. From this observation, we believe that if a system can be enacted to order these interior control points, then structural failure

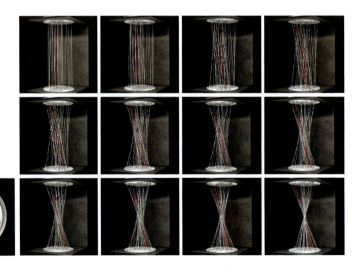

Figure 6.17 Physical iteration derived from digital model; in the sectional control method, fixed points of string attachment allow for variations in weaving within bundles

may be preventable. If ECM bundles can be implemented with a mechanism or switch that maintains the relative location of control points along an individual bundle fiber, then the order of rotational relationships can be maintained and the unordered expansion/densification of the PAH bundle can be prevented. This also presents a series of adaptive structural and material logics for exploring bundled fibrous assemblages at multiple length scales for architectural speculation.

Note

1 K. Ihida-Stansbury, S. Sanyal, C. Navarro, P. Janmey, H Sundararaghavan, J. Burdick, and P.L. Jones, "Lung Vascular Smooth Muscle Cell Differentiation Controlled by Elastic Matrix," *Molecular Biology of the Cell* 21, suppl. (12/2010).

Case Study: BioInspired Skin Systems and Dynamic Boundary Conditions
Joon Hyuk Choe, Danlu Li, and Tzara Peterson

Bioinspired Materials and Design
Instructor: Jenny E. Sabin
Visiting lecturers and critics: Kaori Ihida-Stansbury, Jae-Won Shin and Shu Yang

In observing the bio-cellular behavior of smooth muscle cells in varying extracellular matrix (ECM) conditions (dataset "col TN-C002"), this project focused on the boundary conditions between the two ECMs, pillared and non-pillared. The pillared environment consisted of several geometries and spacing of pillars, while the non-pillared environment was free of elements that influenced cell behavior. We worked

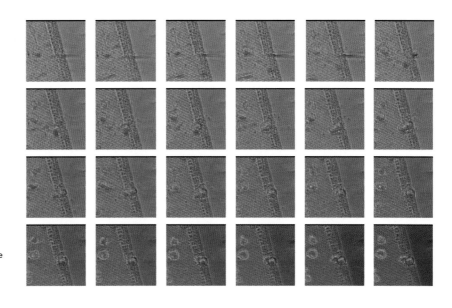

Figure 6.18 Boundary conditions between two extracellular matrix environments, pillared and non-pillared. The pillared environment (left-hand side of each image frame) consists of an array of pillars with regular spacing and height, while the non-pillared environment is free of elements that influence cell behavior.

to discover the diversity of cell behaviors at the boundary conditions and simulated cellular motility and morphology in these two environments.

Our team worked with several premises: (1) that the ECM is able to influence the form of the lamellipodium of the cell; and (2) that the behavior of the cells at given areas will begin to imply the boundary condition in its environment. This is made apparent through the impaired or enhanced motility, fragmentation, and convergence of cells.

In contrast to architectural notions of boundary as hard-lined and well defined, a boundary condition within the context of the ECM is more of a gradient. We propose that the state of a boundary is defined by a change in cell behavior within a given environment.

While the DNA of the cell is outside of our control, its motility and morphology will predictably respond to changes in environment. In relating this concept to architecture, the "ECM environment" may translate into a speculative

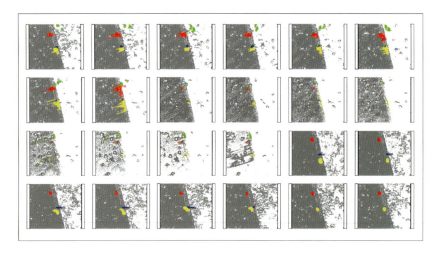

Figure 6.19 Tracing studies of cells in two different ECM environments. Cell behaviors at the boundary conditions change due to changes in geometry, patterning, and compliance. These tracings are used to extract key features and parameters to further inform simulations of cellular motility in the two environments.

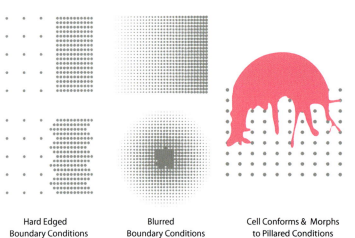

Hard Edged
Boundary Conditions

Blurred
Boundary Conditions

Cell Conforms & Morphs
to Pillared Conditions

Figures 6.20 and 6.21 Parameters for simulations of cellular motility, including pillar spacing, height, pitch, density, pattern, and compliance

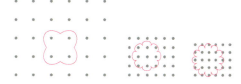

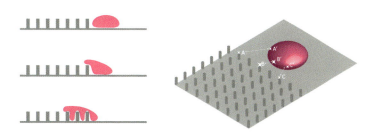

design for a building skin, where the "cell" parallels passive elements such as light, temperature, water, and wind. These elements change, based on variable inputs determined at the architectural scale.

Understanding that the extracellular matrix environment possesses variability that is controllable, we moved forward with speculative processes in the design of new ECMs. The relationship between one ECM and another is implied by cell behavior instead of the hard edge between them, as seen in the given dataset. The cell has the ability to detect and react differently to changes in its environment, which could consist of gradients or patterning of morphing geometries and densities.

In an iterative generation of testing within architectural application, we explored our micro-scaled hypothesis on a macro-scaled test-bed. Specifically applying our findings to a façade condition, we created a developable *skin* environment that begins to react to life-sized "cells." We specifically selected water as our macro-scaled analogous "cell" (passive element) to control within our architectural application.

Conceptually, water behaves in parallel to the cells seen in the dataset. From the dataset, it is apparent that cellular motility is affected by the structure of the ECM. Additionally, the cells experience fragmentation and convergence repeatedly over the entire span of the dataset. The same effect applies to water. The surface properties of water allow for fluid dispersion and grouping, determined by its environment.

At an architectural scale, water is an important element both in terms of function and aesthetic. Functionally, it is used in the cooling and heating system, thus facilitating the exchange between a building's inside and outside conditions, as well as contributing to the creation and control of its interior microenvironment. Aesthetically, an architectural ambience is achieved by the changing of transparency and form. By combining these two aspects, we are aiming to design a performative and adaptable architectural surface that manipulates water simultaneously as part of a building system and an architectural cosmetic.

More specifically, we are exploring a concept from fluid dynamics: the Saffman-Taylor instability. The concept describes the effect of the interface between a low viscosity fluid (for example, water) and a liquid of a higher viscosity, in which a "fingering" effect results. In most lab setups, the experiment is executed using two acrylic plates. The higher viscosity fluid, usually a glycerine base, is then sandwiched between the two plates and held together by an external pressure (clamping, framing, etc.). A small hole is cut into the top plate. When the lower viscosity liquid is injected into the plates through the hole, the "fingering" effect becomes apparent.

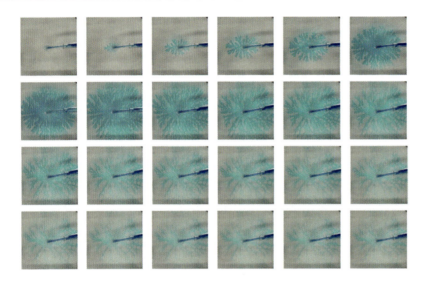

Figure 6.22 The Saffman-Taylor instability. This matrix shows the effect of the interface between a low viscosity fluid (for example, water) and a liquid of a higher viscosity, in which a "'fingering" effect results.

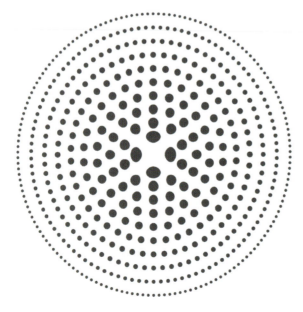

Figure 6.23 Typical pattern for a filter etched in acrylic. In directly translating the concept of a designed ECM for smooth muscle cells to the Saffman-Taylor experiments, a set of acrylic plates were etched using a laser cutter with variable patterns.

BioInspired Materials and Design 255

In our experiments, we used two types of acrylic plates: one smooth, and the other textured with a patterned matrix, laser-scored onto the plate. In directly translating the concept of a designed ECM for cells to the Saffman-Taylor experiments, we saw similarities in the ability to design and pattern the matrices in our experimentations. Our experimentations directly mirrored the setup of the original Saffman-Taylor experiments—we pressurized the two plates after sandwiching glycerine and injected dyed water into the pressurized zone. What may vary from the original experiment is the addition of a consistent pressure flow when injecting the lower viscosity liquid.

In addition, we considered creating permanently sealed plates. Different from the previous setup, in this series of experiments, the dyed water was dripped on top of the glycerine hand soap. Both fluids were then sandwiched between a clear and

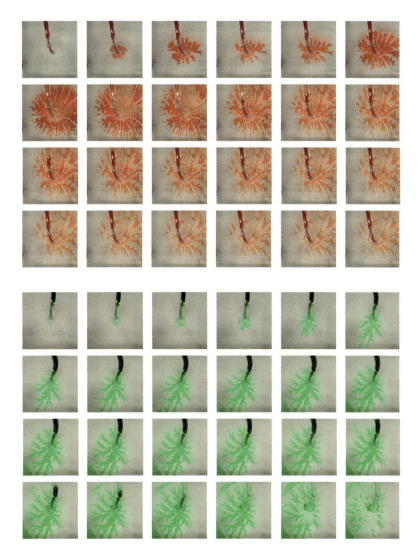

Figure 6.24 Changes to the pattern on the filter directly impact the features or signatures of various outputs displaying the "fingering" effect.

Figure 6.25 Outputs from the gradient filter. In setting up the material experiments, two plates are pressurized after sandwiching a substrate of glycerine and injecting dyed water into the pressurized zone.

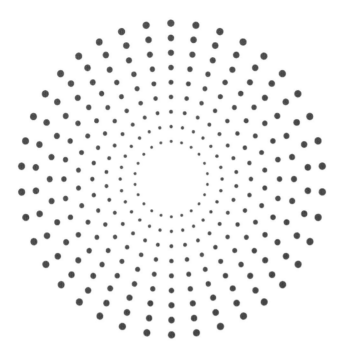

Figure 6.26 Radial pattern. This project opportunistically extracts from cellular motility a set of analogous models or abstract machines for material experimentation at the human scale.

smooth acrylic plate on top, and a textured plate with patterned matrix laser-cut (scored) at the bottom. The plates were first sealed with a clear adhesive tape to maintain water tightness. As a result, any directly applied pressure would produce the "fingering" effect to correspond to the patterned geometries. As varying amounts of pressure were applied to differing areas on the plates, "fingering" appeared and disappeared.

From the testing results of the two setups, we envisioned a bioinspired architectural surface, which responds to changes in pressure. At a large scale, the most straightforward pressure change is applied by human force (walking, poking, pushing, touching). Interaction between the material and the human could come in the form of floor tiles or screens set up in a similar fashion to the second round of experiments. Moreover, besides human force as a way to trigger changes in pressure, the heating and cooling could also contribute to the deformation of the surface and thus trigger unequal pressure points on the surface for it to react to. In this project, cellular motility and a careful study of the role of the extracellular matrix set up a series of analogous drivers or abstract machines for exploring dynamic material behaviors at the human scale. While the original biological datasets are not directly influencing the final design experiments, the data serve as a point of departure for setting up material experiments where changes to a filter (analogous to the ECM) directly influence the changing features and effects within the material substrate. We will see similar models and modes of thinking in the context of the eSkin project described in depth in Part IV.

PART III
Architectural Prototyping

Human-Scale Material Systems and Big Datascapes

Adaptive Materials and Structures: An exploration of the material and ecological potentials of visualization and simulation tools through the production of experimental structures, interventions, and material systems.

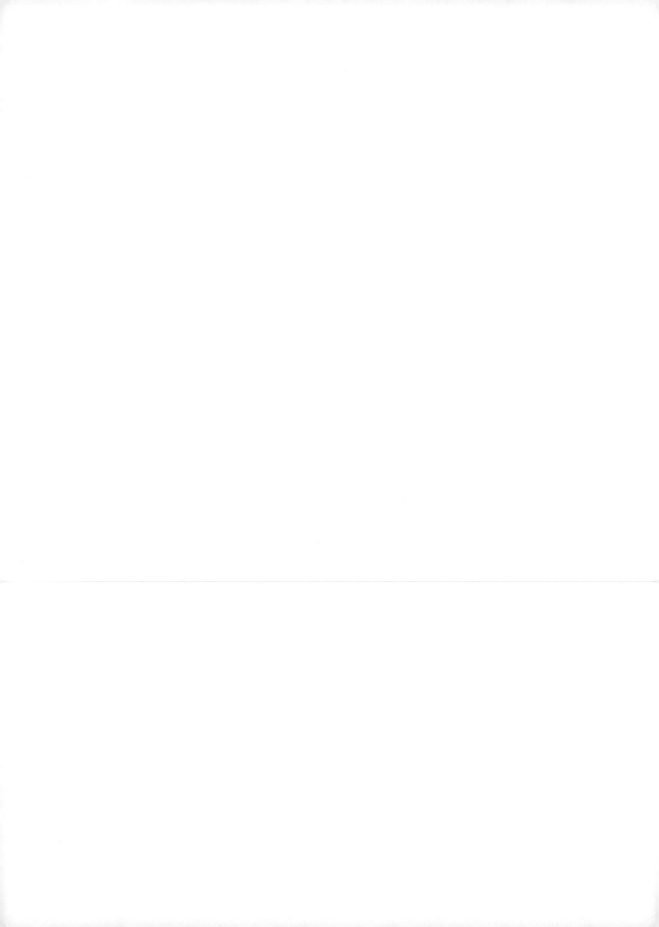

7
The New Science of Making

Mario Carpo

Since the beginning of time, data have been known to be rare and valuable; their collection, recording, transmission, and processing famously laborious and expensive. Sure, the invention of writing, then of print, have increased the amount and reliability of data supply, and diminished its cost over time. Yet, until a few years ago, the cultural economy of data was strangled by a permanent, timeless, and apparently inevitable scarcity of supply: we always needed more data than we had. Today, for the first time ever, we seem to have more data than we need. Moreover, the cost of collecting, storing, and processing data keeps decreasing. So much so that, for many technical purposes, and within the normal ambit of most practical operations, we may safely assume that soon an almost unlimited amount of data will be recorded, transmitted, and processed at almost no cost. Of course a state of zero-cost, unlimited data storage and data mining is, and will always be, technically impossible; but this is the trend, and the trend appears to be going, asymptotically, this way.

As I have discussed elsewhere,[1] one of the most momentous consequences of our newly acquired data opulence is that many traditional scientific strategies, required by and made to measure for our small data past, are being replaced by a "new kind of science"[2] predicated on unlimited data availability. The traditional way to compensate for both human and technical limits in the storage and processing of information was to curtail the intake of data from the start, by culling them, sifting them, then laboriously distilling them by comparison, selection, generalization, formalization, and abstraction. In the history of science, this method is called "inductive," and it leads to formulas or laws that aim at being as general as possible; since Galileo and Newton, these laws have been primarily expressed using a mathematical notation, and to this day many see them as representing a certain order of the universe (which is why they are sometimes called "laws of nature"). By contrast, let's imagine an ideal Big Data environment, where data are so easy to garner, store, and process that all precedents that might be relevant to our understanding of any given event have been recorded, and can be retrieved at will and at no cost, or almost. In these unlikely circumstances, all traditional scientific strategies would be ipso facto unnecessary, because the search for a specific event (as recorded when it actually occurred) would always be more effective than its deduction from abstract mathematical notations, formulas or laws. Scientific formulas condensate a huge corpus of events, from which they derive and to which they refer, into a single short script, which can in turn be used to deduct many more occurrences of a similar kind; but in this sense, formalization (the inference of abstract formulas) is itself a data compression technology, unwarranted in a Big Data techno-cultural environment, and which could be phased out—together with the scientific method of inductive and inferential science in its entirety.

This post-scientific revolution, here too briefly outlined, may sound like the apocalyptic or dystopian view of a post-human intelligence. To the contrary, the new science of data is already deeply embedded in many technologies we use in our daily life; and traces of the same quantitative, heuristic use of data are evident, in some muted, embryonic way, in various branches of the natural sciences (and more openly, for example, in weather forecasting). And sure enough, some historians of science have already started to investigate the matter—with much perplexity and reservations, as one would expect; as the postmodern science of Big Data marks a major shift in the history of the scientific method.[3] As mentioned above, mathematical abstractions such as the laws of mechanics or of gravitation, for example, or any other grand theory of causation, are not only practical tools of prediction, but also, and perhaps first and foremost, ways for the human mind to make sense of the world. But then, if abstraction and formalization (i.e., most of classical and modern science, in the Aristotelian and Galilean tradition) are also seen as contingent and time-specific data-compression technologies, one could argue that in the absence of the technical need to compress data in that particular way, the human mind can find many other ways to relate to, or interpret, nature. Epics, myth, religion, and magic, for example, offer vivid historical examples of alternative, non-scientific methods; and nobody can prove that the human mind is or ever was hard-wired for modern experimental science. Many postmodern philosophers, for example, would strongly object to that notion. And as so many alternatives to modern science existed, historically, in the past, one could argue that plenty of new ones may be equally possible in the future.

As mentioned, the mere technical logic of the new science of data runs counter to core postulates and assumptions of modern science. Additional evidence of an even deeper rift between the two methods is easy to gather. Western science used to apply causality to bigger and bigger groups, or sets, or classes of events—and the bigger the group, the more powerful, the more elegant, the more universal the laws that applied to it. Science, as we knew it, tended to universal laws—laws that bear on as many different cases as possible. Today's new science of data is just the opposite: using information retrieval and the search for precedent, data-driven prediction works best when the sets it refers to are the smallest. Indeed, searches are most effective when they can target and retrieve a specific and individual case—the one we are looking for.[4] In that, too, the new science of data represents a complete reversal of the classical (Aristotelian, Scholastic, and modern) scientific tradition, which always held that individual events cannot be the object of science: with few exceptions, Western science only dealt with what Aristotle called forms (and today we more often call classes, universals, sets, or groups).[5]

In social science and in economics, this novel data-driven granularity means that instead of referring to generic groups, social and economic metrics can and will increasingly relate to individual cases. This points to a brave new world where standards and averages will no longer be either required or warranted: fixed prices, for example, which were introduced during the Industrial Revolution in order to standardize retail transactions, have already ceased to exist, as each retail transaction on a digital marketplace today is a customized one-off delivered at zero processing costs, or almost.[6] Likewise, the cost of medical insurance, calculated as it still is on the basis of actuarial and statistical averages, could become irrelevant, because it will be possible to predict, at the granular level, that some individuals will never have medical expenses, hence they will never need any medical insurance, and some will have too many medical expenses, hence no one will ever sell any medical insurance to any of them. This is a frightening world, because the individual who becomes the object of this new science of granular prediction is no longer a statistical abstraction—it becomes each of us, individually. This may be problematic from a philosophical and religious point of view, as it challenges traditional ideas of determinism and free will; but in more practical terms, it is also incompatible with most principles of a liberal society and of a market economy in the traditional, modern sense of both terms.

In the field of natural sciences, however, the picture is quite different. Indeed, if today's digital style in architecture is already so different from the first digital style of the 1990s, this is due in large part to our newly acquired data affluence, and to the general awareness of this unprecedented techno-cultural development. Twenty years ago, when processing power was rare and costly, digital technologies were designed to skimp on data, and use them sparingly. That was perfectly justified back then, but it is not now. Today's most alert digital designers have already started to explore and interpret the big-data logic of the new tools they use, and the same heuristic method of big-data prediction (or a very similar one) is at the basis of various theories of cellular automata, agent-based design, and form-searching by simulation and optimization, that some of today's digitally intelligent designers already use extensively. As works by Achim Menges, Neri Oxman, Jenny Sabin, and Peter Lloyd Jones among others, have shown, we can now design and fabricate materials with variable properties at minuscule, almost molecular scales; or detect, quantify and take into account the infinite, minute, and accidental variations embedded in all natural materials—a capriciousness that made natural materials unsuitable for industrial use, and almost eliminated them from modern industrial design.[7] Artisans of pre-industrial times did not have much choice: they had to make do with whatever natural materials they could find. So, for example, when farmers in the Alps had to roof a new chalet, they looked high and low (literally) for a tree that would be a good fit for the ridge piece, and sometimes the shape of the roof would be tweaked to match the quirks of the best available trunk. And cabinet-makers could (and the few still extant, still can) skillfully work around almost any irregularity they find in a plank of timber, and make the most out of it. But industrial mass-production does not work that way. To be used in an assembly line, or pieced together by unskilled workers, timber must be processed and converted into a homogeneous material compliant with industrial standards—as plywood, for example, which is a factory-made industrial product, although derived from wood. Artisan masons of old (and few still exist in the so-called industrialized countries) knew very well how to make concrete on site the smart way, making it stronger, for example, in the angles and corner walls (more cement), cheaper in some infill (more rubble), thinner and more polished next to some openings (more sand), etc. But for modern engineers, concrete had to be dumb, homogeneous, and standard, the same all over, because that was the only material they could design with the notational and mathematical tools at their disposal. Even assuming an engineer could calculate the structural behavior of variable property concrete (i.e., concrete with different mechanical characteristics and performances in different parts of the same structure), until recently there would have been no practical way to produce those variations to specs, either by hand or mechanically. After all, reinforced concrete is only an elementary two-properties material, yet it took several decades to learn a consistent, reliable way to design it, calculate it, fabricate it, and deliver it.

Today, in theory and increasingly in practice, digital design and fabrication tools are eliminating many of the constraints that came with the rise of industrial standards. X-ray log scanning, for example, is already used in forestry: trees are scanned prior to felling and the slicing of the boards is customized for each trunk to minimize waste. As Achim Menges remarked recently, the scan is now abandoned by the sawmill after the planks are sold, but there is no reason not to envisage a full design-to-delivery workflow, in this case extended to include the natural production of the source material—from the forest to the end product, perhaps from the day the tree is planted (which would once again curiously emulate ancestral practices of our pre-industrial past).[8] Each tree could then be felled for a specific task: a perfect one-to-one match of supply and demand which would generate economies without the need for scale—which is what digital technologies typically do, when they are used the right way. Likewise, variable property materials can now be designed and fabricated at previously unimaginable levels or resolution, including concrete, which can be extruded and laid by nozzles on robotic arms, so each volumetric unit of material can be made different from all others. This is what artisanal concrete always was anyway—which always scared engineers to death, because they could not design nor calculate it. Today we can.

So it will be seen that, much as modern science tended to more and more general laws, applying to the largest possible sets of events, modern technology tended to the mass-production of standardized materials that were designed to be, as much as possible, homogeneous and isotropic. Industrial standards were meant to generate economies of scale, but also, and crucially, homogeneous materials could be described and modeled using elegant mathematical tools, such as differential equations and calculus. Calculus is a mathematics of continuity, which abhors singularities: it is perfect, for example, to quantify the elastic deformation of any homogeneous chunk of continuous matter—which is why the modern science of engineering could calculate the stress and deformation of the Eiffel Tower, which is made of iron, but until recently the same science could not calculate the resistance of a brick wall.

Today, using digital simulation and data-driven form searching, we can model the structural behavior of each individual part in a hyper-complex, irregular, and discontinuous 3-D mesh. And using digital tools, we can fabricate any unimaginably complex, heteroclite mess precisely to specs, on time and on budget: robots will see to that. Industrial materials were standardized so they could be calculated and mass-produced. Today we can calculate and fabricate variations at all scales, and compose with variations as needed or as found, including natural variations. Used this way, the new science of granular prediction does not constrain but liberates, and almost animates, inorganic matter.[9] And indeed, vitalism is a powerful metaphor, and a source of inspiration, for many experiments, tendencies, and trends that populate today's digitally intelligent design, although a contentious one, and—in today's computational environment, often misleading. Big-Data prediction, as well as cellular automata simulation, may be, or may appear to be, forms of prediction without classical causation; but both posit a universe which is as predictable as Laplace's most classical of all possible worlds.

Notes

1   This article anticipates some arguments from my forthcoming monograph, tentatively called *The Second Digital Turn*, to be published by The MIT Press in the Writing Architecture Series in 2017, and it can be read as an introduction to Chapter 2 in that book, to which the reader is also referred for additional footnotes and bibliographies. See also: "Big Data and the End of History," *Perspecta* 48, *Amnesia* (2015): 46-60, and "The New Science of Form Searching," AD 237 (2015): 22-27.

2   See, in particular, Stephen Wolfram's work on cellular automata, which has been particularly influential in the design community. Stephen Wolfram, *A New Kind of Science* (Champaign, IL: Wolfram Media, 2001).

3   See D. Napoletani, M. Panza, and D.C. Struppa, "Agnostic Science: Towards a Philosophy of Data Analysis," *Foundations of Science* 16(1) (2011): 1–20. In July 2008, a groundbreaking article by Chris Anderson in *Wired* first argued for a scientific revolution brought about by Big Data ("The End of Theory," *Wired* 7 (2008): 108–109); that issue of *Wired* was entitled "The End of Science," although the other essays in the "Feature" section of the magazine did little to corroborate Anderson's vivid arguments. The main point in Anderson's article was that ubiquitous data collection and randomized data mining would enable researchers to discover unsuspected correlations between series of events, and to predict future events without any understanding of their causes (hence without any need for scientific theories). A debate followed and Anderson retracted some of his conclusions.

4   As search always starts with, and aims at, one individual event, the science of search is essentially a science of singularities, but the result of each search is always a cloud of many events, which must be compounded, averaged, and aggregated, using statistical tools. Thus, a lower level of precision, i.e. less intention in a search, will generate more hits, i.e., a larger extension in the definition of the set.

5   The rejection of modern science as a science of universals is central to the postmodern philosophy of Gilles Deleuze and Félix Guattari. See, in particular, Gilles Deleuze and Félix Guattari, "Treatise on Nomadology,"

in *A Thousand Plateaus*, trans. B. Massumi (London: Continuum, 2004), pp. 406–409, 450–451. (First published as *Milles Plateaux*, Paris: Les Editions du Minuit, 1980; English trans. first published Minneapolis: University of Minnesota Press, 1987.)

6 See M. Carpo, "Micro-managing Messiness: Pricing, and the Costs of a Digital Non-Standard Society," *Perspecta 47: Money* (2014): 219–226.

7 See Neri Oxman, "Programming Matter," *AD, Architectural Design* 82(2) (2012). Special issue, "Material Computation: Higher Integration in Morphogenetic Design," guest-edited by Achim Menges, pp. 88–95, on variable property materials; Achim Menges, "Material Resourcefulness: Activating Material Information in Computational Design," *AD, Architectural Design* 82(2) (2012): 34–43, on non-standard structural components in natural wood.

8 Menges, "Material Resourcefulness," p. 42.

9 This vindicates the premonitions of Ilya Prigogine, another postmodern thinker whose ideas were a powerful source of inspiration for the first generation of digital innovators in the 1990s. See, in particular, Ilya Prigogine and Isabelle Stengers, *Order Out of Chaos: Man's New Dialogue with Nature* (New York: Bantam Books, 1984; first published in French as *La Nouvelle Alliance: métamorphose de la science*, Paris : Gallimard, 1979).

8

Matter Design Computation: Biosynthesis and New Paradigms of Making

Jenny E. Sabin

Parts III and IV are dedicated to architectural prototyping where novel tools and methods for modeling biological and material behavior—such as those presented in Part II—are met with material and fabrication constraints. These human-scale material systems and datascapes rigorously negotiate the problem of scale through exploration and simulation of multi-dimensional biological systems. Taking direct inspiration and cues from the dynamic reciprocity between context, code, and form and the adaptive material models presented to us through matrix biology, materials science, mechanical engineering, physics, biological and environmental engineering, and our three primary areas of inquiry: cellular motility, networking, and surface design, we aim to develop prototypes and material assembles that are not only smart and responsive, but are also intelligent, programmable, adaptive, and at times living. We are interested in the coordinated generation of complex form coupled with feedback from environmental, material, geometric, and fabrication constraints alongside the resultant topics of performance, change, and perception in architecture. This interest probes the productive tinkering and misuse of digital fabrication machines informed by issues of craft and making to produce bio-inspired material systems and software design tools that have the capacity to facilitate embedded expressions in our built environment. Finally, this work considers the role of a dynamic matrix in connecting and constantly reforming architecture in context. Similar to Detlef Mertins's description of bioconstructivisms in Chapter 1, the emphasis here is also upon the analogic negotiation of morphological behavior as a dynamic template that is then filtered through material organizations, frequently engaging the human hand and *digital handcraft*. This part will explore approaches to applied research-based architectural design projects across multiple length scales, mediums, and software.

The evolution of digital media has prompted new techniques of fabrication as well as new understandings in the organization of material through its properties and potential for assemblage. The scope of the following projects probes the visualization of complex spatial datasets alongside issues of craft, fabrication, and production in a diverse array of material systems (woven and knitted textiles, Rapid Prototype and 3-D printed ceramics and components, CNC cut plastics and metals). This work exhibits a deep organicity of interrelated parts, material components, and building ecology. Generative design techniques emerge with references to natural systems, not as mimicry but as trans-disciplinary translation of flexibility, adaptation, growth, and complexity into realms of architectural manifestation. In particular, this work includes long-standing traditions of shared relationships between architecture and biology, with sub-topics that include biomimicry, digital fabrication, textile tectonics, responsive architecture, experimental structures, algorithmic design through scripting procedures, and materials science.

Recent advances in computation, visualization, material intelligence, and fabrication technologies have begun to fundamentally alter our theoretical understanding of general design principles as well as our practical approach towards architecture and research. This renewed interest in broadening the discipline has offered alternative methods for investigating the interrelationships of parts to their wholes, and emergent self-organized material systems at multiple scales and applications. The advantages of researching and deploying such methodologies in the field of architecture are immense as they impact aspects of material systems, fabrication and construction, sustainable and ecological design, optimization, and formal aesthetics.

Existing digital fabrication techniques such as CNC milling and cutting are useful tools, but they are also severely limited by their 2-D and reductive constraints. 3-D CNC tools offer a much more adequate and versatile approach to issues of shape, material, and geometry, but these tools are particularly underexplored when it comes to fabrication at the architectural scale. Two of the most promising technologies are 3-D printing and rapid assembly via robotics for manufacturing of individual and continuous component parts or fibrous assemblages. Together, these technologies are geared toward becoming indispensable tools for nonlinear manufacturing as well as complex form-making. How might advances in 3-D printing and rapid assembly in alternate industries and disciplines impact the design and fabrication of building components and skins? What are needed now are rigorous multi-directional and multi-disciplinary investigations that can help shape the future trajectories of these material innovations and technologies for architecture.

Biomimicry casts a wide net. In a general sense, we may define it simply as nature-inspired design. However, if we look a bit closer, biomimicry reveals examples that copy or mimic nature for specific design and engineering applications.[1] In essence, biomimicry is the practice of developing technology and materials inspired by nature to solve applied design and engineering problems. Examples may include energy-efficient buildings inspired by passive cooling in termite mounds, and non-toxic fabric finishes inspired by water-repellent lotus plants. Biomimicry is goal-orientated and therefore fosters goal-based research. In contrast to biomimicry, *biosynthesis* fosters process and process-based research where design solutions and applications emerge along the way in a shared collaborative space. In the following body of work, we hope to articulate not only the differences between biomimicry and biosynthesis, but also more importantly, how biosynthesis may foster new models of thinking in design at the intersection of architecture and science and in the context of variegated materials, digital fabrication, and construction. Biosynthesis has the capacity to maintain powerful modes of bottom-up, materially directed, ecological and systemic design thinking in the long term. In contrast, biomimicry fosters design problems with a start and an end point. It searches for best fit and optimal solutions.

Biosynthesis offers a radical break from our perceived notions of what it means to bring science and architecture together. It is a mode of thinking in design generated through deep immersion within bottom-up processes found in biology and architecture. It is focused upon processes and behavior, not things, images, or static forms. This type of thinking considers biological complexity and formation to emerge through code in context. Here, environment counts in the development of form. It is abstract creative thinking gleaned from productive collisions between science and architecture; it is not a product, style or mimicking of nature. Biosynthesis is a slow process. The shared artifacts within this process are designed tools and methods and new modes of thinking and designing. *Tissueness* and *cellness* therefore are qualities achieved in the 'other', a synthesis of the natural model into something new, an analogic hybrid. Materially directed generative fabrication forms a bridge and point of departure in the successful extraction, transfer, and translation of biological behaviors and processes from the natural model to a synthetic analogic model. Custom design software tools along with material agency are catalysts. What are our next steps and how might we learn from existing and historic models and concepts that operate across disciplinary boundaries in the exploration of dynamic form and structure in context?

Important historical and contemporary protagonists have served as an inspiration to us through their groundbreaking and experimental work at the intersection of architecture and science, offering necessary alternatives to how we negotiate and explore the dynamics of form across disciplines. Popular examples include: D'Arcy Wentworth Thompson, a pioneering mathematical biologist who promoted structuralism over Darwin's survival of the fittest in the formation of species in his seminal book entitled, *On Growth and Form*.[2] Thompson states, "The form of an object is a diagram of the forces acting upon it . . ." and "Morphology is not only the study of the forms of things, but has its dynamical aspect under which we can deal with the interpretation in terms of forces of the operations of energy." R. Buckminster Fuller, a self-proclaimed design scientist, architect, author, industrial designer, and inventor, described nature as conceptual, composed of integrated behaviors and associative potentials. While his applied geometries in a variety of forms—from tents to domes—served highly functional purposes for clients including the U.S. military, and garnered numerous patents, Fuller's dedication to solving the world's problems through functional geometries was driven by a larger obsession with the fleeting relationships between nature and mathematics.[3] Perhaps less known, but equally influential are the contemporary writings by J. Scott Turner who unfolds the dynamic relationships between environment and organism and how context specifies form, function, and structure in his descriptions and comparisons of seemingly unrelated systems such as collagen meshworks and termite mounds. Both are dynamic structures capable of restructuring their environment to regulate changes in force such as tension or wind.[4] How do these concepts and ideas relate to emerging technologies and computation in design and architecture?

The pioneering thinkers at the onset of digital architecture focused primarily upon variation as afforded through simple manipulations of parametric functions.[5] Economist, architect, designer and writer, Bernard Cache discusses variation and variable curvature through his concept called *objectile* and the difference between concavity and convexity or in other words, an inflection.[6] Cache articulates this in the following statement through the lens of geography, "Space is thus no longer a juxtaposition of basins but a surface of variable curvature. We will no longer say that time flows, but that time varies."[7] In his book, *The Alphabet and the Algorithm*, Mario Carpo unfolds how digital technology has altered the role of authorship in architecture and in turn how this has opened up existing notation systems through parametric functions and the possibility of variability. He points to Cache's contribution to digital discourse in the following statement:

> [T]he objectile is not an object but an algorithm—a parametric function which may determine an infinite variety of objects, all different (one for each set of parameters) yet all similar (as the underlying function is the same for all).[8]

The well-established belief that the architect is the singular author of a building through the agency of exact copy and translation of plan, section, and elevation to built form is radically changing through the generation of variegated forms now made possible through the onset of digital technology and digital fabrication in architecture. Carpo writes:

> Indeed, 3-D printing, 3-D scanning and reverse modeling have already made it possible to envisage a continuous design and production process where one or more designers may intervene, seamlessly, on a variety of two-dimensional visualizations and three-dimensional representations (printouts) of the same object, and where all interventions and revisions can be incorporated into the same master file of the project.[9]

No longer privileging the copy through the transaction of the blueprint, notational limitations are opening through variegation and are in fact repositioning the architect as a maker again, where authorship is horizontal and collaborative.[10] As Carpo states:

> By bridging the gap between design and production, this model of digital making also reduces the limits that previously applied under the notational regimes of descriptive and predescriptive geometries, and this may well mean the end of the "notational bottleneck" that was the uninvited guest of architectural design throughout most of its early modern and modern history. Under the former dominion of geometry, what was not measurable in a drawing was not buildable. Now all that is digitally designed is, by definition and from the start, measured, hence geometrically defined and buildable.[11]

However, Carpo warns us that the computer is still a tool which feeds back upon the designer and that "digitally designed or manufactured objects can easily reveal their software bloodline to educated observers."[12]

What if we were to consider dynamic variation where the architecture of context or environment acts upon the *objectile*, and vice versa? This may perhaps get us closer to Waddington's epigenetic landscape so elegantly visualized by painter, John Piper (see "Design Research in Practice: A New Model" in Part I). As Sanford Kwinter writes in his essay, "Landscapes of Change":

> Epigenetic landscape seen from below: The complex relief features of the epigenetic surface are themselves largely the expression of a prodigiously complex network of interactions underlying it. The guy-ropes are tethered not only to random points on the overhead surface, but to points on the other guy-ropes as well, and to pegs in the lower surface that themselves represent only semi-stabilized forms, thus multiplying exponentially the non-linearities flowing through the system. Not to diminish in importance either is the tension surface above as a distinct domain contributing its own forces to the field. No change in any single parameter can fail to be relayed throughout the system and to affect, in turn, conditions all across the event surface.[13]

Here, Kwinter is showing us how biology *is* information theory. The epigenetic landscape is enmeshed with forces embedded in forms. Here, the code creates a developmental landscape not forms. Or in other words, diagrams are now algorithms. It is a space of possibility and of formal potential. As Kwinter points out, biology is the study of hyper-communicative matter.

Giuseppa Di Cristina in her 2001 edited issue of *AD, Architecture and Science*, writes about dynamic variation through the lens of what she calls *topological architecture*, but primarily through geometry and transformation.[14] She states:

> Here, architectural topology means the dynamic variation of form facilitated by computer-based technologies, computer-assisted design and animation software. The topologising of architectural form according to dynamic and complex configurations leads architectural design to a renewed and often spectacular plasticity, in the wake of the Baroque and of organic Expressionism.

She also speaks to the contextual influence of these transformations, "The dynamic system of transformations implies a field of external forces that weigh on the bodies and deform them. The forms of change are an expression of the action of vectors that bring about deformations and bending." But these continuous transformations rarely meet the physical world of material.

Perhaps closer to our methodology and interests is the work of engineer and recent Pritzker Prize-winner, Frei Otto. As historian Detlef Mertins stated:

> Frei Otto also took up the notion of self-generation and the analogy between biology and building, but eschewed the imitation of nature in favor of working directly in materials to produce models that were at once natural and artificial. But unlike other admirers, Otto's group did not take these creatures as models for engineering, but rather sought to explain their self-generation with analogic models.[15]

Why are these concepts and arguments important to architecture now? As Carpo states, "One can discuss, design, and make at the same time—just as pre-modern artisans and pre-Albertian master builders once did."[16] How do we cultivate and generate design methodologies, notational systems, and thinking that are in line with this shift? We think that collaborations across disciplines, specifically in matrix biology and materials science afford new modes of thought and design principles that prepare the architect for the paradigm shift that Carpo describes, a synthesis, a *biosynthesis* where environment, code, geometry, pattern, materiality, structure, and communication inform form; indeed, a *new science of making*.

We are not necessarily preoccupied with bio-inspired design as a primary study, but in cultivating methodological models and design thinking skills that are sensitive to the plasticity, feedback, ecology, organicism and indeed, aesthetics, that the study of nonlinear biological and material systems afford. Rather than opportunistically borrowing models from biology and nature, perhaps we can co-generate—through trans-disciplinary collaboration—tools, methods, prototypes, materials, and eventually adaptive architecture that are truly enmeshed and embedded with feedback. Where generative strategies and new notational systems are not merely about measure, replica, and geometry, but variegated matter that are designed, computed, and informed by bio-inspired design thinking or *biosynthesis* and geared toward the production of what we call *matrix architecture*. This may very well contribute to what Carpo calls "a new, organic, postmodern ecology of Things."[17] Cache's *objectile* gave us variation, inflections in the landscape. *Matrix architecture* brings this variation into the realm of dynamic reciprocity, where matter and form are in communication with context or in other words, architecturalizing and analogically materializing Waddington's epigenetic landscape.

In the following section, networking and cellular motility inform two featured projects: *Branching Morphogenesis*, and our workshop at Smart Geometry, Nonlinear Systems Biology and Design. Other projects such as *Ground Substance* bring together biological principles in surface design with advancements in additive manufacturing and 3-D printing. As with *myThread Pavilion*, the research diversifies into links between computation and the binary natures of weaving and knitting that influenced parallel innovations, including the contemporary computer and digital space. By investigating loops that filter datasets through material organizations, the following work seeks to form a bridge between the human body and technology as an active datascape that influences and contributes to an alternative material practice in architecture. In several of the featured projects, we transform and translate material, geometric, and biological behaviors at the nano- and micro-scales to the design and generative fabrication of adaptable and interactive material systems at the building scale. For example, in the project Branching Morphogenesis, we coupled our expertise in scripting and computational design with cell biologists who generated real-time movies of lung endothelial cells (ECs) networking within a 3-D extracellular matrix, within which they form capillary-like structures. Working collaboratively, we extracted specific rules-sets from this highly dynamic, multi-dimensional system in order to generate a sophisticated catalog of simulation tools, allowing the pre-testing of hypotheses *in silico* regarding the role of specific cell- and matrix-based parameters, including cell number and matrix density, in generating branched networked structures within the lung. Equally important, it was realized that there were more than merely visible parameters governing the emergent structures and networks that eventually give rise to EC networks. More specifically, we were able to map and construct a human-scale "Datascape" of physical forces that ECs exert upon their surroundings as they form capillary networks.

Our design process moves fluidly between analog and advanced digital procedures, often inserting the human hand or *digital handcraft* in the meaningful and rigorous negotiation of scale and complex behavior. As the gap between digital design model, shop drawing, and fabricated result continues to decrease, we seek to learn from fabrication models and natural systems that do not separate code, geometry, pattern, material compliance, communication, and form, but rather operate within dynamic loops of feedback, reciprocity, and generative fabrication. How have these advances impacted material practice in architecture, engineering, and construction at the economic, technological, and cultural levels? How might we address these issues during the design process? The main thrust of this work concerns the integration of material and digital complexity in the built environment through prototypical design experiments that rigorously abstract, extend, and translate dynamic behaviors and models from alternate disciplines with the end goal of generating adaptive architecture and material assemblies that operate at the human scale. On a meta level, this marks a shift away from Cartesian formal orders and toward interiorities, networks, fabrics, and fibrous assemblages that are pliable, plastic, open, and beautiful.

Notes

1. See Janine M. Benyus, *Biomimicry: Innovation Inspired by Nature* (New York: Harper Perennial, 2002).
2. D'Arcy Wentworth Thompson, *On Growth and Form* (Cambridge: Cambridge University Press, 1917). See Chapter XVII, "The Comparison of Related Forms," where Thompson describes differences between related species through mathematical transformations.
3. Jenny Sabin, "Geometry of Pure Motion: Buckminster Fuller's Search for a Coordinate System Employed by Nature," in *The Language of Architecture: 26 Principles Every Architect Should Know*, ed. Andrea Simitch and Val Warke (Beverly, MA: Rockport Publishers), p. 191.
4. J. Scott Turner, *The Tinkerer's Accomplice: How Design Emerges from Life Itself* (Cambridge, MA: Harvard University Press, 2007), p. 34.
5. Mario Carpo, *The Alphabet and the Algorithm* (Cambridge, MA: MIT Press, 2011), p. 39. See Carpo's discussion on the discourse of digital variability in his first chapter, "Variable, Identical, Differential."
6. Bernard Cache and Michael Speaks, "Décrochement," in *Earth Moves: The Furnishing of Territories* (Cambridge, MA: MIT Press, 1995).
7. Ibid., p. 40.
8. Carpo, *The Alphabet*, p. 40.
9. Ibid., p. 33.
10. Ibid.
11. Ibid., pp. 33–34.
12. Ibid., p. 34.
13. Sanford Kwinter and Umberto Boccioni, "Landscapes of Change: Boccioni's 'Stati d'animo' as a General Theory of Models," *Assemblage* 19 (1993): 55–65.
14. Giuseppa Di Cristina, *AD: Architecture and Science* (Chichester: Wiley-Academy, 2001).
15. Detlef Mertins, "Bioconstructivisms," in *Nox: Machining Architecture*, ed. Lars Spuybroek (New York: Thames and Hudson, 2004), pp. 360–369.
16. Carpo, *The Alphabet*, p. 79.
17. Ibid., p. 120.

Project: Branching Morphogenesis, 2008
Jenny E. Sabin, Andrew Lucia, and Peter Lloyd Jones

> What is beautiful in science is the same that is beautiful in Beethoven. There is a fog of events and suddenly you see a connection. It expresses a complex of human concerns that goes deeply to you, that connects things that were always in you that were never put together before.
>
> (Victor Weisskopf)[1]

Originally on view at the Design and Computation Gallery SIGGRAPH 2008 and subsequently at Ars Electronica, Linz, Austria, 2009–2010
A project by Sabin+Jones LabStudio, 2008
Design Team: Jenny E. Sabin and Andrew Lucia
Science Team: Peter Lloyd Jones and Jones Lab members
Special thanks to Annette Fierro for critical commentary
Production Team: Dwight Engel, Matthew Lake, Austin McInerny, Marta Moran, Misako Murata, Jones Lab members
Simulations with Christopher Lee

The project entitled Branching Morphogenesis starts with human data, endothelial cells that compose the vascular lining of the lung. These data are simulated, translated, and abstracted into a new datascape, meaningfully extending data from the human body into a new body, a hybrid artifact that is inhabitable. From the beginning of the project, we were less interested in the direct mimicking or copying of these cellular processes, but rather in the coordinated abstraction, scaling, and materialization of variegated and granular behaviors; as micro exchanges transformed into a new set of potential choreographies at the human scale. As curator and founder of the seminal Archilab exhibition series, Marie-Ange Brayer states:

> [T]he issue today is no longer imitating nature, reproducing its forms following biomorphism, but rather simulating through a generative approach, artificially recreating by renewing ties with the self-organizational dimensions of the living, transformed by principles of mutability and discontinuity.[2]

Further, we posited, how might the insertion of the hand and manual manipulation reveal these hidden and intangible dynamic assemblies and forms? Ultimately, our process involves the embedding of material systems with biological relationships and behavior to generate models and prototypes that are once natural and artificial.

In 2008, Sabin was invited to exhibit work in the Design and Computation gallery at SIGGRAPH, the largest annual conference on computer graphics. We proposed to contribute new work from LabStudio, a project that would explore the analogic materialization of dynamic cellular data. The following questions initiated our starting points for the project. Could we generate a human-scale material system

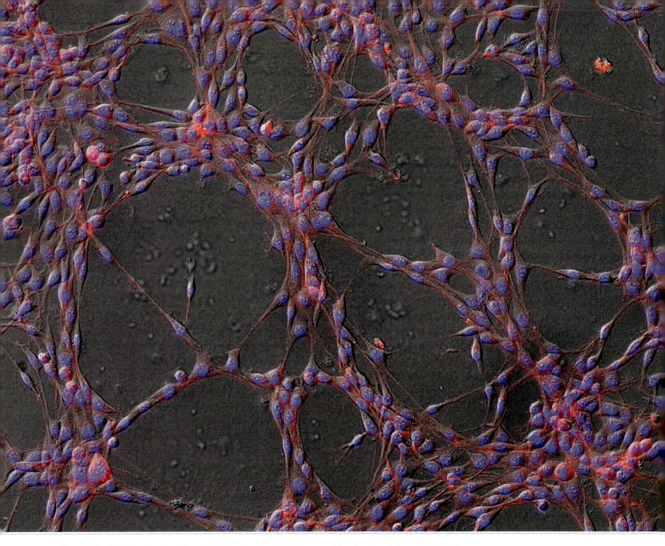

Figure 8.1 Real-time imaging of endothelial cells cultured within a specialized extracellular matrix (ECM) microenvironment, designated the basement membrane, that either suppress or promote networking, formed the basis for this project.

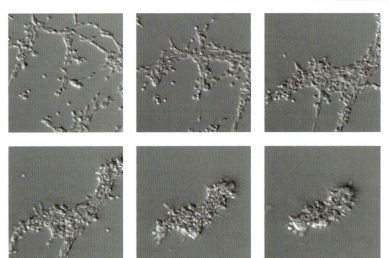

Figure 8.2 Networking endothelial cells. Here, we are interested in the parameters that govern branching morphology in response to the underlying extracellular matrix (ECM), and how this alters cell-cell and cell-ECM interactions during networking.

Figure 8.3 Early concept sketches for Branching Morphogenesis

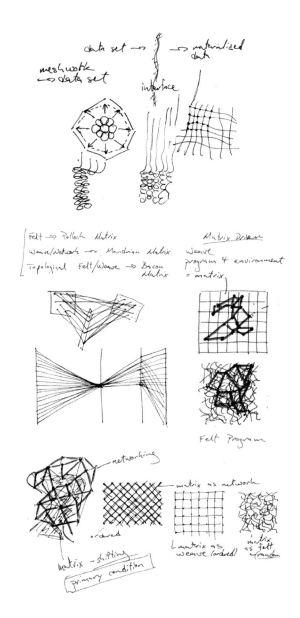

that would visualize these data in a new way? How might slowing down the process of visualization allow for new ways of seeing, predicting, intuiting, and interpreting changes in behavior over time and across multiple length scales? Can collaborative design across disciplines enable new methods and forms of visualization in the context of abundant data? How might filtering data through an abstract material system at the human scale form new habits of thought in both science and design? Could this intermediate step, of abstracting and synthesizing rules for cellular networking behavior and then filtering this dynamic choreography through alternate material forms, give us some clues as to the formation and structure of these exquisite fractal networks that ultimately compose the human lung?

Through the course of the physical manifestation of Branching Morphogenesis, this designed prototype offers the possibility of emergent tectonic and material principles (albeit in their nascent stages) and thinking that may perhaps begin to match the (bio)complexity that many architects have been preoccupied with in the context of digital architecture but frequently in the absence of a consideration of structure and scale.[3]

Antoine Picon argues in his book, *Digital Culture in Architecture*, that, "With the crisis of tectonic, it is not only structural legibility that finds itself jeopardized; the link between architecture and memory is also compromised."[4] Picon takes it a step further in suggesting that this crisis is most likely linked to questions of the body and its senses when he asks,

> How can we reconstruct a dense web of intuitions and analogies between digital architecture and our physical experience of the world, intuitions and analogies that may enable us to rebuild an understanding of how architecture relates to our individual and social destinies?[5]

Picon concludes that the answer may reside within a new notion of materiality. Here, "like nature, materiality is about the way we perceive what is around us with our five senses—and depends on the operations we carry out on our physical environment."[6] One of the central ideas driving the formation of Branching Morphogenesis concerns the inhabitation of the final artifact where simulation organizes a newly constructed material environment with data as agent. Could we generate a human-scale material system that would spatialize these variegated micro cellular processes in a new way, contributing to a "dense web" of ongoing performances at the human scale? Early inspiration for possible spatial arrangements of Branching Morphogenesis and corresponding effects stemmed from an opportune visit to a dance performance where dancers interacted with and were absorbed by a sequence of multiple hanging translucent scrims.

The most recent installment of Archilab, the ninth in the exhibition series, entitled Naturalizing Architecture, curated by Marie-Ange Brayer and Frédéric Migayrou, on view at the Frac in Orléans, France, also explores materiality at the nexus of biology, engineering, design, and computer science and the subsequent impact this is making upon the practice of architecture. They state in their introduction to the exhibition:

> More broadly, the systematization of simulation processes for living organisms marks the advent of definitive mathematization and as such the end of the modern mechanical world. Between nature and technology, the material condition of the "artifact" is henceforth transferable to other materials and other scales.[7]

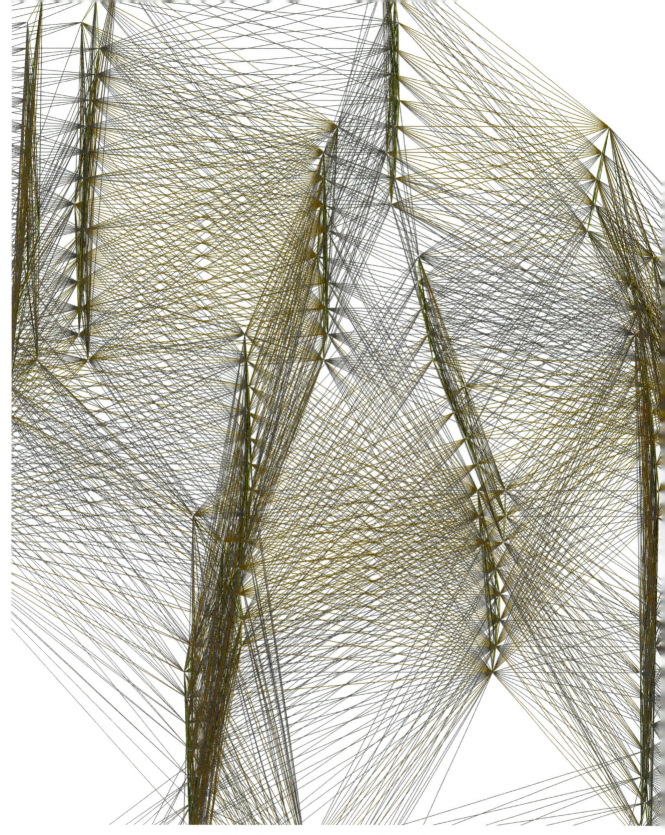

Figure 8.4 Initial networking studies generated in GenerativeComponents

Figure 8.5 View of the installation. Gallery visitors are invited to walk around and in-between the layers of Branching Morphogenesis, and immerse themselves within an organic and newly created "Datascape".

Indeed, Branching Morphogenesis features a complex choreography of interactions, from cell-to-matrix point and component part-to-whole. The final artifact is as Marie-Ange Brayer states, "a proto-prototype working at local and global levels, building bridges between semantic fields, not form, but generative processes."[8] "It is at once software, algorithm, material system, human cell, model, object, building."[9]

Conceptual groundwork, which preceded this project, includes early explorations in the materialization of data, specifically woven systems steered and specified by biodata and harmonic series. One example of this type of material transfer of data through geometry and form is shown in a project that was completed in 2005 by Sabin entitled BodyBlanket.

The intent of this project is to realize interfaces between patients, information, and the hospital setting by giving physical form to patient data, thereby making data directly perceptible. The final woven form is literally

Project: Branching Morphogenesis 279

Figure 8.6 The installation materializes five slices in time that capture the force network exerted by interacting vascular cells upon their surrounding matrix scaffold. Time is manifested as five vertical, interconnected layers made from over 75,000 cable zip ties.

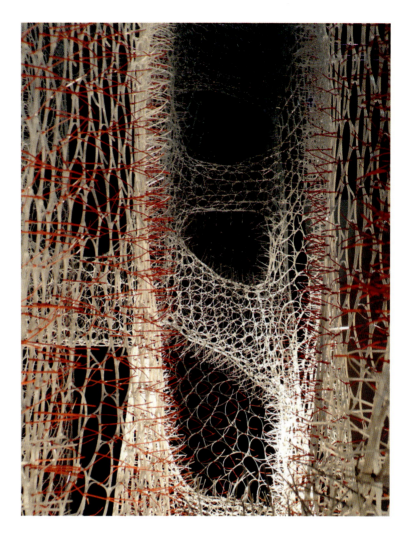

Figure 8.7 BodyBlanket by Sabin, 2005. This project is woven on a digitized Jacquard loom and generated by dynamic biodata.

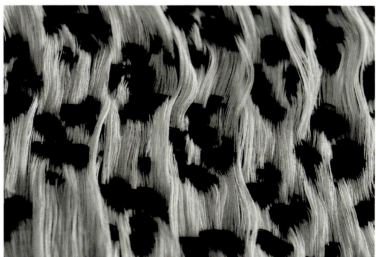

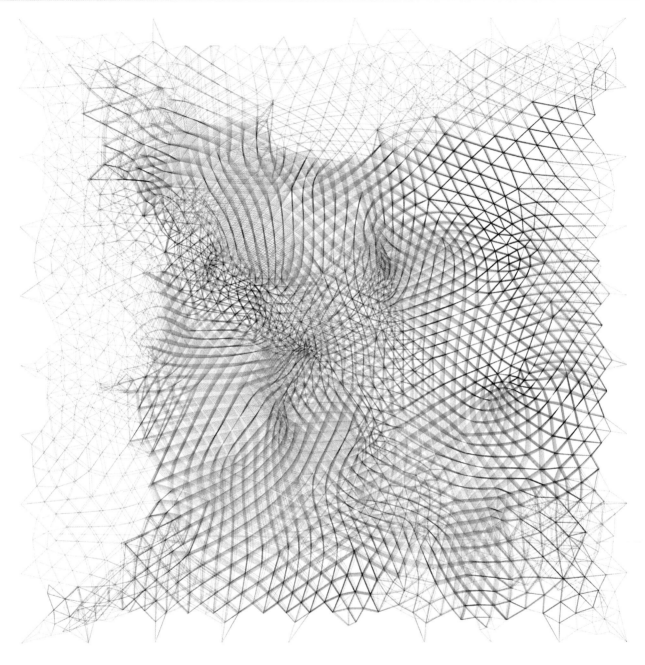

Figure 8.8 Simulations of cellular force networks as a series of variegated and overlapping "time-sheets" or tapestries.

assembled (via a digitized Jacquard loom) through a generic set of patient data. This interest in intuitive pattern recognition and alternative representations of complex datasets that can be seen, heard, and finally fabricated is embedded within all stages of our work. Branching Morphogenesis is yet another example of material and informational transfer of biological data through alternative material systems.

Figure 8.9 Initial simulations of cellular networking behavior. Cell attraction with force exchange through matrix.

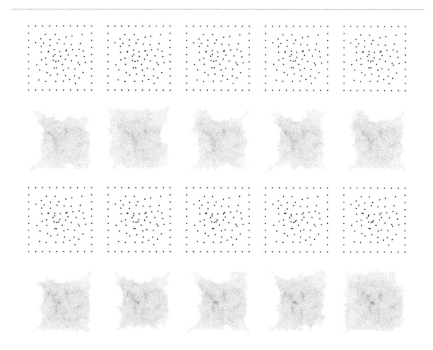

One of our primary aims for the project was to open up new spatial methods for visualizing and interpreting large biological datasets, specifically those being produced in the Jones Lab. In addition to exploration of materialized data, simulations of cellular networking processes initiated by Lucia and Lee in the first installment of Nonlinear Systems Biology and Design taught by Sabin and Jones were further refined to expose the granular and variegated nature of the dynamic networked assemblages.[10] Additionally, the unseen force relationships transferred through the simulated cellular environment were made tangible; rather than studying and depicting the appearance or morphology of the networking cells, a decision was made to make legible and materially translate the invisible force networks exerted throughout the context in which the simulated cells reside. Spatially, this resulted in a series of variegated "time-sheets" or tapestries whose organization, expression, and connectivity were determined by densities of force formally resolved as linkages. This strategy operates counter to pre-rationalized logics of Cartesian assemblies giving preference to deeper structural logics governed through a generative process of dynamic material formation. How might we begin to harness what Mario Carpo calls a "New Science of Making" through the simulation of these data that exhibit exquisite architectures of change? As Carpo points out, "Western science developed absent of big data."[11] The modern scientific method, which is deterministic and causal, favors calculation and repeatable experiments over probabilism and heuristics, or, in other words, searching and retrieving over calculation in the discovery of trends and differences in behavior within large data.[12] In an era where Big Data are challenging

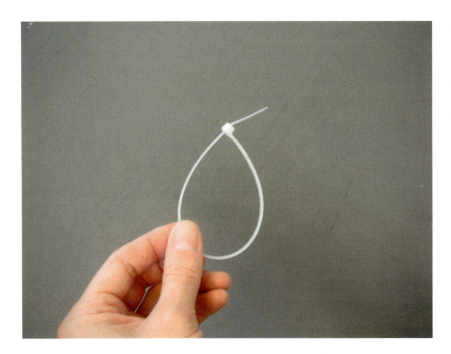

Figure 8.10 The installation is composed of over 75,000 cable zip ties. Organization, expression and connectivity of the force networks are materialized formally as linkages.

research practices in many disciplines including the viability of the scientific method, data-driven methods and thinking that fold information into the creative process are paramount. From cell-to-cell network, to confluent sheet, to tube, to organ, Branching Morphogenesis is a scaled datascape that captures a creative process that encompasses intersections between design and science.

The development of an adjustable material system assembled from simple components and simulation tools based on cellular networking behavior shaped the beginning of the project. 75,000 cable zip ties filter the force network exerted by cells at a micro-scale. Importantly, the use of a ready-made object, a zip tie, plays an important pedagogical role. Its ubiquitous familiarity allows for diverse audiences to engage in complex topics across disciplines that may otherwise be inaccessible. The seemingly banal qualities of these ready-made objects play a significant didactic role as hybrid material systems that by association and recontextualization foster new readings and interpretations of seemingly intangible datasets. Additionally, the inherent flexibility and adjustable nature of each cable unit allow for local and global adjustments within the assembly. At all stages of the filtering and scaling process, the network is adjusted by new material constraints. The final artifact is a synthesis, a biosynthesis, for people to inhabit and experience.

Ultimately, Branching Morphogenesis explores fundamental processes in living systems and their potential application in architecture. The project investigates part-to-whole relationships revealed during the generation of branched structures formed in real time by interacting lung endothelial cells placed within a 3-D matrix

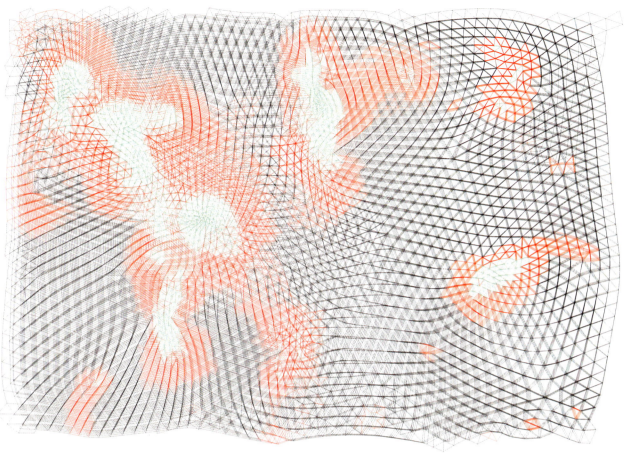

Figure 8.11 Composite construction template depicting five overlapped vertical surface assemblages in *Branching Morphogenesis*.

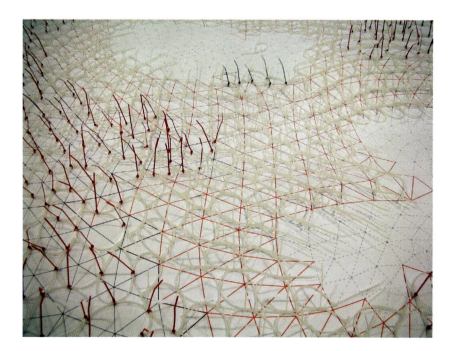

Figures 8.12 and 8.13 Digital simulations inform the analog assembly of linked units. Three differently sized loops are connected with smaller cable ties normal to the generated surface. Density shifts in the field are notated through color change.

Figures 8.12 and 8.13 (Cont.)

Figure 8.14 Detail of the material system.

Figure 8.15 View through a tube linking one sheet or time frame to the next. The densest areas of the force networks connect one sheet to the next with tubes also made of cable tie components.

environment. How does this extracellular matrix environment affect networking behavior, and vice versa? Changes within the architecture of this environment also specify changes within the structures of networking cells. Just as a weave is organized through an invisible background of zeros and ones, of over and under, the communicative landscape that these cells shape also undulates as a dynamic interface steered and specified by both code and environment. The installation captures moments of this changing micro datascape through the materialization of five slices in time that capture the force network exerted by interacting vascular cells upon their matrix environment. The time lapses manifest as five vertical, interconnected layers made from thousands of variable cable zip ties.

    Gallery visitors are invited to walk around and in-between the layers, and immerse themselves within a newly created datascape integrating changes within dynamic cellular networking behavior with human occupation, all through the constraints of a ready-made.

    At the scientific level, the aim of this project was to sequentially model the networking process *in vitro* and *in silico*, and then to abstract this process

Figure 8.16 Frontal view of the overall installation.

Figure 8.17 This image won the AAAS/NSF International Visualization Challenge and was featured on the February 19, 2010 cover of *Science Magazine*.

Project: Branching Morphogenesis 287

into experimental architecture. To approach this, we studied the parameters that govern branching morphology in response to the underlying extracellular matrix (ECM), and how this alters cell-cell and cell-ECM interactions during networking. We have explored potential parameters that prohibit networking behavior, including intercellular communication, environmental instigators, and cellular geometry. Through the investigation of controlled and uncontrolled cell tissue biological models, parallel models work to unfold the parametric logic of these biological and responsive systems revealing their deep interior logic. The result is

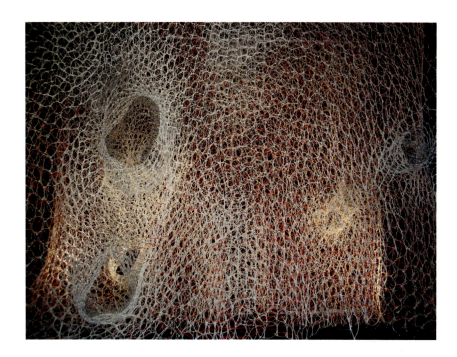

Figure 8.18 Overall installation on view at SIGGRAPH Design and Computation Gallery, Los Angeles, 2008.

Figure 8.19 The project was again on display at Ars Electronica in Linz (Jan. 2009 through Jan. 2011) as part of the exhibition, "New Views of Humankind."

Figure 8.20 From cell-to-cell network, to confluent sheet, to tube, to organ, Branching Morphogenesis is a scaled datascape that captures a creative process that encompasses intersections between design and science.

a component-based surface architecture that abstracts and embeds environment (context) with deeper interior programmed systems.

The project was again on display at Ars Electronica in Linz (January 2009 through January 2011). The curators at Ars Electronica were adept in understanding the technological and social underpinnings of flowing a micro-scale biological dataset through the constraints of a readymade. Their exhibition, "New Views of Humankind," explores the age that we live in and the dominant role that technology and science play in our interpretation of the world that we inhabit. They pair, as can be seen in the gallery shots, Branching Morphogenesis with live demonstrations of Magnetic Resonance Imaging (MRI) and digital models controlled by our own eye-movement. In this juxtaposition, art and science come together to interpret new images, how they are materialized and how we perceive them. Perhaps an installation composed of over 75,000 cable zip ties may tell us just as much about the data constructs at hand and how we interpret them as an image generated through MRI technology. The intent is more than simply translating micro-scale data into formal and often beautiful constructs, but rather to consider pathological interventions through material and affect. At the scientific level, the manifestation of such data as a physical, human-scale architectural installation has the potential to produce and project new hypotheses regarding the role of tissue architecture and environment in controlling morphogenesis within the human lung.

Notes

1. Sean B. Carroll, Jamie W. Carroll, Josh P. Klaiss, and Leanne M. Olds, *Endless Forms Most Beautiful: The New Science of Evo Devo and the Making of the Animal Kingdom* (New York: W.W. Norton and Company, 2005), p. 14.
2. Marie-Ange Brayer and Frédéric Migayrou, *Naturalizing Architecture* (published at the 9th ArchiLab, Orleans, France: FRAC Centre, 2013), p. 16.
3. Antoine Picon, *Digital Culture in Architecture: An Introduction for the Design Professions* (Basel: Birkhäuser, 2010), p. 124.
4. Ibid., p. 133.
5. Ibid., p. 137.
6. Ibid., p. 146.
7. Brayer and Migayrou, *Naturalizing Architecture*, p. 10.
8. Ibid., p. 18.
9. Ibid., p. 17.
10. See Case Study: Modeling a Complex Multi-Variable System: Nonlinear Cellular Networking in Part III, by Asher, Chan, Fukunishi, Lee, and Lucia.
11. Mario Carpo, "Digital Darwinism: Form and Indeterminacy in Contemporary Digital Design Theory," lecture, Bartlett International Lecture Series, UCL, London, January 23, 2013.
12. Ibid.

Project: Ground Substance, 2009
Jenny E. Sabin and Peter Lloyd Jones

Generative Fabrication: Design Experiments in 3-D printing
Sponsored by a University of Pennsylvania Research Development Grant
A project by Sabin+Jones LabStudio, 2009
Design Team: Jenny E. Sabin, Andrew Lucia
Science Team: Peter Lloyd Jones and Jones Lab members.
Production Team: Rebecca Fuchs, Emily Bernstein, Kara Medow
First exhibited at Siggraph 2009, Design and Computation Galleries, New Orleans, USA

In 2009, LabStudio purchased our first large format powder 3-D printer, a ZCorp 510 color printer, which is still in daily use in the Sabin Design Lab at Cornell University. Rather than use the printer for representational purposes solely, we began experimenting with 1:1 rapid prototyping of nonstandard parts and components for larger assemblies. This pioneering strategy informed our project brief for the Smart Geometry workshop described in the next project section. This not only allowed us to work beyond the scale of the print bed, but it also encouraged and inspired a productive medium for translating the nonstandard and nonlinear nature of cellular networks and biological processes into scaled variegated units through 3-D printing that would later be assembled into larger coherent wholes.

Figure 8.21   3-D printed components for Ground Substance. In this project, we began experimenting with 1:1 rapid prototyping of 3-D printed nonstandard parts and components for larger assemblies. We were one of the first groups to do this within the area of additive manufacturing.

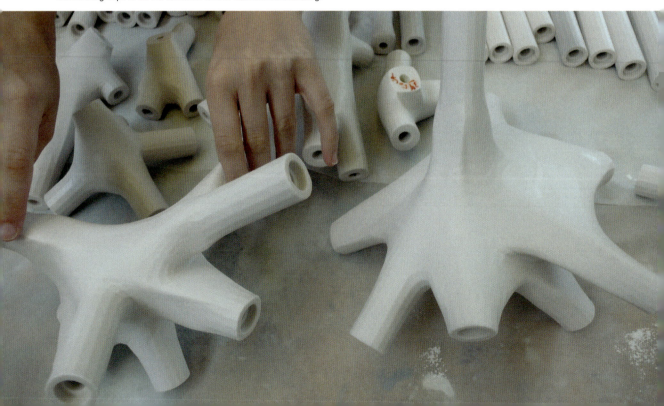

Project: Ground Substance **291**

This project, *Ground Substance*, was our first experiment with component-based 3-D printing. Building upon our work in dynamic cellular surface design as described in the Surface Design Chapter 5, in Part II, the project embeds biological behavior in material systems through the use of advanced technologies in 3-D printing and rapid prototyping. Ground Substance quantifies and spatializes cellular

Figure 8.22 This project examines morphogenesis, lumen formation, and cellular packing behavior as a response to alterations in tissue surface design

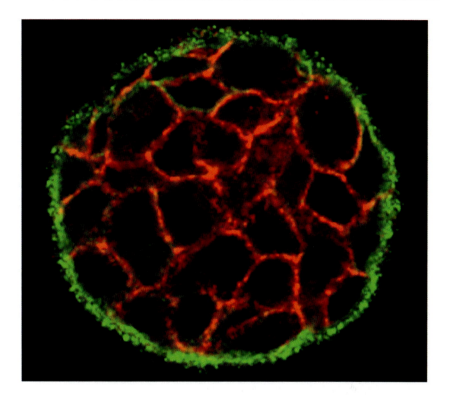

Figure 8.23 Ground Substance quantifies and spatializes cellular and tissue contour information using normal human mammary epithelial cells cultivated within a 3-D normal or tumor-like microenvironment. As can be seen here, the addition or absence of tenascin C (+TN-C or –TN-C), a glycoprotein that is expressed in the extracellular matrix during disease, injury or development, contributes to changes in the normal surface architecture of breast epithelial tissue-at the level of the ECM, which can directly initiate and propagate tumor formation.

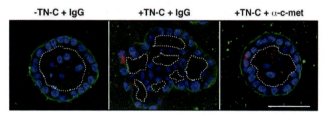

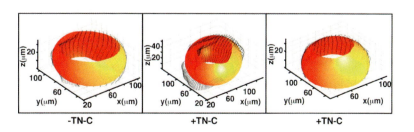

and tissue contour information using normal human mammary epithelial cells cultivated within a 3-D normal or tumor-like microenvironment.

More specifically, when cultured on a non-compliant, i.e., hard, 2-D, chemically inert surface, mammary epithelial cells—that become transformed in breast cancer—fail to achieve a normal form, even though they possess the appropriate genes that should enable them to do this. They form flat monolayers that resemble many other cell types. When cultivated within a compliant, 3-D extracellular matrix within a tissue culture dish, however, cells can be induced to undergo a normal morphogenetic process that resembles the breast *in vivo*. Once they have achieved this state, the breast epithelium will remain quiescent. However, with cancer, the integrity and quality of the external surface change, resulting in an inappropriate growth response (deep within the structure) to this modified surface matrix environment. Understanding precisely how this occurs is a critical step in comprehending this disease, since it is now appreciated that changes in the normal surface architecture of breast epithelial tissue at the level of the ECM can directly initiate and propagate tumor formation. This is a profound discovery, since it suggests that by reversing altered cell-ECM interactions, breast cancer cells can be tricked into normal modes of behavior.

Building upon this scientific foundation, the project examines morphogenesis, lumen formation, and cellular packing behavior as a response to alterations in tissue surface design. We focused our efforts upon simulation of nonlinear behavior in cell biological systems and the translation of this behavior into material systems and fabrication techniques, namely, through 3-D printing. Additionally, we wanted to bridge behaviors abstracted from biology with material

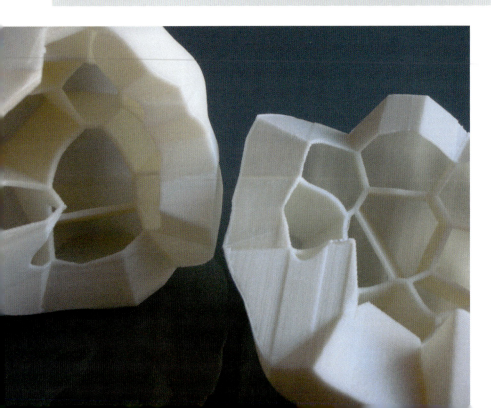

Figure 8.24 3-D print and model of normal human mammary epithelial cells cultivated within a 3-D normal microenvironment; with Voronoi geometric graph

Project: Ground Substance   293

Figure 8.25 3-D print and model of normal human mammary epithelial cells cultivated within a 3-D tumor-like microenvironment inducing tumorigenesis; with Delaunay tessellation

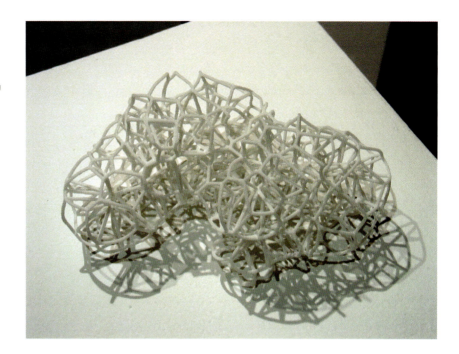

Figure 8.26 Drawing highlighting relationships between cell-to-cell and the basement membrane (extracellular matrix of tissue)

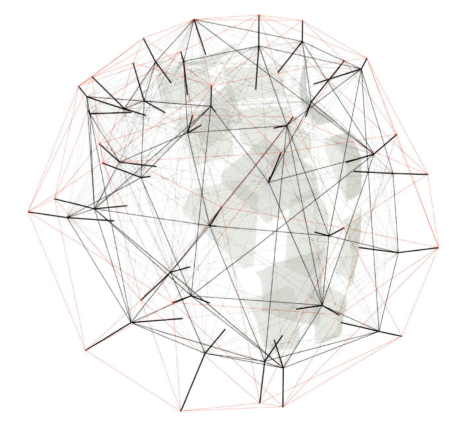

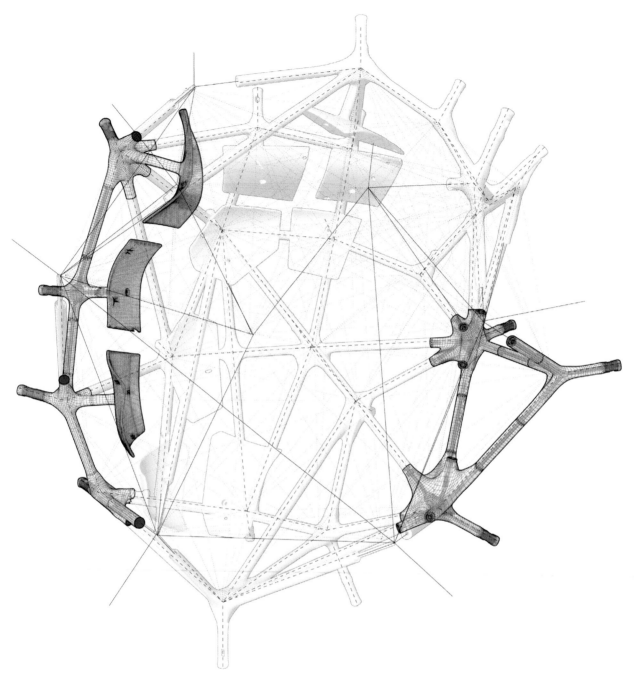

Figure 8.27 Component parts and assembly for Ground Substance, the title of which is an older name for the connective tissue between cells or the extracellular matrix; digital component model abstracted from the original biological system of study

constraints, force transmission, and actual fabrication and assembly as a continuous and unfolding loop. Here, the algorithmic and digital exploration of relationships between interacting cells and their immediate tissue environment gives rise to an abstract, yet deeper understanding of architectural form as it relates to a dynamic boundary condition.

Advanced techniques in additive manufacturing technology influenced the final production of the prototype. In conjunction with the rapid manufacturing of component parts, 3-D printed ceramic modules were also produced. The initial impetus to work with ceramics in architectural projects was driven by an interest in hybridizing the material parameters inherent to the highly plastic nature of clay with basic part to whole relationships present in biological cellular behavior. We intended to move beyond the printing of representational objects in our ZCorp 510 color printer in favor of the rapid manufacturing of parts. This effectively increases the spatial capacity of the build bed through the packing of component parts. The plasticity of clay offers up a useful material interface for projects sitting at the nexus between biology and architecture. Within the Sabin+Jones LabStudio, we started working with 3-D printed ceramics to investigate complex biological phenomena through the visualization of micro-scale datasets embedded in material systems. This was achieved through Sabin's background and expertise in ceramics and direct experimentation with dry clay material recipes and LabStudio's 3-D printer. This was later refined in the context of three seminars on digital ceramics taught by Sabin initially at PennDesign and now in the Department of Architecture at Cornell University;[1] and later through recent spatial prototypes in the Sabin Design Lab (see "PolyBrick: Variegated Additive Ceramic Component Manufacturing [ACCM]", Sabin *et al.*, 2014).[2]

3-D Printed Ceramic Components: Adapted Recipe

During the prototyping phase, we exchanged the proprietary ZCorp powder media for our custom clay body recipe, which is drastically more cost effective. These changes in material require changes in the design in order to adhere to principles that govern

Figure 8.28   3-D printed high fire clay components ready to be excavated from our ZCorp 510 printing bed. This project was our first experiment with component-based 3-D printing of clay material.

ceramics and printing with clay, such as the rounding of edges to reduce the likelihood of chipping. High accuracy and resolution allow for tolerances to be built into the digital model. Parts are printed in a sequence that allows for rapid assembly.

The production of ceramic form includes three distinct phases: greenware, bisque firing, and glaze firing. Greenware is the initial state of the clay form before firing. It is during this phase that the clay may be manipulated through hand forming, throwing, casting, and now 3-D printing. Initial recipes for the dry clay mixture were adapted and later transformed from open source recipes published by Mark Ganter who directs the Solheim Rapid Manufacturing Laboratory located in the Mechanical Engineering Building at the University of Washington in Seattle. The recipes were initially published in *Ceramics Arts Daily* on February 1, 2009.[3]

For the first set of prototypes, we chose to work with a high-fire clay body for durability and strength in parts. Each part was initially fired to cone 06

Figure 8.29 The greenware stage pre-firing. Excavated and cleaned 3-D printed clay parts.

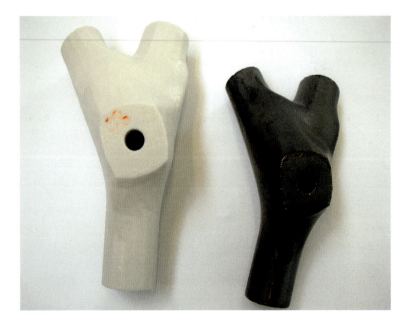

Figure 8.30 Initial set of prototype parts. During the initial firing, the clay body shrinks by approximately 15 percent. The same 3-D printed part is printed in standard starch powder (left) and in clay, then bisque fired, glazed, and glaze fired (right).

Project: Ground Substance 297

Figure 8.31 Close-up view of assembly. The final fabricated model is composed of 146 unique 3-D printed ZCorp prototype parts connected together with an internal system of aluminum rod and an external tertiary system of cable thread.

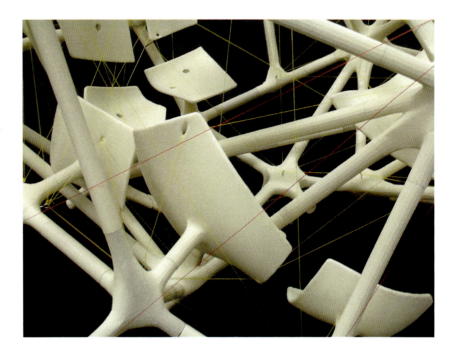

Figure 8.32 Ground Substance on view, originally exhibited at the Design and Computation Galleries, Siggraph 2009, New Orleans, LA.

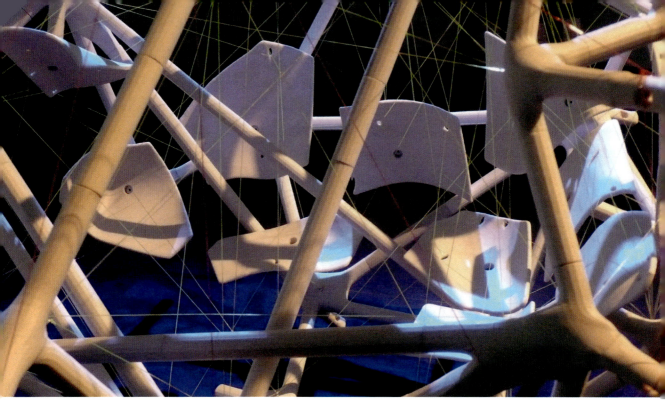

Figure 8.33 Detail of installed prototype. The surface components express and visualize the cells and their connection to the basement membrane.

during bisque fire and then glaze fired to cone 10 or approximately 2300 degrees Fahrenheit. In ceramics, kilns are not fired to temperature, they are fired to a cone level. This incorporates both temperature and overall duration of the firing process, which measures how much heat is absorbed by the clay body. During the initial firing, the clay body shrinks by approximately 15 percent. This shrinkage factor must be incorporated into design parameters when working with low tolerance models and construction techniques.

The final fabricated model is composed of 146 unique 3-D printed ZCorp Powder and ceramic prototype parts connected together with an internal system of aluminum rod and an external tertiary system of cable thread.

Notes

1. Jenny E. Sabin, "Digital Ceramics: Crafts-based Media for Novel Material Expression and Information Mediation at the Architectural Scale," in *ACADIA 10: LIFE in:formation, On Responsive Information and Variations in Architecture, Proceedings of the 30th Annual Conference of the Association for Computer Aided Design in Architecture (ACADIA)* (New York, 2010), pp. 174–182.
2. Jenny Sabin, Martin Miller, Nicholas Cassab, and Andrew Lucia, "PolyBrick: Variegated Additive Ceramic Component Manufacturing (ACCM)," *3-D Printing and Additive Manufacturing* 1(2) (2014): 78–84.
3. Mark Ganter, Duane Storti, and Ben Utela, "The Printed Pot: Ceramic Arts Daily," *Ceramic Arts Daily* 2009. Available at: http://ceramicartsdaily.org/ceramic-supplies/pottery-clay/the-printed-pot/.

Workshop: Nonlinear Systems Biology and Design, 2010

Jenny E. Sabin and Peter Lloyd Jones
[UL]Smart Geometry Workshop, 2010
The Institute for Advanced Architecture of Catalonia (IAAC), Barcelona, Spain
Later on view at DHUB, Barcelona, as part of the Working Prototypes exhibition, 2010
Cluster Leaders: Jenny E. Sabin, Peter Lloyd Jones, Andrew Lucia, Erica Swesey Savig
Participants: Jacob Riiber, Sahar Fikouhi, Mario Guidoux, Mania Aghaei Meibodi, Johannes Beck, Barcu Arkut, Madhi Alibakhshian, Jeffrey Halstead, Wendy Teo, Christy Widjaja, Pavel Hladik, and James Patterson

Building upon our research and exploration of cellular networking processes and behavior (see Part II, Chapter 3), LabStudio directed a unit at the 2010 Smart Geometry Workshop in Barcelona, Spain, at IAAC. As one of ten workshops within the Working Prototypes session, we focused our efforts upon simulation of nonlinear behavior in cell biological systems and the translation of this behavior into material systems and fabrication techniques, namely, through 3-D printing. Fundamental to this work is the integration of the human body as a primary point of departure in the conception, materialization, and making of complex form.

Our cluster was entitled *Nonlinear Systems Biology and Design*. This workshop situated itself at the nexus between architecture and systems biology to gain insight into dynamic living systems for the development of novel computational design tools and material systems. This approach examines the nature of nonlinearities, emergent properties and loosely coupled modules that are cardinal features of complexity. As with our previous student projects,

Figure 8.34 Nonlinear Systems Biology and Design; The team from left to right: Peter Lloyd Jones, Jenny E. Sabin, Erica Savig, and Andrew Lucia.

students were exposed to new modes of thinking about design ecology through an understanding of how dynamic and environmental feedback specifies structure, function, and form.

This workshop consisted of two parts: (1) Scripting and Simulation; and (2) Fabrication and Production of a 1:1 physical structure composed of 3-D printed components and connections with variable stiffness. For the first stage of the workshop, our efforts were focused upon abstracting and extracting the underlying rules and criteria for cellular networking behavior.

We are interested in abstracting behaviors as opposed to static form and cellular shapes. Additionally, we taught the students about the interdependence

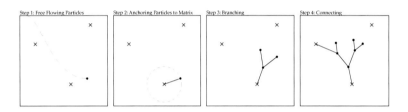

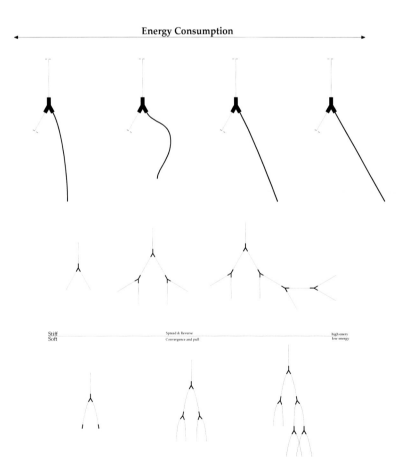

Figures 8.35–8.40 Nonlinear Systems Biology and Design with SG cluster champions Jenny E. Sabin, Peter Lloyd Jones, Andrew Lucia and Erica Savig. The generative rules by participant, Jacob Riiber, 2010. For the first stage of the workshop, our efforts were focused upon abstracting and extracting the underlying rules and criteria for cellular networking behavior.

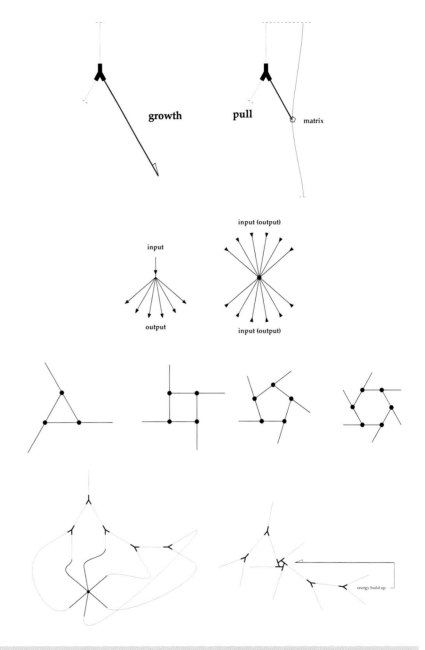

between environment, code, and algorithmic processes. Before the start of the workshop, our testing ground included heavy use of our ZCorp 510 color 3-D printer. We were interested in prototyping components, not wholes. We were also interested in using the powder-based printer to rapid prototype full-scale nonstandard parts instead of representational forms.

ZCorp, now 3-DSystems, provided generous support in the printing and production of the components before and during the actual workshop. We were interested in engaging an analogic mode of designing; bridging

Figure 8.41 As one of ten workshops within the Working Prototypes session, we focused our efforts upon simulation of nonlinear behavior in cell biological systems and the translation of this behavior into material systems and fabrication techniques, namely, through 3-D printing.

Figure 8.42 Families of 3-D printed nodes produced by the workshop participants. Each node produces a different network behavior within the analogic assembly.

behaviors and nonstandard geometries abstracted from biology with material, structural, fabrication and assembly constraints as a continuous and unfolding loop.

In the context of the workshop, there was a constant reengagement with feedback, where constraints from the actual physical environment informed our virtual computational models.

The final Working Prototype engaged a process of nonlinear fabrication informed by biological behavior and novel fabrication techniques that incorporated 3-D printing.

We started the workshop several months in advance by providing specific biological concepts (i.e. vascular networking within the lung) to our group of international students. The initial intent was for the students to build some

Figure 8.43 *Nonlinear Systems Biology and Design*, with SG cluster champions Jenny E. Sabin, Peter Lloyd Jones, Andrew Lucia, and Erica Savig. Installation with participants at IAAC. This workshop situated itself at the nexus between architecture and systems biology to gain insight into dynamic living systems for the development of novel computational design tools and material systems that are at once natural and artificial.
Source: Photo by Shane Burger

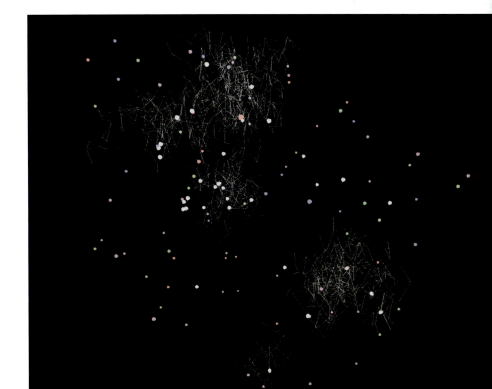

Figure 8.44 Simulation of networking behavior by workshop participant, Jacob Riiber. Built in Processing.

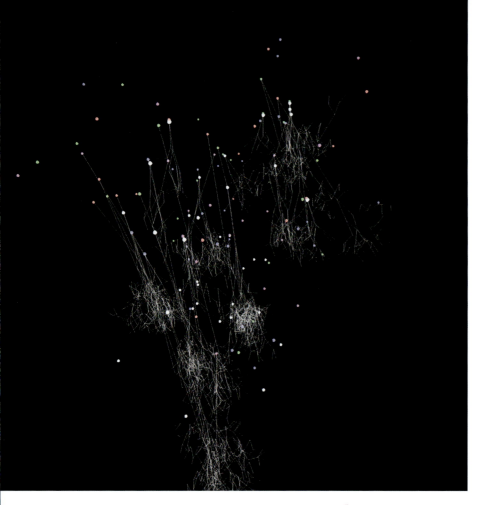

Figure 8.45 Simulation of external forces acting upon the overall network. Simulations were used as a guide for the design and production of the analog network assembly.

Figure 8.46 Cluster leaders, Jenny E. Sabin and Peter Lloyd Jones, describing dynamic cellular surface modeling through the lens of deployable structures.

Figure 8.47 Workshop participants assembling the final prototype at IAAC

basic rule sets that could be applied and developed at the workshop. Deviation, abstraction, and bending of the actual rules governing these systems were highly encouraged. Our cluster relied upon understanding certain rules that promote the development and branching of networked structures (i.e. blood vessels) within the developing lung, and specifically focused upon the role of the microenvironment, which includes the extracellular matrix and multiple cell types, including those that make up the lining of the blood vessel wall (also known as endothelial cells). Within this research trajectory, we aimed to place genetic code in context, without which, proper form and function within the body fail to exist and operate. Some of the questions that we posed: what does context mean in the setting of lung vascular development, and how might it influence code? Why do we need a vascular network, and what are the rules of engagement between code and context that allow branching morphogenesis to occur in the lung, ultimately giving rise to an exquisite fractal network upon which our life depends? How can these rules be extracted to give rise to novel and abstract forms at that the macro-scale?

We then presented a few basic questions and concepts in contemporary cell and matrix biology for the students to consider, which included the following. First, if all the (somatic) cells in our body possess the same genetic code (which they do), how does the body give rise to so many distinct forms with so many different functions, and how do these differentiated structures remember to be what they are, and to remain where they are? In other words, why doesn't your eye turn into a foot, and your arms into lungs? How is the linear DNA code transformed into a dynamic, nonlinear, multi-dimensional system? A simple answer to this is that combinations of specific gene products (i.e. proteins) give rise to specific structures, which means that certain genes are sometimes on, while others remain off. It is the combinatorial nature of the expressed DNA code that gives rise to specificity in form and function in living systems.

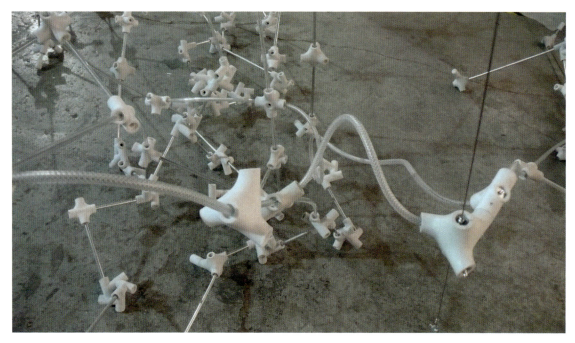

Figure 8.48 The material system for the network is composed of 3D printed components and three connector components featuring floppy, semi-rigid, and rigid materials. The assembly starts at two locations as a set of seeds: on the ground and near the ceiling with tensioned cable. Here, changes in force transmission through the materials are considered as active design parameters.

So how might context or microenvironment turn genes on or off? Imagine a cell. A membrane encompasses the cell, and within that membrane there are many structures, including the nucleus where the DNA code is contained. However, cells are not merely water-filled balloons with a nuclear core, they have structure that is provided by an internal skeletal structure, called the cytoskeleton. The cytoskeleton provides a tension network within the cell, and it runs all the way from the cell membrane to the nucleus. What is more, cells don't merely float around the body at random during development; they stick to and interact with each other and with proteins that exist outside the cell. As described in the previous section, this protein network is the extracellular matrix or ECM. When a cell encounters an ECM protein, it forms a physical connection with the ECM through receptors on its cell surface. In essence, the cells dock with the ECM. Once docked, the ECM and the cell become a single unit, underscored by a structural continuum from the ECM on the outside, and the cytoskeleton on the inside, which in turn is coupled to DNA. So it is relatively easy to imagine that if the structure of the ECM differs, then the connections between the cytoskeleton and the DNA will also differ. These differences in connectivity, at the structural and chemical levels specify which parts of the DNA (genes) are turned on or off.

In simpler terms, imagine the ECM and cytoskeleton as an interconnected cable network under tension, and at the end of this network is Gene A, which, when expressed, will ultimately give rise to Form A. Next to Gene A is Gene B, which has the potential to give rise to Form B. When interacting with ECM Type I, cytoskeletal tension changes and it pulls on the folded DNA code, allowing Gene A to be exposed and expressed. On the other hand, with this specific tension network, Gene B is folded and hidden, and so it is not expressed. However, if the nature of the ECM changes to Type II, the tension network also changes, and this may lead to pulling Gene B out of its cryptic state, unfolding it, allowing it to be expressed. So in this scenario, and depending upon the type of ECM, and thus the tension network that cells experience, the cell may have Gene A on and Gene B off, Gene A off and Gene B on, Gene A on and Gene B on, or Gene A off and Gene B off. Each of these different combinations will give rise to a different type of form and function. Now imagine the possible permutations when considering all the approximately 30,000 identical genes in each of your cells, and given the fact that cells may simultaneously engage scores of different types of ECM protein networks!

Possible Conditions and Details That We Used to Evolve the Students' Notion of Code in Context

(1) There are three different types of cytoskeleton within the cell. Only one type (the microfilament network) directly couples to the ECM via cell surface receptors, and it is this network that also interacts with DNA at the nuclear membrane, yet all three cytoskeletons do interact with one another inside the cell. These other types are called microtubules and intermediate filaments. Each type is made up of different proteins, and so their physical properties, including density, size, and elasticity also differ. How might a change in the microfilament network in response to alterations in context lead to alteration of the other cytoskeletal elements? Importantly, tensegrity models have been

Figure 8.49 Nonlinear Systems Biology and Design by SG cluster champions Jenny E. Sabin, Peter Lloyd Jones, Andrew Lucia, and Erica Savig for the exhibition "Working Prototypes."
Source: Photo by Jenny E. Sabin, DHUB, Barcelona, 2010.

Figures 8.50 and 8.51 Final installation at DHUB on opening night, Barcelona, Spain.

used in this research area, yet these do not account for all the nonlinear responses adapted by these networks as they sense and modify in response to changes in ECM context.

(2) Although context specifies code usage, context would not exist without code. What this means is that the ECM is derived from a genetic code, and cannot exist without it, but the development of higher levels of organization at the

genetic level, must be dependent upon an increasingly complex context. So what all of this means is that form, specified by context, regulates code usage, and that code usage feeds back to specify context and this form, and so on. The system is in a state of constant and dynamic feedback.

(3) Speaking of dynamics, once the ECM is laid down, it does not stay necessarily in one state. It can be embellished or ornamented by the addition of other protein modules, or it can be modified by enzymes that chew away at it, thereby editing or removing it, leading to another type of context. This is particularly evident in disease, where the ECM changes, and gene behavior can then follow suit, giving rise to abnormal forms and function, the result of which is a pathology. In fact, pathologists look at changes in the ECM as a way to predict the severity of certain diseases, including cancer.

Having introduced the ECM and its ability to control the behavior of genes, we then turned the students' attention to specific examples of how context specifies form and function via the appropriate use of code in the context of blood vessel development within the lung. This provided the starting point for the students' initial simulations and nodal experiments with 3-D printing. This workshop marked our second investigation into the use of 3-D printing for the rapid manufacturing of component parts, *Ground Substance* being the first. The following projects build upon these fabrication and part-to-whole relationships with entirely new materials and spatial structures including human scaled adaptive materials and environments.

PART IV
Personalized Architecture and Medicine

Adaptive Environments: The generation of scientifically-based, design-oriented applications in contemporary architecture practice for adaptive building systems, protocols, and material assemblies.

9

eSkin: BioInspired Adaptive Materials[1]

Jenny E. Sabin and Andrew Lucia with Jan Van der Spiegel, Nader Engheta, Kaori Ihida-Stansbury, Peter Lloyd Jones, and Shu Yang

While the previous Parts highlight the development of tools, methods, and prototypes for modeling and materializing biological behavior and processes for both architecture and science, Part IV describes advanced research supported by the National Science Foundation in the context of topics such as sustainability, building ecology, and adaptive materials and architectural assemblies. Here, we apply select methods and tools to model and simulate cellular processes, in particular, cellular motility. As was briefly described in Part I, in 2010, the National Science Foundation (NSF) within the Emerging Frontiers for Research Innovation (EFRI) Science in Energy and Environmental Design (SEED) umbrella solicited proposals for trans-disciplinary research teams that would engage the problem of sustainability concerning building energy and its associated impacts upon our built environment. In an unprecedented occurrence, the teams were to also include architects. Importantly, the program manager for EFRI SEED did not require AIA licensure as a requisite for architects to submit. This opened up opportunities for both licensed architects and architectural designers engaged in practice and core design research to apply with their collaborative teams across academia, practice, and industry.[2] At the time that the call was made public, the Sabin and Jones LabStudio was in its fourth year of collaborative research and

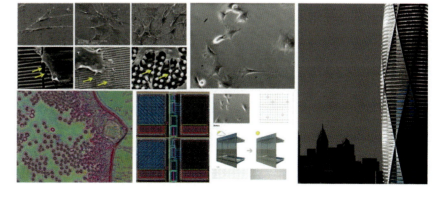

Figure 9.1 eSkin summary with team contributions by: Shu Yang (PI) (materials science), Jan Van der Spiegel & Nader Engheta (Co-PIs) (electrical and systems engineering), Kaori Ihida-Stansbury, Peter Lloyd Jones (Co-PIs) (cell biology), University of Pennsylvania; Jenny E. Sabin (Co-PI) and Andrew Lucia (Senior Personnel) (architecture), Cornell University.

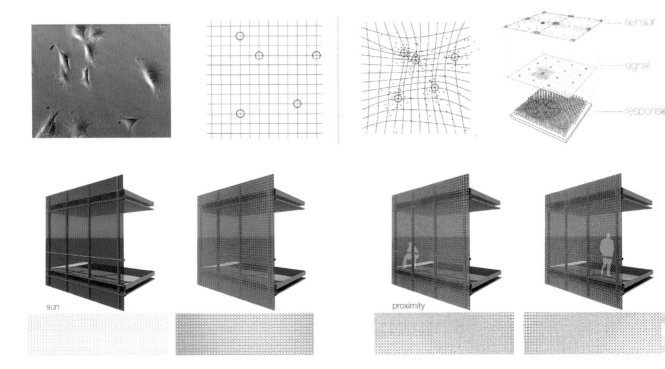

Figure 9.2 eSkin inputs: cell-matrix interface and architectural speculation as adaptive wall assembly

project work. We had also engaged in a collaborative exchange with Dr. Shu Yang, a materials scientist also based at the University of Pennsylvania and who had obtained her PhD in Chemistry and Chemical Biology from Cornell University. We shared a graduate architecture thesis student, Erica Swesey Savig, and engaged in joint seminar lectures, lab meetings, and studio reviews. This initial foundation helped pave the way for the successful procurement of one of ten $2 million grants awarded by the NSF.

As was briefly highlighted in Part I, since the official public launch in the fall of 2010 of our NSF EFRI SEED project entitled, *Energy Minimization via Multi-Scalar Architectures: From Cell Contractility to Sensing Materials to Adaptive Building Skins*, we, Jenny E. Sabin (Co-Principal Investigator) and Andrew Lucia (Senior Personnel), have led a team of architects, graduate architecture students, and researchers in the investigation of biologically-informed design through the visualization of complex data sets, digital fabrication, and the production of experimental material systems for prototype speculations of adaptive building skins, designated eSkin, at the building scale. The full team, including Nader Engheta, Jan Van der Spiegel, Kaori Ihida-Stansbury, Peter Lloyd Jones, and led by Dr. Shu Yang (Principal Investigator), is actively engaged in rigorous scientific research at the core of ecological building materials and design. We ask, how might architecture respond ecologically and sustainably whereby buildings behave more like organisms in their built environments? We are interested in probing the human body for design models that give rise to new ways of thinking about issues of adaptation, change, and performance in architecture. These biological models, based on human cellular processes and behavior, engage biomimetic principles in the synthesis and design of new materials that are adaptive to external inputs, such as light. The eSkin project starts with these fundamental questions and applies them to the design and engineering of responsive materials and sensors. The work presented here, entitled eSkin, is one subset of ongoing trans-disciplinary research spanning the fields of cell biology, materials science, electrical and systems engineering, and architecture. The goal of the eSkin project is to explore materiality from nano- to macro-scales based upon an understanding of nonlinear, dynamic human cell behaviors on geometrically defined substrates. Sensors and imagers are then being designed and engineered to capture material and environmental transformations based on manipulations made by the cells, such as changes in color, transparency, polarity, and pattern. Through the eSkin project, insights as to how cells can modify their immediate extracellular microenvironment are being investigated and applied to the design and engineering of highly aesthetic passive materials, sensors and imagers that will be integrated into responsive building skins at the architectural scale.

Comprised of a field of low-cost sensors and passively responsive materials, eSkin is conceived to be generic and homogeneously structured upon installation (i.e. laden with the full potential) but readily adaptable to local heterogeneous spatiotemporal conditions, thereby reducing the overall functioning demands upon it and ultimately lowering overall energy consumption. In this regard, a "learning" and adaptive second skin would forgo the need for lengthy, costly, and one-time site analysis relegating ever changing environmental analysis and response of the local and global spatiotemporal environments to its own internal/local functionality. This manner of operation not only maximizes immediate performative efficiency, but also allows for ongoing contextual adaptation, which is not possible by glass fritting alone, an established architectural treatment for passive solar control in the

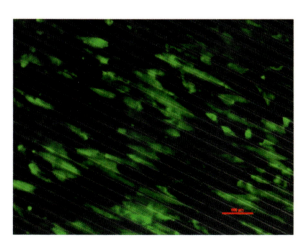 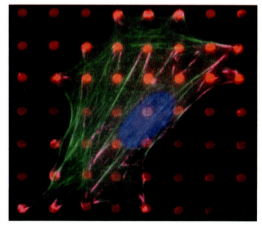

Figure 9.3 Human smooth muscle cells plated on wrinkled and pillared samples of eSkin material, an organic polymer known as PDMS (polydimethylsiloxane). PDMS produced in Dr. Shu Yang's lab; Cell data originally produced in Dr. Peter Lloyd Jones lab, University of Pennsylvania.

building industry. We posit that sustainable building practice should not simply be a technical endeavor, but one that also includes the transformation of existing built fabric into sustainable models that inspire both positive socio-cultural change and innovation in science, technology, and the material arts. In the context of the eSkin project, we hope that the transformation of existing buildings in post-industrial cities with the application of adaptive building skins will not only address the dire need for greener building practices, but will also generate the necessary equivalent effects that may foster and inspire urban ecosystems, beauty, collective levity, and playfulness at the scale of the city.

Our architectural research with the scientific team operates within a multi-year and multi-phase research plan. Currently, the project is broken down into three phases including: (1) the production of catalogs of visualization and simulation tools that are then used to discover new behaviors in geometry and matter; (2) an exploration of the material and ecological potentials of these tools through the production of experimental prototypes and material systems, and (3) the generation of scientifically-based, design-oriented applications in contemporary architecture practice for adaptive building skins.

Additionally, our group is carrying out fundamental and applied research to develop novel materials synthesis and fabrication methods not yet available on the market. Our applied design offers additions to the paradigm of responsive and adaptive architecture through architectural treatments in the form of an adaptive building skin, which modulates passive solar, light, and moisture control with embedded sensors that ultimately (re)configure their own performance based upon local criteria.

Our role as architects involves generating tools to visualize and simulate cell attraction forces and cell behavior such as forces distributed via a virtual extracellular matrix environment, over multiple time-states while also incorporating micro-scale material constraints. Beyond visualization, we also direct the architectural intent of the project by constantly speculating on how results at the nano and micro-level will potentially look, feel, and assemble at a building scale. For example, one of the featured interactive prototypes integrates actual simulation data from micro-scale material substrates to test how these complex behaviors and effects may translate to the building scale and initiate human interactivity. Based upon these nonlinear and dynamic responses that

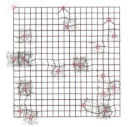
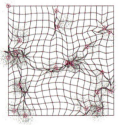
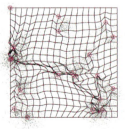

Figure 9.4 Custom digital tools to visualize and simulate cell attraction forces and cell behavior such as forces distributed via a virtual extracellular matrix environment, over multiple time-states

Figure 9.5 Architectural speculation of eSkin effects simulated as an interior partition assembly

human cells generate, we are redesigning and re-engineering interfaces between living and engineered systems with the ultimate goal of implementing some of the key features and functions revealed by cells on a chip for sensing and control at the building scale.

Background on Materials Research and Importance of Scaled Prototypes

The particular material research presented here focuses on the latest prototypes and applications within the eSkin project, the optical simulation and application of geometrically defined nano-/micro-scale substrates that display the effects of nonlinear structural color change when deployed at the building scale. Given the groundbreaking nature of this work, many of the tools we use to simulate, visualize, and model nonlinear nano-to-micro scaled material properties and effects at the architectural scale must be custom-written and designed. Through the development of our own simulation tools that work directly with these optical data while also understanding the state of the art in optical engineering and design simulation, it is possible to generate a thinking space and design intuition for materiality not yet realized at the architectural scale. We are currently limited to a 4-inch by 4-inch maximum swath of the eSkin material due to high material costs and fabrication time. This requires that we develop and fabricate prototypes that exhibit the same material effects of eSkin, but that can be fabricated at the human scale. Specifically, nano-/micro-scale pillar substrates, designed in the Yang lab, form the basis of this investigation. These substrates are fabricated via microlithography and soft

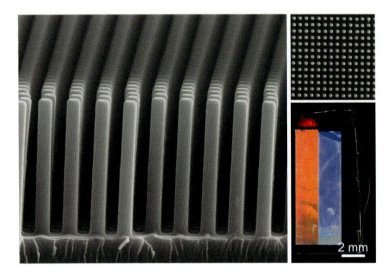

Figure 9.6 SEM image of square micro pillar array (left). SEM (upper right) and optical (lower right) images of micro pillar arrays. Two different colors (lower right) result from Bragg diffraction of micro pillar arrays with different periodicities.

Source: Photo credit: Lo, Zhang, and Yang, University of Pennsylvania.

lithography, first requiring a negative nano/micro pattern to be etched into a substrate in which an organic polymer known as PDMS (polydimethylsiloxane) is subsequently cast, cured, and removed, thus producing a positive relief of nano/micro pillars (see, for example, Thompson *et al.*, 1994, Xia & Whitesides, 1998, and Zhang *et al.*, 2006). Though these positive substrates may be cast using an array of polymers, compounds and mixtures, for the purpose of this study we are exclusively interrogating the properties of PDMS as a proof of principle.

Demonstrating unique angle-dependent and wavelength filtering optical properties of interest, these periodic pillar arrays act as passive filters of light given the specific nano or micro scale periodicity of their structures and the angle at which they are viewed. Because of the particular periodic spacing and geometry of these arrays existing at the nano-to-micro scale, light is absorbed via the PDMS material, but also filtered as a property of the specific wavelength of light that is allowed to pass through a particular pillar array. Depending on factors such as the diameter and the periodic distance between each pillar in an array, the visible spectrum of light, which exists between 390–750nm, will be filtered out, absorbed, or scattered and reflected or refracted from the material. Through changes in pattern, compliance, geometry, and structure, we can manipulate material features including color, transparency, and opacity. Here, color change is generated by an optical effect such as refraction or interference as opposed to a change in pigment. This is known as structural color. In our case, physical structures in the form of pillars interact with light to produce a particular color. These colors are also dependent upon angle of view or one's orientation to the given materials. There are many examples of structural color change found in nature such as the wings of the Blue Morpho butterfly or the feathers of hummingbirds. We are interested in harnessing these material features and effects and translating them into scalable building skins. Imagine dynamically blocking sunlight throughout

Figure 9.7 Left, schematic geometry of nano/micro pillar arrays. Right and below, schematics of nano/micro pillar units.

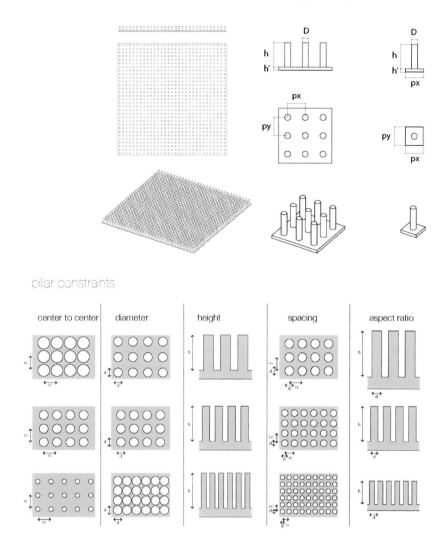

the day through simple mechanical changes of the eSkin film via stretching or compressing or in other words, creating and tuning your own window!

Though these specific types of optical qualities can be seen by the naked eye, extracting their performance quantitatively for speculation at larger architectural scale applications is necessary given (1) the current limitations in which these substrates can be fabricated (currently 4 inches maximum); and (2) the need to speculate on large-scale deployments of potential materials without the need to actually fabricate. Thus, simulating the effects of larger swathes of these materials has been the goal and focus of this research; to speculate as to the larger-scale application and effect of these substrates in an architectural context.

As actual fabrication of the desired nano material is still unattainable, we demonstrate in principle two distinct types of speculation for the architectural scale. The first is a simulated, digital, real-time and interactive prototype that

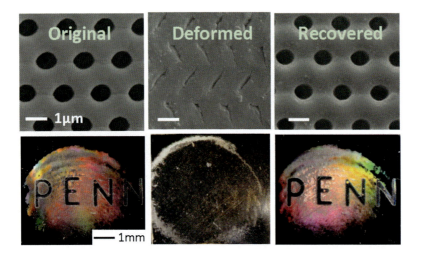

Figure 9.8 Example of a predefined geometric pattern embedded within a shape memory polymer material displaying structural color change under deformation and recovery. Here we exhibit SEM images of membranes consisting of a hexagonal array of micron-sized circular holes (top row) and demonstrated dramatic color switching as a result of pattern transformation (bottom row).

harnesses the optical properties and attributes of the particular desired nano materials being designed and tested within the Yang Lab for the eSkin project. Here, through the use of video input, a viewer's proximity and movement trigger a simulated response in the virtual substrate's appearance.

The second output advances speculative design trajectories within the eSkin project. It is a physical, interactive, and scaled component prototype whose properties behave in a comparable manner to those sought, but which can be fabricated at a larger scale. Importantly, this second scaled prototype provides a test bed for another fundamental element in the conception of eSkin, that of local adaptation to environmental stimuli. Here, we consider the role of the individual sensing node and its effects upon a local region of change within a global surface substrate. To this end, our scaled prototype makes use of sensing and control technologies developed in the labs of Van der Speigel and Engheta, which then influence local regions of change within the eSkin prototype. While the adaptive materials, colloidal particles, used in this prototype are not the intended final material output for eSkin, they allow for a rapidly responsive testing ground for the local sensors and their adaptive technologies. The thrust of this section demonstrates these two distinct trajectories, both of which offer insights and allow for development of further architectural speculations as we traverse scalar-dependent properties generated at the nano and micro levels, yet applied at the macro human scale.

Prototyping Optical Properties at the Nano/Micro Scale

Previously (Wang et al., 2013), we have demonstrated the extraction of simulated optical properties and their deployment through a suite of off-the-shelf and custom written software, enabling the speculation of nano-/micro-scaled material properties at an architectural scale.

The unique physical and angle-dependent optical properties of a small portion of these periodic geometric substrates were simulated in the labs of Van der Spiegel and Engheta through the use of material simulation software. These simulations, which derive the angle-dependent optical properties of the material substrates, ultimately form the basis for larger-scale simulations of potential material applications within the eSkin system. The angle-dependent property can be formatted into a function of reflection coefficient and transmission coefficient versus wavelength of light. As illustrated in Figures 9.9 and 9.10, the structure of the untreated pillar substrate is 2-D periodic. Therefore, the optical properties of the pillar array may be obtained by simulation of a single unit in a matrix. While neighboring units must be considered, immediate neighbors in the simulation are taken to be the same geometry in order to increase efficiency. To develop a digital material that reproduces the optical properties of the selected sample faithfully, its bidirectional light distribution function must first be determined in transmission and reflection, so that the spatial distribution of emerging light can be identified for varying incident directions. For this simulation we used software specifically designed to determine nano-scale optical effects.

Since the PDMS material is wavelength-sensitive, it is necessary to obtain a spectral power distribution for both reflection and transmission light at each light

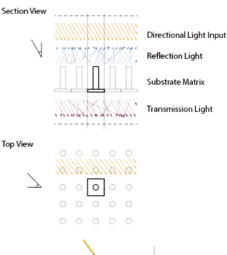

Figure 9.9 Two sub directions of incident light and reflection and transmission light

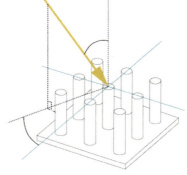

Figure 9.10 Incidental angles

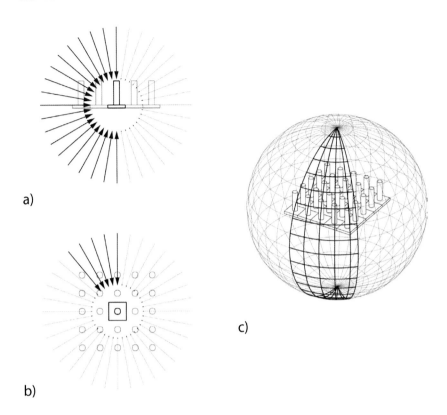

Figure 9.11 Simulation of the incident angle about a sphere

angle. As opposed to conventional architectural material simulation and rendering software, the direction of reflected and transmitted light of these unique nonlinear nano/micro pillar surfaces and substrates do not perform like a conventional specular surface, or a diffuse material. To this end, our method requires integrated spherical measurements rather than a total summation of reflection light and transmission light intensity in all directions.

Interpreting the Simulation Data

Methods for interpreting large amounts of simulation data are key for simulations operating at the building scale and in understanding the optical properties of the nano/micro material in architectural contexts. Light in the natural world is almost never purely monochromatic. For the particular material tested in this proof of principle, given an increasing incident angle, the material tends to be more transparent or translucent. Having interpreted the nano-/micro-scale simulation data, it is now possible to deploy these optical and material properties at the architectural scale.

Initially at the architectural scale, speculations as to the extracted performative and aesthetic qualities of these nano/micro materials were deployed using custom-written algorithms in conjunction with 3-D modeling software. We would later rewrite these algorithms in alternate software to facilitate a smoother real-time

Figures 9.12 and 9.13
Incident light and normals of a surface; rendered optical properties of eSkin

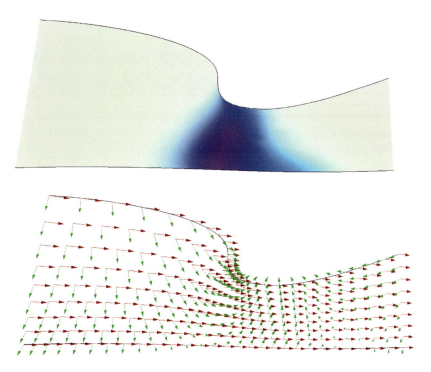

simulation. To maximize the color-changing effect and efficiency calculation, the simulated geometry was approached as a tessellation of panels that respond to people's movement in front of a video camera serving as a sensor. Each panel can be treated as an individual surface with a single light source of uniform distribution, with each point on the surface having the same optical properties. Therefore, evaluation of the center point upon each discrete tessellated surface represents properties of the whole. With one single light, the two sides of the surface can be called the reflection side and transmission side. After calculating the incident angle, a matrix at different wavelengths can be obtained by interpreting the measured data.

Interface Design for Simulation

While using the prior studies as a framework, open-source scripting environments were selected for computational efficiency. Custom-written scripts and algorithms were developed to easily import geometry and optical data from the prior simulations. This was then interfaced with real-time interactive inputs via a camera installed in the prototype, thus integrating a viewer's experience as part of the simulation. For the purposes of this simulation, the new application retains much of the original functionality of the initial studies (i.e. rotation and color change), querying simulated angle-dependent optical data arrays as a user moves about the environment, ultimately affecting the viewers' virtual angular difference with respect to the simulated substrate.

This interactive simulation features a built-in camera that detects motion through a custom script. When one waves their hand or moves in front of the screen—to the left, right or center—the input adjusts the virtually simulated eSkin in real time. The simulation incorporates real optical data in the form of color and geometric transformation from micro-scale material substrates to speculate upon how these nano-to-micro-scale material effects and related geometries may be applied at the architectural scale. Here, visitors are able to interact with the actual material effects of eSkin in real time and in a scaled way.

In order to achieve this, a faceted digital mesh "wall" model composed of 1,728 individual faces was generated. The number and geometry of mesh subdivisions were based on empirical design decisions governed by spatial fidelity (impacting computational speed) and aesthetics. Each mesh facet within the sketch is embedded with the identical angle dependent optical color data array extracted from simulated nano/micro materials fabricated in the Yang Lab. At its inception, this color data appears generic, but is given specificity per simulated viewing angle. Within the simulation, this color data is then queried through time as a list array of color values corresponding to virtual incident-dependent viewing angles initially established according to actual material testing in the Yang Lab.

In order to engage with viewers in an environment external to the simulation, a video camera captures a real-time image feed and a resultant low fidelity image difference is taken. This image difference is achieved by simply comparing corresponding pixel brightness at each pixel of a possible 256 values through time and at each frame of image capture. If the difference in brightness between two spatially coincident pixels has changed sufficiently through time, the resultant pixel is said to have changed and is added to a counter for further inquiry. In order to increase computational speed for real-time interaction, not every pixel in the initial image array is considered for calculation. For the purposes of this study and simulation, the optimal fidelity of the differenced pixel array was set to empirically satisfying outcomes based upon speed and the amount of sensitivity to environmental stimuli deemed sufficient. Depending on the amount of difference at a given moment within the underlying grid of circular bins, varying degrees of rotation about each mesh centroid are triggered within the corresponding meshes. Finally, based upon the amount of rotation and each corresponding mesh normal to the viewing plane of the simulation screen, the array of simulated angle-dependent color data from the Yang Lab is accessed, assigning appropriate color data to the mesh at each instant in the simulation. Thus, within the overall field of change, as a single mesh facet is allowed to rotate within the model, a resultant color variation is expressed within each facet, corresponding to the optical color data array.

While the primary aim of this prototype is to simulate macro-scale possibilities of nano-scale technologies and the effects of eSkin, it was also important to create a provocative representation of the technology through

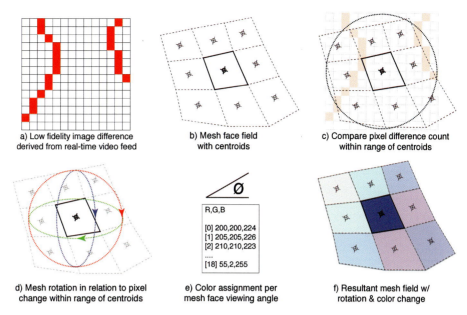

Figure 9.14 Schematic diagram of real-time simulation. A real-time image feed is subject to low-fidelity image differencing. The virtual resultant difference map (a) is compared to the actual mesh face field visible in the simulation (b). Within a given range of every unique mesh centroid, the number of changed pixels is quantified (c). Each mesh is then rotated about its centroid corresponding to the given pixel count (d), while a corresponding viewing angle is calculated and coupled with a change in color (e).

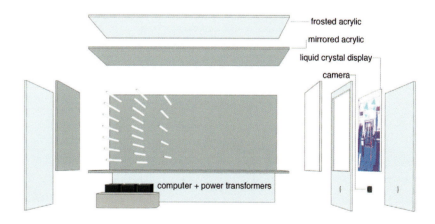

Figure 9.15 Diagram of simplified installation assembly. An array of cold cathode fluorescent bulbs is housed within a mirrored box. The face of this box is a sandwich of a liquid crystal display and clear acrylic. The mirrored box provides a reflective depth that is much deeper than the actual volume of the box.

an interesting and dynamic simulation that exemplifies the capabilities of the technology and suggests further real-world applications. Therefore, we approached this installation as an interface that alludes to the possible spatial characteristics of a thin film laminate in relation to the tangible built environment. In order to distance the installation from a typical screen displaying an interactive simulation, we disassembled a contemporary Liquid Crystal Display monitor, removed everything nonessential to the display, and applied the LCD panel to a box of our own creation. The liquid crystal screen serves as a façade for the box, constructed of frosted

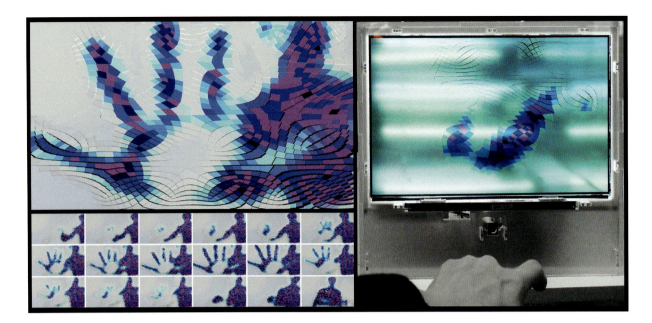

acrylic glass, mirrored acrylic glass, and an array of high-voltage cold-cathode fluorescent bulbs, 2 mm in diameter. The pattern of the bulb array within the box, in conjunction with the mirrored surface, provides a seemingly deep space that expands within the box and behind the screen, lit at a standard interval by bulbs that resemble typical office fluorescent lights. Along with the screen, at the face of the box is an embedded camera. This camera is the input for the simulation run on the machine hidden in the underside of the box that is in turn displayed upon the front of the box as a mock interactive building façade unit.

Interface Design for Scaled eSkin Components

The second prototype, originally exhibited as part of the 9th Archilab, Naturalizing Architecture, at the Frac in Orleans, France, aims to advance speculative design trajectories within the eSkin project as a physical, interactive, and scaled component prototype whose properties behave in a comparable manner to those observed at a nano-to-micro scale, but which can be fabricated at a human scale. Specifically, we are working with structural color where physical structures in the form of particles interact with light to produce a particular color. Silica colloidal nanoparticles dispersed in an organic medium (solvent) are sandwiched between two transparent conductively treated indium tin oxide (ITO) pieces of glass, housed within an assembly of three laser-cut acrylic glass frames. These colloidal particles are typically found in nature in the form of seashells where opalescent color gradients are present. The light reflected from the ordered structure (depending on the particle size, distance, and reflective index contrast between the silica nanoparticles and the organic medium) provides a specific wavelength of light. When a voltage is applied to the particulate solution, the surface charge of the particles is altered, thus changing the distance between the particles, leading to the change of color

Figure 9.16 Still image (top left) and sequential still images (bottom left) taken from real-time simulation built with Processing, harnessing geometric variation, optical data arrays, and user input as parameters. Simulation presented on custom LCD monitor, backlit with cold-cathode fluorescent bulbs, and embedded in custom-built acrylic glass housing (right).

Figure 9.17 Rendering of eSkin material prototype demonstrating user interaction as an active input with resultant speculative transformation of the material substrate (top). Schematic diagram of circuit design interfacing with nano-colloidal particle solutions through voltage control. Individual sensing nodes interact with the material substrates locally through voltage control via the sensing of changes in ambient light, ultimately affecting the appearance of the prototype components (bottom).

appearance. At each intersection between the color cells, a sensor based on shifts in light intensity levels actuates a voltage change between the adjacent color cells. Thus, when a finger, hand, or figure passes by a sensor, a detected shift in light intensity level triggers a small voltage shift across the ITO component, reorganizing the distribution of particles in the solution, ultimately affecting the reflected appearance of color from the nanoparticle solution.

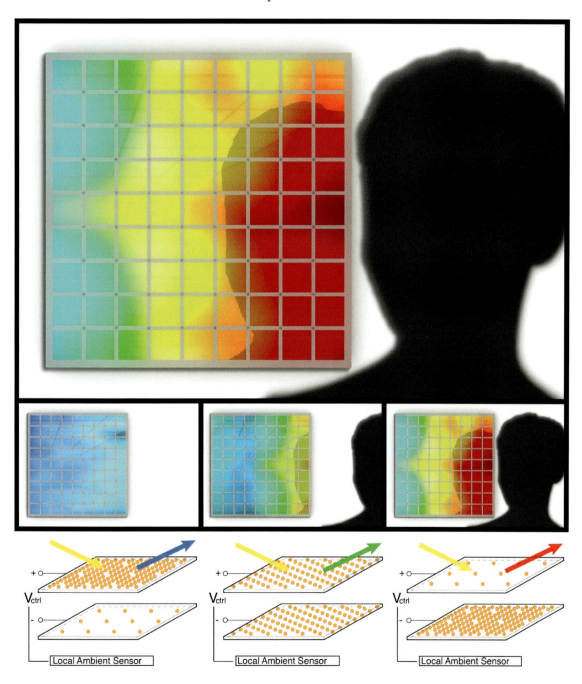

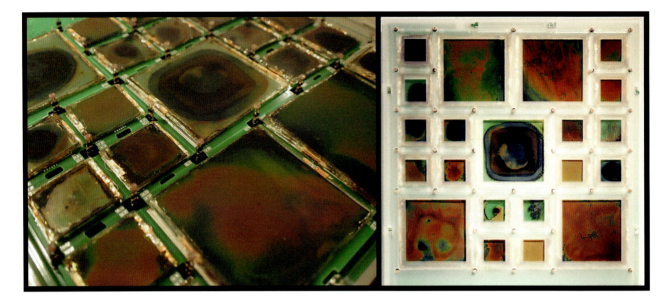

Rather than purely display a simulation based upon actual data as recorded from the original nano-pillar material simulation, we were able to create a panel that displayed the dynamism and control achieved with an actual nanoparticle solution. In collaboration with the labs of Yang, Van der Spiegel and Engheta, we present a modular component design that exhibits the structural color adaptation of the nanoparticles in relation to variable voltage.

Using glass treated with ITO, acting both as the frame and the conduit for the nanoparticles within, we were able to create a mosaic of a variety of individual units, of multiple sizes. Each of these components is then mounted onto a PCB, with its own control circuit that adjusts the charge across the glass according to the dynamic impedance of an arrangement of photo-resistors. These photo-resistors,

Figure 9.18 ITO-treated glass cells with voltage-controlled nanoparticle solution within, housed on a custom-built PCB substrate, and controlled locally via ambient sensing nodes. Note, front face of prototype housing removed to reveal component arrays (left). Component material prototype with local sensing nodes affecting component cells, harnessing user interaction as an active input and resultant transformation of the material substrate (right).

Figure 9.19 Detail photo of the eSkin prototype featuring structural color change

acting as sensing nodes, are placed at each corner of the cells, and react to the presence of the participant or spectator.

Within the context of the two prototypes presented, the practical inability to fabricate nano-/micro-patterned materials at a human scale in turn inspired a speculation as to the behavior of these materials and their corresponding effects as scale increases were taken into consideration. We found the scalability of effects and material fabrication to be the greatest challenge to the group at hand. Specifically, the micro and nano lithography arrays we intend to utilize cannot currently be fabricated above roughly a 2 cm × 2 cm feasible sample size with an absolute maximum of 4 in × 4 in. The two prototypes presented here can thus be seen as a designed response and speculation to this limitation as well as a test for more advanced human interaction, including local and regional environmental stimuli.

Through simulation, modeling, and prototyping, unique nonlinear aspects of the desired eSkin materials, and projected use of these materials *in situ*, were extracted and redeployed through a series of prototypes. Through each of these prototypes (the simulation and the scaled components), we were limited to the exploration of aspects of the nano-to-micro material behavior and its application in part, but never as a whole. Thus, for each investigation, we took this limitation as an opportunity to interrogate and expand upon particular behavioral aspects of the eSkin materials, while not trying to merely mimic the total functionality within a single prototype.

By their very nature, the eSkin materials in question demonstrate optical variability based upon the relative location between a viewer and the surface. In an ideal large-scale application of an eSkin prototype, material effects, specifically color change, transparency, and transmission, would be the result of geometric variation within a surface, a viewer's relative position with respect to that surface, and the source of illumination upon or from behind the surface. Naturally, it is impossible to take all of these parameters into consideration in a single virtual modeled environment, especially one in which a viewer's interaction with the simulation space is transmitted to an entire audience via a flat screen display. In the case of the simulation, multiple angle-dependent observation points cannot be accounted for, given the singular nature of the flat screen display technology deployed here. However, this optical simulation space is essential given the inadequate scalability of the material at hand. While a single virtual simulation allows for a robust overall speculation of material behavior and effect, it does not truly allow for the simulation of a curved geometry in actual space from multiple observational vantages. In order to overcome this limitation of a virtually variegated geometric interactive surface, the use of a gently rotating faceted surface geometry was deployed, one that could approximate multiple viewers' interaction and involvement with the virtual environment.

In the case of the second prototype, we were able to test changes in color and to some degree, transparency, through nuanced environmental stimuli including multiple participants and changes in light intensity (day-to-night light shifts in a gallery room) that in turn affected features of the prototype in local, regional, and global ways. The ability to 'tune' the second prototype to specific environments was made possible through the incorporation of individual potentiometers located in the back of the control board of the prototype. This enabled adjustments of local thresholds of illumination relative to each component for the control of the materials across the entire prototype. While the selected material for the second prototype, silica colloidal nanoparticles, is not the actual material being tested currently at the nano-to-micro scale, these interim material assemblies behave in a comparable manner to those observed at a nano-to-micro scale, but which can be fabricated at a human scale, thus providing a useful testing ground for eSkin.

Furthermore, through an understanding of the nano/micro material behavior and their properties at an architectural scale via the two prototyping methods presented here, we are thus able to further intuitively speculate as to the broader implications of the material in given scenarios within the built environment. The speculations shown here demonstrate these potentials in the context of the eSkin system deployed on the interior surface of a double skin building façade. These are presented from interior and exterior frontal vantage points as well as an exterior perspective in order to demonstrate the unique and relative angle-dependent nature of the materials. Here, Figure 9.17 takes into account the global ambient light-based properties of angle-dependent color change, reflection, and transmission while taking direct cues from the material simulation properties and Prototype I. Figures 9.21 and 9.22 build upon these variable optical properties while including local responsiveness via environmental stimuli taking impetus from behavioral properties demonstrated in Prototype II.

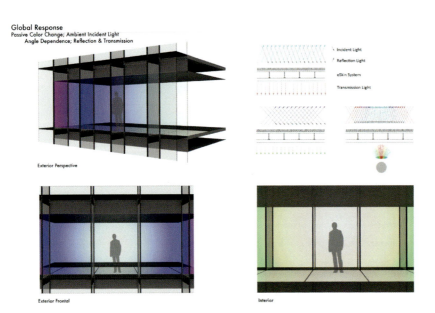

Figure 9.20 Speculative large-scale deployment of eSkin presented in the context of an interior application within a double-skin façade system demonstrating relative global optical features including angle dependent color change, reflection, and transmission.

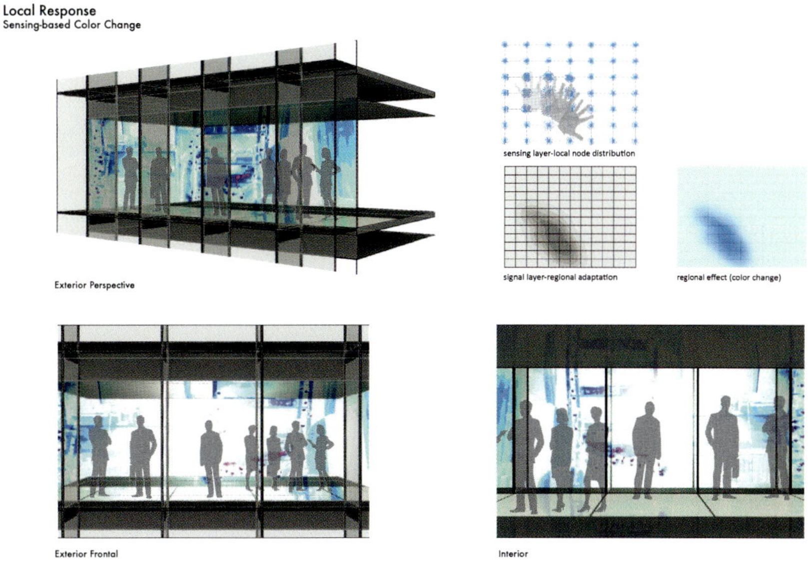

eSkin: BioInspired Adaptive Materials   331

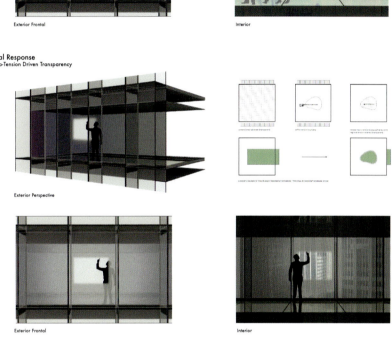

Figure 9.21 Speculative large-scale deployment of eSkin presented in the context of an interior application within a double-skin façade system. Here relative local optical features including angle-dependent color change, reflection, and transmission respond to environmental stimuli, in this case the presence of change within the interior environment.

Figure 9.22 Here, we present a tunable window through micro-tension-driven transparency. In this scenario, occupants may tune and adapt their environment to generate personalized and responsive windows.

Having learned from these two prototypes, we continue to explore and apply these features and effects in human-scale architectural assemblies. For example, *ColorFolds*, by Sabin Design Lab at Cornell University, is a large-scale prototype whereby the material effects sought are simulated in a manner in which multiple viewers' unique experiences are the consequence of variegated geometry and relative viewing relationship with a tessellated and folded surface. ColorFolds is a large-scale prototype for an adaptive architecture system, which uses a combination of novel surface materials and Nitinol linear actuators to react and respond to its environment and inhabitants. As part of two projects funded by the National

Figure 9.23 Schematic proposal for eSkin at the scale of a building façade

Science Foundation in the Sabin Design Lab at Cornell University, KATS: Cutting and Pasting—Kirigami in Architecture, Technology, and Science, ColorFolds is one product of ongoing trans-disciplinary research spanning the fields of cell biology, materials science, physics, electrical and systems engineering, and architecture. ColorFolds incorporates two parameters that the team is investigating: optical color and transparency change at the human scale, based upon principles of structural color at a nano-to-micro scale. Though again this prototype does not incorporate the specific nano-to-micro materials fabricated in the Yang Lab, our aim is to architecturally prototype the behaviors of the intended materials. Through development and installation, various customized digital tools, algorithms, and processes were explored and created for the generation of deployable geometries and unique architectural surface assemblies with a rudimentary responsive control system. Together, this synthesis of design, material, and programming allows a small set of simple actions to manifest complex emergent behaviors. Accentuating the manipulation of kirigami-inspired folding patterns is the implementation of surface materials, which inherently change their appearance dependent upon their orientation. From the virtual simulation, we maintain the modeled material

Figure 9.24 Final installation view of ColorFolds, seen from below

properties of angle-dependent color change, coupled with a regioned notion of triggered event space within the folded surface. This event surface is further controlled and developed by the movement of people in space relative to their unique viewing of the prototype. To this end, we are integrating the material substrate's transparency and reflectivity with respect to the unique viewing angle via geometry. Taking cues from the second eSkin physical prototype, ColorFolds maintains an aggregate of faceted surfaces whose variation is the result of nodal characteristics, or regions of change. These prototypes serve as an interactive testing ground for selecting and refining nonlinear nano-to-micro-scaled material effects at the human and architectural scale, specifically engaging structural color. Currently, the eSkin team is seeking industry partners to move our fundamental research into a viable building product.

Notes

1   Adapted from conference paper: Jenny Sabin, Andrew Lucia, Giffen Ott, and Simin Wang, "Prototyping Interactive Nonlinear Nano-to-Micro Scaled Material Properties and Effects at the Human Scale," in *Symposium on Simulation for Architecture and Urban Design (SimAUD 2014) 2014 Spring Simulation Multi-Conference (SpringSim'14); Tampa, Florida, USA, 13–16 April 2014*, ed. David Jason Gerber (Red Hook, NY: Curran, 2014): Article 11 (Best Paper Award).
2   Adapted from Jenny Sabin, "Transformative Research Practice: Architectural Affordances and Crisis," *Journal of Architectural Education* 69(1) (2015): 63–71.

Bibliography

Chandra, Dinesh, Shu Yang, Andre A. Soshinsky, and Robert J. Gambogi, "Biomimetic Ultrathin Whitening by Capillary-Force-Induced Random Clustering of Hydrogel Micropillar Arrays," *ACS Applied Materials and Interfaces* 1(8) (2009): 1698–1704.
Fairchild, Mark D., *Color Appearance Models*. 3rd ed. Chichester.: John Wiley and Sons, Ltd, 2013.
Kretzer, Manuel, Jessica In, Joel Letkemann, and Tomasz Jaskiewicz, "Resinance: A (SMART) Material Ecology" in *ACADIA 2013 Adaptive Architecture*, ed. Phillip Beesley, Omar Kahn, and Michael Stacey, pp. 137–146. Cambridge, ON: Riverside Architectural Press, 2014.
Li, Jie, Jongmin Shim, Justin Deng, Johannes T.B. Overvelde, Xuelian Zhu, Katia Bertoldi, and Shu Yang, "Switching Periodic Membranes via Pattern Transformation and Shape Memory Effect," *Soft Matter* 8(40) (2012): 10322.
Sabin, Jenny, Andrew Lucia, Giffen Ott, and Simin Wang, "Prototyping Interactive Nonlinear Nano-to-Micro Scaled Material Properties and Effects at the Human Scale," in *Symposium on Simulation for Architecture and Urban Design (SimAUD 2014) 2014 Spring Simulation Multi-Conference (SpringSim'14); Tampa, Florida, USA, 13–16 April 2014*, ed. David Jason Gerber (Red Hook, NY: Curran, 2014).
Saito, T. and M. Ditton, "OBJ Loader for Processing." Available at: http://code.google.com/p/saitoobjloader/.
Thompson, L.F., *Introduction to Microlithography*. 2nd ed. Washington, DC: American Chemical Society, 1994.
Vos, J.J., "Colorimetric and Photometric Properties of a 2-Deg Fundamental Observer," *Color Research and Application* 3(3) (1978): 125–128.
Wang, S., A. Lucia, and J. Sabin, "Simulating Nonlinear Nano-to-Micro Scaled Material Properties and Effects at the Architectural Scale," Simulation for Architecture and Urban Design (SimAUD), April 7–13, 2013.
Xia, Younan and George M. Whitesides, "Soft Lithography," *Annual Review of Materials Science*, 28 (1998): 153–184.
Zhang, Ying, Chi-Wei Lo, J. Ashley Taylor, and Shu Yang, "Replica Molding of High-Aspect-Ratio Polymeric Nanopillar Arrays with High Fidelity," *Langmuir* (2006): 8595–8601.

10

myThread Pavilion[1]

Jenny E. Sabin

Temporary Pavilion Structure
Client: Nike Inc. for Nike FlyKnit Collective
Location: Nike Stadium, New York, New York, 2012
A project by Jenny Sabin Studio, 2012
Architectural Designer: Jenny E. Sabin
Design and Production Team: James Blair, Simin Wang, Martin Miller, Meagan Whetstone, Brian Heller, Nicola McElroy
Consulting Engineer: Daniel Bosia, AKT Engineers
Consulting Textile Designer: Anne Emlein
Knit Fabrication: Shima Seiki WHOLEGARMENT
Lighting Design: Kayne Live

Building upon the foundations laid by earlier data-generated projects such as BodyBlanket and Branching Morphogenesis, the *myThread Pavilion* by Jenny Sabin Studio integrates data from the human body with knitted lightweight, high-performing, form-fitting and adaptive materials. Seeking to develop novel textile-based assemblages steered and specified by the hidden organizations and structures of biological data, the myThread pavilion features a generative design strategy that hybridizes the simplicity of knitting processes and fabrication technologies with the flexibility and sensitivity of the human body as a biodynamic model for new adaptive forms.

  Advances in weaving, knitting, and braiding technologies have brought to the surface high-tech and high-performance composite fabrics. These products have historically infiltrated the aerospace, automobile, sports, and marine industries, but architecture has not yet fully benefited from these lightweight freeform surface structures. myThread, a commission from the Nike FlyKnit Collective, features knitted textile structures at the scale of a pavilion. Textiles offer architecture a robust design process whereby computational techniques, pattern manipulation, material production, and fabrication are explored as an interconnected loop. The myThread Pavilion integrates emerging technologies in design through the materialization of dynamic datasets generated by the human body engaged in sport and movement activities in the city.

  How do you knit and braid a building? Could a building be as lightweight as air? Can sport, in terms of the measurable performance of the human body, influence both design and fabrication and inspire the next generation of buildings?

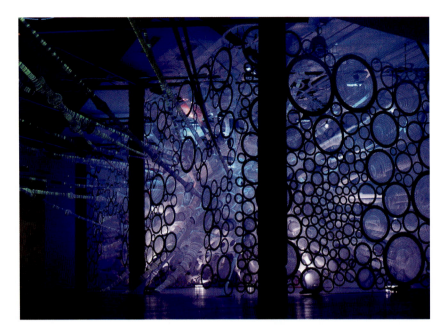

Figure 10.1 myThread Pavilion for Nike FlyKnit Collective, NYC

Source: Sabin & Nike Inc., 2012.

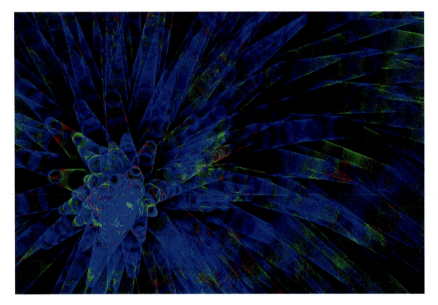

Figure 10.2 Visualization of participant data (bio-data) from Workshop 1 mapped as changing striation patterns of material and knit parameters across the population of cones

What if we could form-fit and enhance architecture with bio-architecture and performance of our own bodies? Turning performance into structure for the Nike Flyknit Collective, Jenny Sabin Studio works at the intersection of art, architecture, design, and science. There are instant similarities in this approach to the work of Nike's Innovation Kitchen, where disciplines from different fields are brought together with a view to re-thinking basic principles and approaches to design challenges.

Figure 10.3 Interior view; elliptical tension ring visible at base

myThread features novel formal expressions that adapt to changes in the environment through form-fitting and high-performance lightweight structures. By digitally crafting these new form-fitting material systems at architectural scales, beautiful formal possibilities emerge, allowing for the construction of novel spaces and immersive adaptive environments that ultimately advance textile tectonics in architecture. Importantly, this project seeks to communicate, document, and make public advances in tooling and textile fabrication toward the design and production of nonlinear systems via complex geometries. Central to this is the integration of fields and industries outside of our own with the promise of advancing the functional, adaptive, and form-fitting nature of knitted material alongside provocative emergent forms and spaces. This is achieved through visualization models of human bio-data—motion data from a large group of runners—transformed and realized as a new choreography of performances in the context of a fully knitted pavilion. Simply put, the generative design strategy is based on prior performances that are translated into present-tense performance through a finely tuned material assembly of knitted threads that respond and adapt to the presence or absence of light.

As described previously, we often start from a molecular point of view where the singularity of a single unit such as a cell or zip-tie becomes a building block for structures of greater complexity. Like Nike FlyKnit, which uses simple threads to create a complex form-fitting structure on a performance-enhancing shoe, the fusion of science, architecture, art, and technology opens the door to new ways of thinking about structure and the relationship of the body to technology. In this project, the human body is used as a biodynamic model to inspire new ways of thinking about issues of performance and adaptation at the architectural

Figure 10.4 Generative model featuring schematic form-finding for the pavilion

Figure 10.5 Digital knitting simulations featuring changes in knit behavior due to the fluctuating nature of the data streams

scale. Performance, lightness, form-fitting, and sustainability become immediately relevant architectural criteria.

The body, or more specifically the body in motion—pure performance itself—is the starting point of our New York collaboration for this project. Using Nike+ FuelBand technology to collect motion data from a community of runners

Figure 10.6 Single cone prototype with highly dense hole patterns due to increased intensity of participant bio-data; as the data change, so too does the material and structural behavior of the knit assembly

Figure 10.7 Workshop 1 in New York City. Using Nike+ FuelBand technology, motion and bio-data are collected from a community of runners, which are then used as design drivers for the knitted pavilion.

during an earlier Nike FlyKnit workshop, the Sabin Studio transformed the patterns of this biological data into the geometry and material of a knitted structure, based on prototypes developed during workshop sessions. The motion data, collected and organized in Excel files, then linked to geometric features in a 3-D modeling environment, forms a material construct for a unique response to the form-fitting question delivered in the original Nike FlyKnit Collective brief.

The pavilion consists of a harder outside construction and softer, organic inner material. Composed of adaptive knitted, solar-active, reflective photo-luminescent threads and a steel cable net holding hundreds of aluminum rings, the simplicity of knitted geometries meets the complexity of a body in motion. An inner structure of soft, textile-based whole-garment knit elements absorbs, collects,

and delivers light as the materials react to variegated light sources and the presence of people through embedded shadows. The material's response to sunlight as well as physical participation is an integral part of our exploratory approach to the subjects of performance and form-fitting.

The myThread Pavilion is the result of collaboration across disciplines and industries including architecture, textiles, sportswear, and engineering. Linking biology and innovation, technology and tradition, this is an analog manifestation

Figure 10.8 Gradient patterns of participant activity mapped as changes in material within each shaped and knitted cone

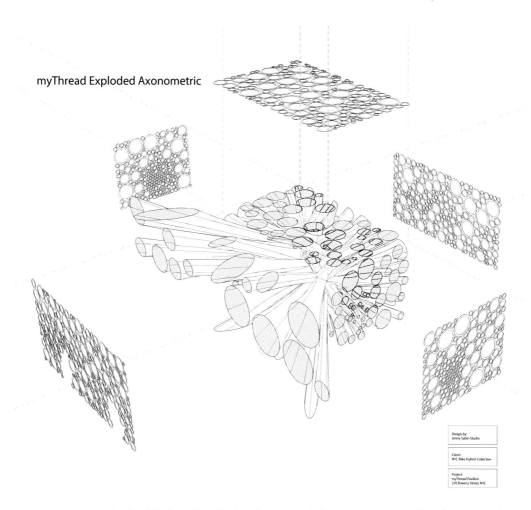

Figure 10.9 Axon drawing of the myThread Pavilion; knitted fabric structure composed of WHOLEGARMENT elements fabricated by Shima Seiki and laser-cut aluminum ring network

both of the benefits of Nike FlyKnit, and also the activities and performance of the individuals that went into its making. This installation's adaptable sensitivity and flexibility mirror the human form. It is its own environment, its own community, and its own energy. While interpreting and visualizing human data, the pavilion also becomes a body itself by virtue of a dynamic, spatial interiority and the presence of a multitude of actual human interactions.

Upon receiving an invitation from Nike Inc. to participate as one of six international innovators from around the world in the Nike FlyKnit Collective, the project began with a visit to the Nike Campus in Beaverton, Oregon. All six innovators were able to spend time with Ben Shaffer, director of the Nike Innovation Kitchen, and to see first-hand the design and fabrication process for the Nike FlyKnit shoe. The Innovation Kitchen is composed of labs and creative studios similar to what an architectural designer may engage in, but with altogether different directives and resources.

Recognizing the promise of a material fabrication process wedded to the nonlinear characteristics of the human form—albeit at the scale of a shoe—the

Figure 10.10 Interior shot on opening night, photoluminescent threads activated

Figure 10.11 myThread Pavilion, exterior view through the front, projected ring panel; solar active threads activated

Sabin Studio project for New York City took shape. The first task was to assemble a framework for a series of workshops composed of invited participants with diverse backgrounds, including extreme athletes, artists, scientists, fashion designers, architects, and textile designers that would feed into the final built legacy project in New York City. The starting point for the New York City project by Sabin was simple: link the complexity of the human body in motion with the simplicity, performance, and materiality of knitting. Using the Nike FlyKnit Collective as a platform for design research and experimentation, the myThread Pavilion incorporates data from the human body to explore nonlinear fabrication processes alongside practical issues such as sustainability and performance in architecture through groundbreaking form-finding and knitting techniques that link sport with architecture.

Knitting may be described as a line affected by force and, more specifically, as a loop affected by force. Differing from a weave, a system of discrete strings or members, the knit is a continuous system. One single member navigates the system, and in each instance looping its current self through its former self. Knitting takes into account parameters such as stiffness, end conditions, loop length, material thickness, and stitch density as well as more complex variables such as dropped stitches, crossovers, and clusters. Given these links between material, structure, and construction, the knit is rich in architectural potential, both as a literal translation and as one that works well with biologically informed design strategies that demand generative fabrication techniques. Much like 3-D printing, knitting is an additive process, where one row is added to the next to form larger wholes. Our first task in the context of the project was to understand the fundamentals of knitting in order to begin rigorously translating these features into a computational environment that may then interface digitized knitting machines. The Sabin Studio initiated a dialogue and collaboration with Anne Emlein, a practicing textile designer and artist who is the Chair of the Textiles, Apparel and Fashion program at the Main College of Art, New York. With Emlein, we developed a series of test swatches and design experiments exploring different types of materials and knit patterns, including parameters such as holes, ladders, changes in tension, and tapering. We produced prototypes on hand-knitting machines and on a Stoll Flatbed digital-knitting machine at RISD. This catalogue of material swatches formed the basis for our digital modeling exercises and eventually a template for attaching data points collected from human motion data to actual knit parameters. This work simultaneously investigates the potential of the knit system as a whole and the geometry of a single knit loop. The work explored material thickness and elasticity, scale, nesting, and overlay. The process was intentionally open-ended and unrestrained, allowing for experimentation of these intricate relationships.

This also framed the sequence of public workshops leading up to the final legacy pavilion in NYC. The first workshop focused on data collection from participants engaged in various movement activities in the city. Each participant was fitted with sensing technology to collect movement activity with GPS tags. These data were partitioned into various parameters and dropped into an Excel environment, which were eventually linked directly with our three-dimensional Rhino models. The second workshop focused on making sense of the human motion data through fabrication and modeling exercises geared toward making the intangible nature of the data—its structure, rhythm, and form—into organized and tangible material systems. Yarn materials and stations of expertise were carefully orchestrated to opportunistically use the workshop format as a test bed for the final pavilion project.

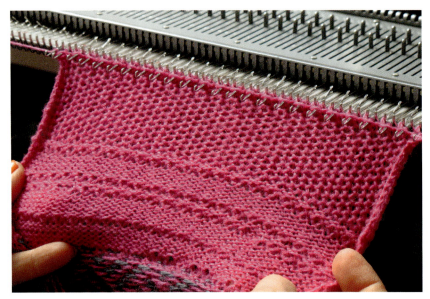

Figures 10.12–10.14 Knit prototypes exploring changes in knit parameters including density, transparency, material, scale and geometry; with collaborator Anne Emlein

With an interest in amplifying the hidden structures and unseen beauty of the data collected from the participants, responsive and adaptive yarn materials were selected. Here, changes in pattern and geometry were linked and choreographed with the dynamics of the material. The aim was to develop a soft material palette that would reflect the complexity of the human data and amplify it through changing conditions reflected in its environment. Solar-active, reflective, and photo-luminescent yarns were chosen for their integrated responsive behavior and ability to change in the presence of active environmental inputs. For example,

Figures 10.15 Interior view with photoluminescent yarns activated

the photo-luminescent yarn absorbs UV light throughout the day and glows at night, while the solar-active yarns change their color immediately in the presence of bright sun. They also hold shadows in fabric form. Finally, the reflective thread provides immediate bursts of light through a micro-scale bead structure on marled yarns when activated by a flash or beam of light.

The design experiments produced during phase two were used as model prototypes for moving into the final design and production phases of the pavilion. Schematic strategies began early on in the project, linking the initial knitted material swatches with formal investigations into data visualization. Working closely within the parameters of knit fabrication and the considerable formal restrictions that were presented, our generative strategies followed several linked trajectories. One focused on form-finding techniques through minimal surfaces and relaxation methods in Kangaroo for Grasshopper. The second explored algorithmic strategies for organizing populations of data produced by the individual participants. These individual agents populated the surface studies and were organized based on nearest neighbor relationships. Meanwhile, analog tension tests were conducted in the studio to understand how much the knitted fabric would stretch under considerable stress. Additional structural analysis and form-finding modeling were conducted by the engineer on the project, Daniel Bosia of AKT Engineering, but due to the unprecedented scale and nature of the knit forms, we found that hands-on testing with 1:1 scale material elements produced the most accurate results. These results fed back into our digital form-finding models.

With a keen understanding of the dynamics of the knit structure, we began to explore ways to distribute forces and "pull" the pavilion into shape. The lines normal to the interior surface—each attached to one unique agent—were then

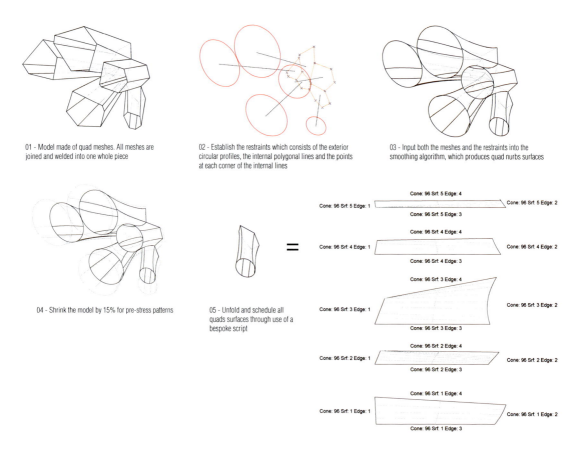

connected to fields of circles, packed in orthogonal planes to form a rectilinear exterior volume. This wireframe geometry describes a collection of cones or conoids, each one representing a knitted volume pulled into tension as well as a collection of data. In order for the pavilion to be inhabitable, the overall structure would need to behave like an inverted basket with a hovering tension ring at the base, thus allowing visitors to enter the interior of the knitted volume. This required that each individual knitted element be seamed together with its neighbor to form one cohesive surface on the interior. To achieve this, a tessellation strategy called Laguerre tessellation was incorporated, which is different from a Voronoi tessellation where inter-distances between nuclei define the geometry of cells only. Compared to a Voronoi, the Euclidean distance between cells is replaced by a power distance in the definition of each cell. The Laguerre allows for the assignment of weights to nuclei. Thus, we were able to use the circles at the edges of the volume as input, which then defined cells on the interior and the edges of the final seam pattern. At the same time, we were able to make adjustments to the input parameters as new material and fabrication constraints emerged. For example, the knitting machines have a limited base diameter and due to cost we were not able to implement unlimited cone diameters. Thus, the circles on the exterior were

Figures 10.16 Topology of the cone family rendered in fine and crude meshes; generative form-finding techniques included the generation of minimal surfaces subjected to relaxation methods

optimized to three diameters only. The circle-packing algorithm was run again, as was the tessellation. Overall, the final pavilion form integrates external datasets from the human body as a population of active external forces that integrate with the inherent performance of knit geometries. This generative fabrication process drives the form-finding techniques that describe emergent geometries, where force, material, and form are understood to co-evolve with human data.

Locating a knitting manufacturer for production was not an easy task as most knitting factories have ceased operation in the United States. Time restrictions and the need to work closely with the manufacturer prohibited working with a knitting manufacturer overseas. Eventually, collaboration with the Japanese-based company Shima Seiki was formed. Shima is at the cutting edge of what they call WHOLEGARMENT knitting. This process enables the production of seamless three-dimensional knitted forms as the knitting process is done in continuous tension as shaped elements. The process is similar to additive 3-D printing, but in rows of knit stitches. One could actually argue that knitting is the first example of 3-D printing! However, Shima had never done a project on the scale of a pavilion before, nor had they knitted responsive and adaptive threads, which tend to have thread counts and stiffness factors that are difficult to knit in tension. The cone elements for the pavilion range in length between 2 and 50 feet. Working side by side in their factory, we innovated an entirely new process whereby the complexity of the data populations determined material striations and hole patterning in each cone, which then informed the file input for the digitized knitting machine. The knitting process took just under two weeks to complete, at which point all the cone elements were transferred to a fabric finisher, Dazian Fabrics, for assembly and final seaming. The tagging scheme for each cone remained the same from digital model to knitted cone to the final seam pattern, which was used at 1:1 by the finishers to accurately label and sew each edge of the elliptical forms to their neighbors. The final fabric structure weighs less than 170 pounds and fits within a single large canvas bag.

The assembly and construction process operated much like tuning a drum. The first task entailed laying out each circle-packing pattern with the various sizes of laser-cut aluminum rings, linking them together and then hoisting them in place. Each ring panel was pulled into tension, using the actual space of the gallery as a frame. The gallery, called Nike Stadium NYC, was located at Bowery and East Houston Streets in New York. Although the project had originally been specified to be outside, the final pavilion installation was housed inside due to permit restrictions. Once the rigging was complete, the knitted structure was slowly pulled into tension, each cone receiving its own unique ring panel, with one wall of cones projecting beyond the front panel and anchored to the opposite wall. We moved concentrically from bottom to top to evenly distribute the changes in tension across the overall form. Once all the cones were fully installed, a tight tension ring emerged at the base of the structure allowing visitors to pass through the door in the front ring panel and pop up into the interior.

348 Personalized Architecture and Medicine

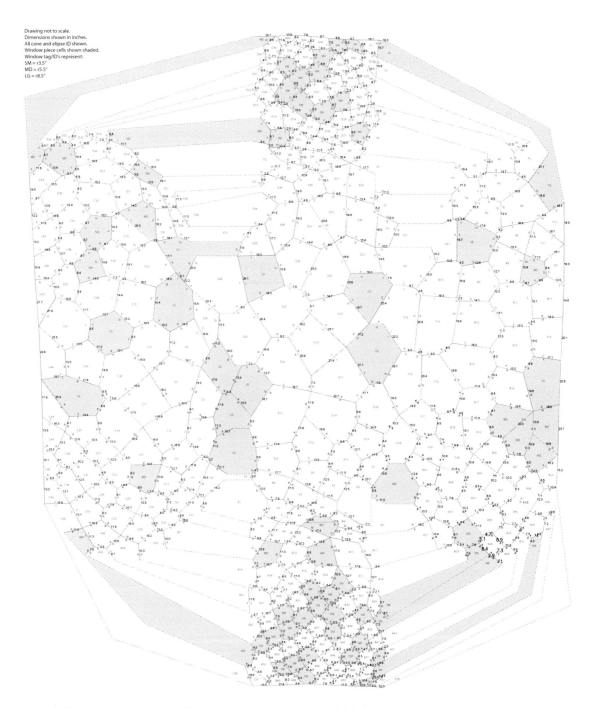

Figures 10.17 Seam pattern drawing used by the finishers to connect and sew each knit element to its neighbor to form a large fabric structure

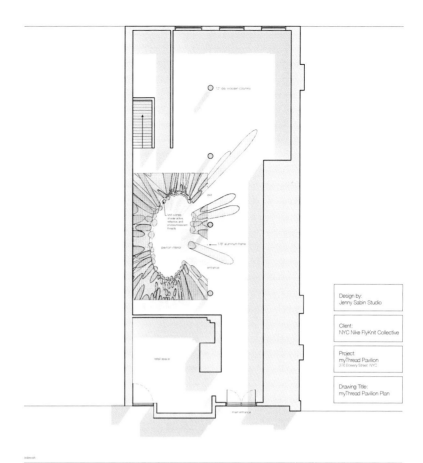

Figures 10.18 myThread plan sited at Nike Stadium, NYC

Figures 10.19 myThread section sited at Nike Stadium, NYC

The Sabin Studio worked closely with Benji Kayne, a lighting designer with expertise in theatrical lighting, to choreograph a lighting scheme that would emphasize the performance of the responsive and adaptive yarns in the context of the overall form and user interaction. UV and blacklights were placed strategically around the pavilion and placed on digital timers to simulate an accelerated day-to-night sequence. The affected knit structure changes dynamically to shifts in its environment.

Figures 10.20 Installation process for the pavilion; each knitted cone is received and tensioned by its unique home within the aluminum ring network.

Visitors were subsequently inspired to inhabit and use the pavilion in multiple ways from lying down to stretching to simply watching the fabric change over time.

Through the productive misuse and tinkering with fabrication technologies in alternate disciplines, namely fashion and sport, this project pushes the boundaries of current scaled applications of knitted surfaces. We believe that the myThread Pavilion opens up potential for soft textile-based architectures. Knitting processes provide material production and generative fabrication processes that work in tandem with nonlinear design concepts, thus closing the gap between design intent, making, fabricating, and final built form. Recognizing the successful and seamless design process that Nike innovated for the FlyKnit shoe that links the nonstandard and complex nature of the human foot with knitted material and formal production, the myThread Pavilion also seeks to explore this example of generative fabrication, but at an altogether different scale. The project opportunistically uses the flexibility and sensitivity of the human body as a biodynamic model for pioneering pavilion forms. myThread features formal expressions that adapt to changes in the environment while engaging material performance through form-fitting and lightweight structures. WHOLEGARMENT knitted, solar-active, reflective, and photo-luminescent threads in combination with aluminum laser cut rings and a steel cable net compose the overall structure of the pavilion. While the technical and formal features of the pavilion are new to architecture, at least in scale, these are not the most convincing aspects of the pavilion in terms of future promise, but rather, point toward a design process engaged in generative fabrication. This may bring us closer to realizing forms and landscapes that are the products of emerging diagrams over static representations of discrete moments in time. Perhaps, this is what is most

Figures 10.21 Detail, knitted fabric structure composed of individual knit cones, tensioned

Figures 10.22 Interior shot on opening night. The entire structure is pulled into tension, cone by cone, until the bottom forms a taut elliptical tension ring.

intriguing: the promise of a design process that is adaptive in real time. Like cells weaving their own extracellular matrices in order to signal through their newly constructed environments, which in turn affects the form, function, and structure of the cell, myThread seeks to engage a similar active environment of making and designing analogically. Architectural features of the human body are erased and then found again in newly constructed architectural surfaces that also inspire and delight, shelter, and embody.

Note

1   Sections of this text have been adapted from Jenny Sabin, "myThread Pavilion: Generative Fabrication in Knitting Processes," in *ACADIA 2013 Adaptive Architecture: Proceedings of the 33rd Annual Conference of the Association for Computer Aided Design in Architecture, October 21–27, 2013,* ed. Philip Beesley, Michael Stacey, and Omar Kahn (Cambridge, ON: Riverside Architectural Press, 2013), pp. 347–54.

Interview: RE(IN)FORM(ULAT)ING Health Care via Medicine + Creativity

Mark L. Tykocinski, Peter Lloyd Jones, and Jenny E. Sabin

The previous sections demonstrate applied design research started in LabStudio and ongoing in Jenny Sabin Studio and the Sabin Design Lab at Cornell University. The following interview delves into advances currently being made at MEDstudio@JEFF, located at Sidney Kimmel Medical College within Thomas Jefferson University. MEDstudio was founded by Peter Lloyd Jones in 2013 at the invitation of Mark Tykocinski, and was established as a post-disciplinary creative agency and catalyst promoting health, wellness, and dignity for all, co-serving health professionals and the public alike. Projects include embedding "Creativity" within the medical curriculum, the design or large urban spaces, and the invention of smart wearable fabrics and devices that are radically re-shaping the future of medicine. This 2015 interview sets the stage.

> *Jones*: Hi, Mark, many thanks for joining us from your retreat in Woods Hole. As an early supporter of LabStudio, you are already aware that a specific interest and pursuit are to couple, re-define and then catalyze collaborations in science, medicine, engineering and technology in the context of advanced design, the arts and humanities and vice versa. Could you spend a moment or two describing where your interest in this type of convergence began, and how it has influenced, or has been incorporated into your own life and work, intentionally or otherwise?
>
> *Tykocinski*: My interest in converging seemingly diverse and unrelated fields began as an undergraduate while studying at Yale within the directed studies program, which surveys a broad swathe of the roots of Western civilization. In addition to an interest in medicine, I was also a would-be philosophy major in those early days. So, in a certain sense, I was programmed from my time at college to see things through a broad humanities lens, a sort of meta-physics before physics, if you will. And I distinctly remember asking for, and then getting permission to combine my final term papers for two separate courses: One was under the directed studies program in political science and the other was a philosophy course in existentialism. This allowed me to write one final large paper in which I viewed Nietzsche from both political science and philosophy perspectives. Now that I look back, decades later, maybe that was an early sign that I was somehow mentally primed in trying to bridge concepts, subjects, and fields.
>
> Perhaps another very telling indication in this regard is that over the years I've always gone out of my way to sit in on lectures that are outside of my own field. For almost twenty years, and as is the case today, I spend a month or so each summer here at the Marine Biological Laboratory (MBL) at Woods Hole. Mostly, I'm here conducting scientific writing, sitting up in the

library stacks, looking out at the sailboats and the old pond and everything else around. But while here, I'll also wander into lectures from time to time, and almost always I make it a point of going to ones that had absolutely nothing to do with my own field. And as might be expected, this type of conscious meandering across subjects has had a positive effect on my own research work. For a decade, I had acquired a National Institutes of Health (NIH)-funded line of investigation, a multimillion-dollar grant that actually emerged from that chancing on a lecture in the neurosciences, during one of those summers at the MBL. The lecture touched on a concept that was foundational for neuroscientists; it was really just "Neuroscience 101," on receptor triggering thresholds . . . but guess what? That idea hadn't really made its way into my chosen field of immunology. So I co-opted this basic neuroscience concept to understand anew receptor triggering within the field of immunology. One gene that I cloned encoded a protein that has unusual properties in modulating T-cell function, and having heard the neuroscience lecture, and the function of the neuroreceptor, it became clear that the same mechanism was relevant to immunoreceptor function, thus explaining how this protein works. Thinking back, this was a reinforcement of the notion that wandering beyond one's own field, and touching other areas can really lead to some useful and novel ideas and applications.

Jones: Given your current role as Dean of the Sidney Kimmel Medical College and Provost at Thomas Jefferson University (aka JEFF), it's enthralling to hear about your previous scientific work as an experimental pathologist, since it has clearly played a significant role in generating many of the avant-garde educational and research programs and tools that you have encouraged and supported, including MEDstudio@JEFF. But before discussing these initiatives, I'd like to talk more about your past, and to ask you about other influences in your life and work, including the artist Tobi Kahn, who lectured within our LabStudio program when Jenny and I were both at UPenn.

Tykocinski: So you've perfectly anticipated my next set of points! I have a handful of close lifetime friends, some dating back to high school days, including Tobi Kahn. Interestingly, I always seem to go for friends who are anchored outside of the hard sciences, mostly out of medicine. Tobi, who is a world-renowned visual artist, and whose paintings are in many significant private and museum collections, is one great example. One of the distinguishing things in Tobi's abstract landscape paintings is the richness and visual depth where image components meet up; there's a lot going on in these border areas. And for him, this is by design as he sees borders as generative, and within these he sees depth. In fact, Tobi would come and visit us with his family here on the Cape for summer after summer. We'd go bike riding and he'd stop me along the way and he'd look out at the ocean and say, "You see that area where the ocean meets up with the sky?" I'd say, "Yeah, there's ocean and there's sky." He

says, "No, look more carefully at the border, there are more colors, almost an overlaid image that's emerging out of that border."

So it's hard to say how this influenced me, but I'm pretty sure that it had a subliminal effect, as I focused on my own science, which largely was about bringing protein parts and cells together in different kinds of ways, and changing the way in which cell networks function. So when fusing a piece of protein "A" together with a piece of protein "B" to form a single chimera protein, new functions appear that are neither in "A" nor "B," the union is generative. Suddenly, when you deploy these fusion proteins, to change the way a cell "A" interacts with cell "B," they create a whole new molecular display which changes the ways they interact with cells within their network. You can elicit an entirely new cell state within a cellular network by changing that border condition. I think a word that captures the notion of individual entities transforming when they come together is *emergence*, generally defined as the way complex systems and patterns arise out of a multiplicity of relatively simple interactions.

Emergence speaks to how novel and coherent patterns arise during the process of self-organization in complex systems. That emergent state is generally greater than the sum of its parts. In this instance, what Tobi was looking at from his perspective as an artist, on the one hand, and what I

Figures 10.23 Tobi Kahn, Sky & Water installation, first shown in the exhibition, *Aligned: Paintings by Tobi Kahn* at the University of Maryland Art Gallery, 2011. This installation has been exhibited in other museums since 2011, including the National Museum of American Jewish History in Philadelphia, PA, in 2012.

was looking at from the biomedical side, on the other, were emergent states manifest from things happening at junctures. Again, it's the borders that are generative. They are wellsprings of creativity. This concept has been central to my thinking, and it goes well beyond physical things to idea generation itself. Where concepts and knowledge domains touch, that's where flashes of insight occur. Obviously, this is nothing particularly unique to me. You see a lot of that in the history of science and other things, but often people don't pay attention to that.

The other person I was going to raise as an example of someone who fuses subjects to bring about profound change, is another high school friend, Steve Winter. It turns out that, much like yourself, he's another of these brilliant creative-type people. He's a law professor who's pioneering linguistics in legal thinking. But that's its own discussion, it's own form of fusion. He's taking linguistic theory principles and interpreting all of law essentially around that. But that's not why I'm bringing him up. In his instance, his work is reinforced and provoked by navigation through another notion of the nonlinear, *complexity*. In fact, the title of the first book he wrote was "A Clearing through the Forest" about finding one's way through dense forest of complexity towards simplification, thereby clarifying insight. There's a phrase from naturalist John Muir who said the clearest way into the universe is through a forest wilderness. I just love that! Really what it's saying is that you can't just bypass complexity. You have to embrace it, the metaphor of the forest, embrace the forest ecosystem. It is that system which is a prerequisite for clarity.

Along these lines, there is a magazine article that I clipped out a number of years ago which caught me, and it said, "Keep it complex, stupid." And I thought, how wise! Embrace complexity, that's the magic formula. Complexity equals nonlinear thinking, a predicate for creativity. We're accustomed to celebrating simple, elegant solutions, but the science of complexity is on the rise. And it is by delving into the complex, by absorbing the richness of the canvas, that you'll find your way back to the simple unifying concept. My motto, as I said in a recent graduation speech, was, "Resist the urge to oversimplify."

Jones: It's abundantly clear that you are on a major mission to transform contemporary medical education. Why do you think this needs to happen now? How are you accomplishing this?

Tykocinski: I see this as one central question. To put it into context, twentieth-century medical education, to this day, is framed by the Flexner Report of 1910. One of the cornerstones of that report was bringing fundamental science into medical education to complement what had, until then, been largely almost purely an art, albeit it was a voodoo art. Bloodletting, for example, a lot of that was done in the art of medicine of those days. It was unsubstantiated. Having brought science into medicine with the Flexner Report benefited because of its perfect timing. It was fully in sync with the twentieth-century ascendance of science in all areas of human endeavors. Science became king in the twentieth century and here was the Flexner Report saying that medicine, at its heart, was science and that's what medical education should be about. That revolutionary Flexnerian introduction of science in medical education was in fact so impactful that it almost crowded out any other further change. It's almost as if the last word had been said. So, for the past century in our med schools, it's all been about science, with all the rest as maybe just filler. We like to talk a little bit about humanities. We like to talk a little bit about professionalism. We talk about all these other qualities, but they're just sort of decoration around the edges.

But now it becomes clearer that there's another revolution that has to happen. I'd frame it as bringing the art back into medicine. Not to displace the science, but to complement it, and I have very specific things in mind when I'm talking about science and art in this context. I can define science as algorithmic thinking—all the stuff that's right up the computer's alley, if you will. But I define art, and the art that we need to reintroduce into medicine, as being in more tune with the nonlinear. Three things come to mind for me around that: one is creativity, another is complexity, we've talked about both of those, whereas the third is *emotivity*, the ability to show an emotion and to connect through emotion. These are areas that the human mind still holds the trump card, at least for the foreseeable future. It's people like Kurzweil at Google, one could call the Googleplectic artificial intelligence assault, all of this will be done by robots. I'm not quite sure it's going to ever reach that point, but at least for now those are the human domains. And that's what medical education must key in on . . . the ability, the creativity, the complexity, the emotivity.

As I see it, medical education, in part, will be about cross-cutting all the knowledge domains. There are two things that are going to have to happen. One is the position of the physician of the future who will have to justify themselves by having deeper domain knowledge, probably through some kind of tracking of education, but the other is to have cross-cutting knowledge domain—higher-order thinking modalities, if you will. And in no small part what we've been talking about as the nonlinear. I've come up with sort of phrase, I don't know if I'll stick with it for now, but it's *Medicine Plus*.

Some examples are very near and dear to your heart, which is medicine *plus* design, or design thinking through MEDstudio@JEFF. And one of its ramifications is in shaping environments, cultivating creativity through design. But there are a series of others that we're going to start incubating. One is medicine *plus* humanities, and through that kind of union you can foster visual and reflective thinking. Medicine *plus* policy, which will train the physicians who are going to drive and shape health policy in our nation. Medicine *plus* computational science, cultivating what I would call the digital nexus, with medicine in its myriad of dimensions. Another example is medicine *plus* entrepreneurship, the people who will create value through medical discovery, translating discovery into the marketplace. These are the sort of cross-cutting domains, abilities, fields that will distinguish the physician and will be things that the robots likely are not going to be all that great at in our lifetimes.

Jones: Do you know what strikes me as well? I've been talking with people like Professor George Brainard, a world-renowned neuroscientist at Jefferson, whose research explores the power of color and light on human health, and he's already acting as both scientist and designer. I think that many other scientists are practicing and having a major influence on design, and I bet if we allowed people to re-frame what they're already doing, which is often intuitive and highly creative, they could be defined as designers, per se. Much of this is just about reframing language.

Tykocinski: That would be a fascinating exercise, actually, now that you mention it. But in re-engineering medical education you have to think about how you get there. We need to stop thinking about medical education as a series of discrete, disconnected educational phases. What we're looking at now is a lack of continuum between college, medical school, residency, fellowship, and later medical education; each one sort of happens as if nothing happened before and nothing will happen later, in effect. Instead, our concept here is to re-think medical education in the world. We have to be thinking of a lifetime continuum of medical education, which not only cultivates the specialties and deep domain knowledge needed in medical practice, but one that also promotes a hefty dose of the non-medical throughout. The kinds of things we've been discussing. What we've been doing around Medicine+Design, in a sense, showcases what we mean by continuum of medical education, because we set out to cultivate the most creative medical physicians.

Jones: Along the lines of creating these continuums, when you arrived as Dean at the medical college, you created a series of new elective courses for med students, including a design track under the umbrella of MEDStudio. Can you describe that initiative?

Tykocinski: Yes, the idea to institute cross-cutting knowledge domains in a more intense kind of way, almost a sort of master's degree equivalent, and yet not extending med school, given the length of medical training already. We coined a term, *College within the College*. The first two tracts were translational research and global and population health. Then, upon your arrival, we began discussions around a design track for the *College within the College*.

The idea for each of these endeavors is to have a certain amount of class group discussion and formal lecture-didactics that are complemented with real, hands-on experience through projects and exercises. That is where MEDStudio@JEFF looms large, because it creates those kinds of assignments that direct toward particular questions, on one hand, but then remain open-ended in terms of the creativity that goes with them, on the other.

Jones: What are the main obstacles?

Tykocinski: Well, you know, the obstacle for all of medical education right now is accreditation and the accrediting bodies. One can come up with incredibly creative ideas of what you would want to accomplish during med school, but how do you actually do that when there are exams that require certain content to be tested in certain timeframes? This won't necessarily fit with what you want to do. The second impediment of accreditation, a definition if you're talking about continuum medical education, is that there are different accrediting bodies and exams that govern each of the stages and none of them are coordinated. So, if you're really going to bring together a continuum of medical education, you would have to somehow break down those barriers and have everyone working off of the same new playbook that looks at how we're going to generate that physician in the future. The conclusion that I've come to is that the only way to do it is through a revolution. And a revolution requires bringing together a consortium of leading medical schools that, together, issue a new report. It's a sort of beyond-Flexner Report for the next century that says, "No this is the way we're going to do it! And by the way, you've got to play with our report and our way of seeing things rather than the other way!" People forget the Flexner Report led to the folding of, I can't remember the number, it could be as many as half the med schools in existence at the turn of the twentieth century because they didn't have a prayer of getting the sciences taught. So these revolutions can sort of demand a whole lot and if you're ahead of the curve, you're very well positioned for when the revolution actually happens.

Sabin: Western science and medicine developed in the absence of big data. How is abundant data impacting how science and medicine are done?

Tykocinski: I approached this question from a number of standpoints. There is an informative graph that a speaker at Jefferson showed a year or two ago that I thought was very telling, it charts information against time. There were two curves, one curve said what is the medical information needed to practice medicine optimally. You can obviously break that down by field, but just generically, what you would need to know. That curve goes up quite steeply over the last fifty years. The second curve, which is simply a straight line, shows the capacity of the human brain in terms of memory. And that is what it is. Some people have a little more, a little less, but that doesn't change with time. That is hard-wired to the human being.

Importantly the first curve, the amount you need to know to practice, has crossed the capacity of the human mind. All to say that, accessing stored data now becomes indispensable to the caregiver. There is no way once you have seven hundred genetic markers for different diseases or pharmacogenomics with the endless, endless pharmacogenomics with endless compilations . . . you can't deal with that as an individual. All to say that for the practice of medicine, big data and accessing big data are here to stay whether we want it or not.

The second thought in response to your question addresses how science is done. Until recently, the limiting factor was the laborious and costly acquisition of data. You got an NIH grant and you spent it gathering bits of data—so data was king. Because data was so expensive to acquire, you had to have it very targeted, so therefore it was generally linked to specific well-formulated hypotheses. But in many areas of science this has now fundamentally changed. Data has become a mere commodity, and we've reached the point that within the next five years for the cost of a latte you'll be able to sequence your own genome. It really now becomes that data interpretation is king. Those who can deal with big data and think about it in different ways, will rule the realm. Data is information, information per se is not the value, the premium is knowledge generation from this information base. That, I think, is where we're headed.

In part, this book speaks to this ascendency of data interpretation. Thinking isn't about memorizing or generating a million facts. It's how you take this vastly growing repertoire of facts and data points and approach them through a lens that challenges linearity and looks for unique paths through that information maze. That is one of the implications.

Jones: That's very interesting because there's a sort of paradox. The research that Jenny and I do takes big data through a creative process, humanizing it, and that is alien to many in science and medicine. But it also speaks to a question that you have raised many times and that is: is the doctor of the future going to just be merely a data-deliverer? How do you humanize that process? I think, again, that revisiting the idea of combining big data acquisition with the creative process that designers bring to the table is a potentially important step, because they can humanize that process. That is what they do, they make human occupation possible.

Tykocinski: I love this concept. I would take it in another direction just a tad here, it is something which I tell the medical students, on the very first day in medical school during orientation. Going into a patient's room is an act of creativity, if it's done right. If you walk in there and your only goal is to pigeonhole a patient into a disease category, then, guess what, the computer will ultimately do it far better than you. They will get all the lab data, they will generate the differential, and [the patient] will be pigeonholed. The physician of the future is, in a sense, the patient's consultant in navigating the baseline. When the physician walks in, the computer already told us this is the differential and these are the diagnoses, but now let's really work with you. I call this *N of One Medicine*. Each individual patient is not just part of a huge denominator. All of the patients with prostate cancer are lookalikes, but only you have this one particular combination of 32,000 genes that interact in this way with this life experience within this socioeconomic setting. With all the things that make you an individual genetically, environmentally, and sociologically, let's figure out what's wrong. The computer tells you that there are five anti-inflammatories for your condition, but which one is right for you? It's impossible to imagine, if done right, that the person that is going to make that decision is going to be an AI robot.

Sabin: What are your thoughts on the differences and similarities among architecture, science and medicine, specifically in how research is conducted? What needs to change in either?

Tykocinski: Well, I can give you an initial thought. My sense at the most general level is that both architecture and medicine had their roots as arts and then each imported science, mathematics, and computer analytics in its own unique way. There is probably something in comparing how that was done, and I'm sure *LabStudio: Design Research between Architecture and Biology* will address that. That analogy between the two and the differences that emerge and how that comes about, I'd have to give that a lot more thought.

Jones: In regards to this question, I think there's an interesting moment happening where if you visit Jenny at Cornell, she now has a sizable lab space as well as studio space. So, a few designers have begun crossing into our sphere and we're tackling it from the opposite direction, i.e. creating a studio culture within medicine. I think that those extremes are necessary to generate new languages and ideas that will be mutually beneficial as well as unique to each of our respective fields. Hopefully, LabStudio has contributed to this, and will continue to alter the cultural landscapes of medicine and design.

Tykocinski: It's exciting! The substantiation of this dream that we shared back when we were at Penn and that gave rise to LabStudio, together with Jenny and her own brilliance.

August 15, 2015

Conclusion

Jenny E. Sabin and Peter Lloyd Jones

This book is about collaborative design research across disciplines. It is also about new models of thought in the creative process at the nexus of architecture, science, and medicine. More specifically, it is about the generation of methods, tools, materials, prototypes, and scaled applications in the co-production of a new model for collaborative design research across architecture, science, and medicine. It is also a book about deep relationships. The arc of a *project* across science and architecture spans a lifetime where collaboration is as much about friendship and daily life as it is about professional discoveries, patents, papers, and new materials. This book is about people taking risks together with the positive aim of innovative applications in architecture, science, and medicine.

LabStudio is a model for teaching, research, and practice, where participants continue to engage in individual and collaborative projects based at Cornell University, Thomas Jefferson University, the University of Pennsylvania, Stanford University, the University of Minnesota and beyond. LabStudio and its current offspring, including Sabin Design Lab at Cornell Architecture and MEDstudio@JEFF at Sidney Kimmel Medical College, are now viewed as important models for thinking and teaching in architecture and biology, and have since inspired other departments and schools of architecture and medicine. Architects, cell biologists, pathologists, medical doctors, mathematicians, materials scientists, textile designers, engineers, and physicists continue to actively collaborate to develop, analyze, and abstract dynamic systems through the generation and design of new tools, methods, prototypes, materials, and architectural interventions. These new approaches for modeling complexity, visualizing large datasets, and teaching across disciplines are subsequently applied to both architectural and scientific research. The real and virtual world that LabStudio continues to occupy has already offered radical new insights into generative and ecological design within architecture, and it is providing new ways of seeing and measuring how dynamic living systems are formed and operate during development and in disease. Overall, the mission of LabStudio is to produce new modes of thinking, working, and creating in design, medicine, and the sciences through the modeling of dynamic, multi-dimensional systems with experiments in biology, applied mathematics, fabrication, and material construction.

Our work operates within a multi-year and multi-phase research plan. Currently, each project is broken down into three phases including: (1) the production of catalogs of visualization and simulation tools that are then used to discover new behaviors in geometry and matter within specific dynamic biological systems; (2) an exploration of the material and ecological potentials of these tools through the production of experimental structures and material systems; and (3) the generation of scientifically-based, design-oriented applications in contemporary architecture practice and biomedicine. In part, these applications range from new concepts of materiality to adaptive structures with complex geometries to avant-garde pedagogical models for medical education and diagnostic tools for patients with lung disease. Overall, our research has already had a significant impact upon hypothesis- and design-driven research in architecture, biomedicine, and education.

At the bench side, our research has included the study of the extracellular matrix protein network that surrounds cells within tissues, and asking how this dynamic network affects tissue behavior at the genotypic and phenotypic levels. Jones has studied the extracellular network described herein since 1992, and his work has helped form a new paradigm in biology that is underscored by the idea that 3-D tissue architecture, specified by the matrix, exerts a dominant effect over genes. Essentially, function follows form since much of the secret of life exists outside the cell within the matrix. Through Sabin's expertise in design computation and digital fabrication of alternative materials, we aim to understand—through simulation and analogic modeling—the dynamic and reciprocal relationships that exist between cells (i.e., a component), their extracellular matrix (i.e., an environment), and code (i.e., genetic information). This exchange of information between the outside of the cell and its interior environment and back again presents a potent ecological model for architecture, based upon an understanding of how context or environment shares a reciprocal relationship with code, geometry, and matter. The advanced digital procedures and generative design strategies implemented in LabStudio to study these relationships at multiple scales have been fostered by Sabin's expertise in design and emerging technologies, and by Jones' expertise in building 2-D and 3-D environments at the micro scale. This includes the systematic study and production of formal and contextual relationships through generative processes, which may include custom-designed algorithms and parametric models. Our work and collaboration have contributed substantially to these new trajectories within architecture and research, including material technology and fabrication.

The projects described in this book underscore our mission to probe the ecological and material import of our collaborative research in biology and architecture. For example, our research on cellular *motility* has led to several unique research paths in collaboration with Shu Yang at the University of Pennsylvania and others within adaptive and responsive materials, as well as forthcoming applications at the scale of building skins. For this project, visualization tools have been designed and developed to assess how human smooth muscle cells differentially respond to a normal or diseased 3-D tissue microenvironment. Co-investigators in the fields of architecture, cell biology, materials science and mechanical and electrical engineering investigated the role of the microenvironment in controlling cell motility and the shifting geometries inherent in this movement. This study has influenced research development of a rapid, accurate and cost-effective diagnostic tool for patients with lung disease. Concurrently and with the eSkin project, this work explores materiality from the nano to the macro scales, based upon a quantitative understanding of dynamic cell behaviors cultivated on custom-designed geometric surfaces. This approach to comprehending cell sensing and responsivity to a specific surface condition is now being investigated for the rational design of materials that will form the basis of a building skin that responds passively (i.e. without the need for an external energy input) to changes in climate and user input. This is but one research outcome that also creates a significant opportunity to excite the general public about dynamic biomimicry for energy harvesting and sensing, thereby provoking and engaging interest in science, technology, engineering, and mathematics.

As with Branching Morphogenesis and the myThread Pavilion, fibrous structures are steered by dynamic cellular and bio datasets and explored through hierarchy, heterogeneity, anisotropy, and redundancy. Here, loops, holes, and ladders generate knots and openings in the material systems, a continuous deposition of material, one link and row to the next. Productive failures and unconventional computation become opportunities as the assemblages adapt and design intuition is honed through the agency of the material and process. Perhaps one of the most important outcomes of the work concerns the development of a design process saturated in material computation and making, as steered and specified by a biological model or dynamic template. The intention is not merely about mimicking biology, but learning how to design like nature: synthetically, opportunistically extracting principles and processes for novel tectonic systems and formal expressions in our built environment. Geometry, materiality, pattern, structure, and form are inextricably linked. We purposely resist the post-rationalization of complex form through an approach that engages materially directed generative design. Here, architectural affordances reveal themselves as evolving flows of force through geometry and matter that are computed, designed, and fabricated. As we have discussed previously, these transformative models may in parallel provide potent contributions to issues of sustainability, construction, digital fabrication, and material ecologies in architecture. Critical to this approach in architecture are design processes rooted in experimentation without predetermination of form. Emphasis is placed upon the dynamics of natural systems in context, of behavior and process in the material formation of difference and heterogeneous entities.

In summary, we have demonstrated a radical departure from traditional research and design models in both architecture and biomedical science, with a move toward hybrid, trans-disciplinary concepts encompassing new modes of thinking and collaborating. We believe that this type of environment will continue to provide research models and applications in each of our specialties that addresses a diverse array of pressing issues. Examples include the development of customized digital tools to visualize large and complex datasets to address how we perceive and use information, lowering our carbon footprint for buildings through dynamic interfaces, and novel diagnostics for lung disease.

Overall, we suggest that any future investigation between architecture and biology consider models that capture and cultivate the dynamic reciprocity of the less obvious organic systems of architecture, and the more obvious living complexities of biological systems. To address this, architecture can take its cue from biology in matching the complexity of its generative design models to the very dynamic features of the living environment and organic milieu in which the architecture is a part. Architects might learn from these biological models imbuing their design processes with a systemic understanding of context and form.

The exploration of biological and nano-to-micro-scaled material properties and effects at the human scale form the starting points for many of the featured projects. Engaging with the disciplinary hurdles that we encounter through the co-production of projects across scales culminates in what is perhaps the most potent deliverable: a new model for trans-disciplinary collaboration. The scalar constraints span materials science, cell biology, textile engineering, fashion, sport, electrical and systems engineering, and architecture, which in turn challenge the differences between fundamental and applied research. Through the collaborative production of these applications, we encounter key differences between the conceptualization and materialization of the projects whose success demands that science, engineering, and design meet. The creative navigation of this ambiguous line between science and architecture in turn offers a unique model for collaboration across disciplines that defines a new future for architecture and the role of the architect, where authorship is horizontal, giving way to interiorities, elastic networks, fabrics, and topological meanders that are pliable, plastic, ecological, and open—where geometry and matter are steered and specified by the flexibility and sensitivity of the human body. Perhaps the most important deliverable to date is this new model for collaboration across disciplines where the featured research and projects conducted in LabStudio, Sabin Design Lab, MedStudio@JEFF and Jenny Sabin Studio form a bridge and a point of departure.

Notes on Contributors

Editors and Authors

Jenny E. Sabin is an architectural designer whose work is at the forefront of a new direction for twenty-first-century architectural practice—one that investigates the intersections of architecture and science, and applies insights and theories from biology and mathematics to the design of material structures. Sabin is the Arthur L. and Isabel B. Wiesenberger Assistant Professor in the area of Design and Emerging Technologies and the newly appointed Director of Graduate Studies in the Department of Architecture at Cornell University, New York, where she is also establishing a new advanced research degree in Architectural Science with concentration in *Matter Design Computation*. She is the principal of Jenny Sabin Studio, an experimental architectural design studio based in Ithaca, and Director of the Sabin Design Lab at Cornell AAP, a trans-disciplinary design research lab specializing in computational design, data visualization, and digital fabrication. In 2006, she co-founded the Sabin+Jones LabStudio, a hybrid research and design unit, together with Peter Lloyd Jones. Sabin is also a founding member of the Nonlinear Systems Organization (NSO), a research group started by Cecil Balmond at PennDesign, where she was Senior Researcher and Director of Research.

Sabin's collaborative research, including bioinspired adaptive materials and 3-D geometric assemblies, has been funded substantially by the National Science Foundation with applied projects commissioned by diverse clients including Nike Inc., Autodesk, the Cooper Hewitt Smithsonian Design Museum, the American Philosophical Society Museum, the Museum of Craft and Design, the Philadelphia Redevelopment Authority, and the Exploratorium. Sabin holds degrees in ceramics and interdisciplinary visual art from the University of Washington and a master of architecture from the University of Pennsylvania where she was awarded the AIA Henry Adams first prize medal and the Arthur Spayd Brooke gold medal for distinguished work in architectural design, 2005. Sabin was awarded a Pew Fellowship in the Arts 2010 and was named a USA Knight Fellow in Architecture, one of 50 artists and designers awarded nationally by US Artists. She was recently awarded the prestigious Architectural League Prize for Young Architects and was named the 2015 national IVY Innovator in design.

She has exhibited nationally and internationally, including in the acclaimed 9th ArchiLab entitled *Naturalizing Architecture*, at the FRAC Centre, Orleans,

France, and most recently as part of *Beauty*, the 5th Cooper Hewitt Design Triennial. Her work has been published extensively including in the *NY Times*, *The Architectural Review*, *Azure*, *A+U*, *Metropolis*, *Mark Magazine*, *306090*, *American Journal of Pathology*, *Science* and *Wired Magazine*. She co-authored *Meander: Variegating Architecture* with Ferda Kolatan, 2010. In 2017, she won the prestigious MoMA PS1 Young Architects Program competition.

Peter Lloyd Jones is a biologist, whose discoveries have uncovered fundamental mechanisms in embryogenesis and human disease, including breast cancer and pulmonary hypertension. Jones's work actively seeks and finds new solutions to complex problems via extreme collaborations within seemingly unrelated fields, including industrial, textile, and architectural design. Following completion of his PhD at Cambridge University, Jones conducted post-doctoral fellowships with Mina Bissell and Marlene Rabinovitch at UC Berkeley and the University of Toronto, respectively. Currently, he is Professor of Emergency Medicine, and the first Associate Dean of Emergent Design and Creative Technologies at the Sidney Kimmel Medical College at Thomas Jefferson University, Philadelphia, where, in 2013, he founded MEDstudio@JEFF; a research and education space focused on discovering new and dignified solutions in health care, using approaches rooted in human-centered design. In 2014, MEDstudio@JEFF partnered with Design Philadelphia and Friends of the Philadelphia Rail Park to explore how design could be deployed to benefit health at an urban scale.

Prior to this, Jones was a tenured Associate Professor of Pathology at the University of Pennsylvania, where he established a national center for the study of pulmonary hypertension, and co-founded the Sabin+Jones LabStudio with Jenny E. Sabin. As a lecturer teaching in the Graduate School of Design at UPenn, he was appointed Fellow of the NonLinear Systems Organization. Jones has been funded by the National Institutes of Health, the American Heart Association, the National Science Foundation, and the American Physiological Society, from whom he received the Giley F. Filley Memorial Award for Excellence in Respiratory Medicine. Within science, he has published more than 60 peer-reviewed articles, book chapters, and reviews. Together with Jenny Sabin, he has also widely published and exhibited work, including within *Science, National Geographic, Wired UK, Metropolis, Ars Electronica*, Museu del Disseny de Barcelona, the Slought Foundation, and the Esther Klein Gallery. LabStudio have received a number of honors and awards including first prize in the 2010 NSF/AAAS International Visualization Challenge, and were the inaugural Artists in Residence at Breadboard gallery in Philadelphia, generating new advances in 3-D printing. In addition, Jones's ideas on contemporary relationships between biology and design have been featured in the catalog accompanying the *Gen(H)ome* exhibition at the MAK Center in L.A., and in an issue of *306090* dedicated to models. In 2015, Jones was nominated for Scientist of the Year at the Philly Geek Awards, and in 2016 he will be making his TV acting debut as a master-spy on the Emmy Award-winning National Geographic science series, *Brain Games*.

Contributors

Mario Carpo is the Reyner Banham Professor of Architectural Theory and History, the Bartlett School of Architecture, UCL London. After studying architecture and history in Italy, Dr. Carpo was an Assistant Professor at the University of Geneva in Switzerland, and in 1993 received tenure in France, where he was first assigned to the École d'Architecture de Saint-Etienne, then to the École d'Architecture de Paris-La Villette. He was the Head of the Study Centre at the Canadian Centre for Architecture in Montréal from 2002 to 2006, and the Vincent Scully Visiting Professor of Architectural History at the Yale School of Architecture from 2010 to 2014.

Dr. Carpo's research and publications focus on the relationship among architectural theory, cultural history, and the history of media and information technology. His award-winning *Architecture in the Age of Printing* (MIT Press, 2001) has been translated into several languages. His most recent books are *The Alphabet and the Algorithm* (MIT Press, 2011) and *The Digital Turn in Architecture, 1992–2012* (Wiley, 2012). Dr. Carpo's recent essays and articles have been published in *Log*, *The Journal of the Society of Architectural Historians*, *Grey Room*, *L'Architecture d'aujourd'hui*, *Arquitectura Viva*, *AD/Architectural Design*, *Perspecta*, *Harvard Design Magazine*, *Cornell Journal of Architecture*, *Abitare*, *Lotus International*, *Domus*, *Artforum*, and *Arch+*.

Andrew Lucia is a designer and artist specializing in a computational approach to the analysis and production of data and form in relation to aesthetics and perception. Through a systems approach to design and materiality, Lucia's research agenda focuses on the structure of ambient light, material organization, surface, and the environments in which they are produced.

Prior to his current academic appointment as Cass Gilbert Visiting Assistant Professor in the School of Architecture at the University of Minnesota, Lucia was a visiting lecturer and critic at Cornell University, AAP (2011–2015), faculty in Visual Studies at the University of Pennsylvania, School of Design (2008–2011), and notably a founding member and Senior Researcher within LabStudio from 2008–2015.

Lucia is principal of the design and research practice ALDeR, previously having practiced architecture as designer and project manager in the offices of Yunker+Asmus Architects and M+A Architecture; as a designer in the offices of su11 architecture+design, Ruy Klein, and Barkow Leibinger Architects; and as project team member for Keller Easterling Architect.

Lucia's solo and collaborative works and writings have been published and exhibited internationally, including *Leonardo Music Journal*, *Science*, *American*

Journal of Pathology, the *New York Times*, *Wired Magazine*, *SimAud*, *Cornell Journal of Architecture*, ACSA and CENTER, Ars Electronica and FRAC.

Detlef Mertins was an architect, historian, and educator with extensive experience in teaching, scholarship, and academic administration, as well as architectural practice. He completed his BArch at the University of Toronto (1980) and his PhD at Princeton University (1996). He was Professor of Architecture and former Chair of the Department of Architecture at the University of Pennsylvania. Prior to joining Penn in 2003, he was Associate Professor at the University of Toronto in the Faculty of Architecture, Landscape, and Design where he held a Canada Research Chair in Architecture.

After completing his Bachelor of Architecture at the University of Toronto in 1980, Mertins practiced in Toronto, first with Baird/Sampson and Jones and Kirkland, then in his own firm for a number of years. At that time, he served as Professional Advisor for several important Canadian competitions. Later, in 2000, he was professional advisor for the Downsview Park Competition, Toronto, the results of which were subsequently exhibited at the Van Alen Institute and published by Harvard University.

His revisionist interpretations of Modernism have appeared in publications such as *The Presence of Mies*, the English edition of Walter Curt Behrendt's *The Victory of the New Building Style*, and essays in *Architecture and Cubism*, *Autonomy and Ideology*, *Mies in Berlin*, *Mies in America*, *NOX: Machining Architecture*, and *Phylogenesis: FOA's Ark*. Recent publications include "The Modernity of Zaha Hadid" (Zaha Hadid exhibition catalogue, Guggenheim Museum), "Mies's Event Space" (*Grey Room* 20), "Interview with Natalie de Blois" (*SOM Journal* 4), and "Walter Benjamin and the Tectonic Unconscious" (*Walter Benjamin and Art*, edited Andrew Benjamin). He co-edited the English edition of *G: An Avant-Garde Journal in Art, Architecture, Design, and Film, 1923–1926* (Getty, 2010) and a collection of his essays was published as *Modernity Unbound* (AA, 2011). Mertins's landmark monograph *Mies* (Phaidon, 2014) was published posthumously.

Keith Neeves is an Associate Professor in the Chemical and Biological Engineering Department at the Colorado School of Mines. He obtained his BS in chemical engineering at the University of Colorado, Boulder, a PhD in chemical and biomolecular engineering at Cornell University, and was an NIH NRSA postdoctoral fellow at the University of Pennsylvania. His lab at the Colorado School of Mines investigates transport phenomena in biological and geological media. His research has been recognized with NSF CAREER, Colorado Bioscience Association Educator of the Year, and Biomedical Engineering Society Young Innovator awards and is supported by the National Institutes of Health, the Department of Energy, the National Science Foundation, and the American Heart Association.

Antoine Picon is the G. Ware Travelstead Professor of the History of Architecture and Technology and Director of Research at the Harvard Graduate School of Design. He teaches courses in the history and theory of architecture and technology. Trained as an engineer, architect, and historian, Picon works on the history of architectural and urban technologies from the eighteenth century to the present. He has published extensively on this subject. Four of his books are devoted to the transition from early-modern societies to the industrial era: *French Architects and Engineers in the Age of Enlightenment* (1988, English translation 1992), *Claude Perrault (1613–1688) ou la curiosité d'un classique* (1988), *L'Invention de l'ingénieur moderne: L'École des Ponts et Chaussées, 1747–1851* (1992), and *Les Saint-Simoniens: Raison, Imaginaire, et Utopie* (2002).

With *La Ville territoire des cyborgs* (1998), Picon began to investigate the changes brought to cities and architecture by the development of digital tools and digital culture. His three most recent books are dealing extensively with this question. *Digital Culture in Architecture: An Introduction for the Design Profession* (2010) offers a comprehensive overview of this important transition. *Ornament: The Politics of Architecture and Subjectivity* (2013) focuses on the "return" of ornament in digital architecture to further the investigation. Finally, *Smart Cities: A Spatialised Intelligence* (2015) discusses its impact on cities.

Picon has received a number of awards for his writings, including the Médaille de la Ville de Paris and twice has won the Prix du Livre d'Architecture de la Ville de Briey, as well as the Georges Sarton Medal of the University of Gand. In 2010, he was elected a member of the French Académie des Technologies, and in 2015 became an associate member of the French Académie d'Architecture. He has been a Chevalier des Arts et Lettres since 2014. He is also Chairman of the Fondation Le Corbusier.

Erica Swesey Savig is founder and lead designer at weLeap in Menlo Park, CA. Her work has spanned the fields of architecture, biomedicine, analytics, and health care delivery, where she uses an exploratory process of investigation into complex problems to innovate new tools, techniques, and product concepts. Erica's design solutions are influenced by her motivation to strengthen emotional well-being, connectedness and engagement. To that end, her designs have included an empathetic responsive architectural surface, hyper-dimensional data visualizations that help intuit complex biological data, and a clinical patient engagement model that facilitates emotionally difficult conversations.

In 2016, Erica earned a PhD in Cancer Biology from Stanford University's School of Medicine and Stanford's Center for Design Research. She was a National Science Foundation Graduate Research Fellow and Gabilon Stanford Graduate Research Fellow. She has a Master of Architecture from the University of Pennsylvania and a BS in Management Science and Engineering from Stanford University.

Mark Tykocinski serves as Provost and Executive Vice President for Academic Affairs, Thomas Jefferson University, Philadelphia, and is the Anthony F. and Gertrude M. DePalma Dean of its Sidney Kimmel Medical College. For over five years, he also served as President of Jefferson University Physicians. Before joining Jefferson in 2008, he was Professor and Chair of the Department of Pathology and Laboratory Medicine at the University of Pennsylvania for a decade. Dr. Tykocinski's research contributions have been in the fields of molecular and cellular immunology, for which he holds a series of patents. His scientific focus has been on the design of novel fusion proteins with therapeutic potential, as well as unique cellular engineering strategies that invoke these proteins. He serves as Scientific Advisory Board Chair for KAHR-Medical, the Israeli biotech company he founded in 2007 for fusion protein pharmaceuticals. He is also a member of the SAB of BioCancell, a cancer gene therapy company, based in Jerusalem.

Dr. Tykocinski is Past President of the Association of Pathology Chairs and the American Society for Investigative Pathology/FASEB. He is a recipient of the NYU Distinguished Alumnus Award, and was inducted as a Fellow into the National Academy of Inventors. He has been on the editorial boards of the *American Journal of Pathology*, *Molecular Oncology*, and *Genetics Research*, and was honored with the Warner-Lambert/Parke-Davis Award for Outstanding Research from the ASIP/FASEB.

Dr. Tykocinski earned a BA in biology magna cum laude from Yale University and his MD from New York University. He completed an internal medicine internship at Columbia-Presbyterian Medical Center, residency training in anatomic pathology at New York University, and a research fellowship at NIAID/NIH in Bethesda. His first faculty position was at Case Western Reserve University, where he founded its Gene Therapy Program.

Image Credits

Figures 1.1 and 1.2 Ernst Haeckel, *Kunstformen der Natur* (1904). Published with permission from Thames and Hudson, Architectural Association Publications and Lars Spuybroek. Originally published in "Bioconstructivisms," by Detlef Mertins in *NOX: Machining Architecture* by Lars Spuybroek. Copyright 2004, Lars Spuybroek. Later published in *Modernity Unbound* by Detlef Mertins as part of Architecture Words 7. Reproduced by kind permission of Thames and Hudson Ltd, London, and the Architectural Association Publications, London.

Figures 1.3 and 1.4 Epiphyte Chamber; Caption and Copyright: © PBAI; Photograph by Philip Beesley, 2013.

Figure 1.5 eSkin inputs; Jenny E. Sabin and Mark Nicol, Sabin Design Lab and Sabin+Jones LabStudio, Cornell University and University of Pennsylvania; cell data by Kaori Ihida-Stansbury, University of Pennsylvania; PDMS substrates by Shu Yang Group, University of Pennsylvania.

Figure 1.6 ColorFolds; Photographs by Jenny E. Sabin, Sabin Design Lab, Cornell University, 2014.

Figure 1.7 Jie Li, Guanquan Liang, Xuelian Zhu, and Shu Yang. "Exploiting Nanoroughness on Holographically Patterned Three-Dimensional Photonic Crystals," *Advanced Functional Materials* 22(14) (2012): 2980–2986. Image was rendered by Felice Macera. Copyright: © Shu Yang Group University of Pennsylvania.

Figure 1.8 M.P. Gutierrez (BIOMS director/lead) with L.P. Lee (BioPoets director), UC Berkeley; BIOMS team (Charles Irby, Katia Sobolski, Pablo Hernandez, David Campbell, and Peter Suen); B. Kim (BioPoets team). Copyright: © BIOMS UC Berkeley.

Figure 1.9 Copyright: © BIOMS UC Berkeley.

Figure 1.10 Copyright: © ICD/ITKE University of Stuttgart.

Figure 1.11 Caption and Copyright: © ICD/ITKE University of Stuttgart.

Figures 1.12 and 1.13 Copyright: © ICD/ITKE University of Stuttgart.

Figures 2.1 and 2.2 Charter-Sphere Dome by Buckminster Fuller and Thomas C. Howard, Vitra Museum; Photograph by Jenny E. Sabin, 2010.

Figure 2.3 "Entrance of the Seed Cathedral," 2010. Photograph, UK Pavilion at the Shanghai Expo 2010, Heatherwick Studio, London. Accessed July, 15 2015. Photographer Daniele Mattioli.

Figure 2.4 Robert Le Ricolais, "Trihex, Three Lattice Structure," in Robert Le Ricolais, *Visiones y Paradojas [Visions and Paradoxes]*. Madrid: Fundación Cultural COAM, 1997.

Figure 2.5 First day of Nonlinear Systems Biology and Design, University of Pennsylvania; Photograph by Jenny E. Sabin; pictured, left to right: Christopher Lee, Andrew Ruggles, Misako Murata, Andrew Lucia, Kenta Fukanishi, Wei Wang, Erica Savig, 2007.

Figure 2.6 Drawing by John Piper illustrating the epigenetic landscape described by C.H. Waddington, a leading embryologist and geneticist from the 1930s–1950s. Piper's painting was first published in Waddington's book, *Organizers and Genes*.

Figure 2.7 Courtesy Jones Lab, Institute for Medicine and Engineering, University of Pennsylvania.

Part II, Introduction Figure 1 Pictured left to right: Peter Lloyd Jones, Allison Schue, Wei Wang, Erica Savig; Image courtesy of Sabin+Jones LabStudio, 2007. Photograph by Jenny E. Sabin.

Part II, Introduction Figure 2 Image courtesy of Sabin+Jones LabStudio, 2007.

Part II, Introduction Figure 3 Diagram courtesy of Sabin+Jones LabStudio

Figure 3.1 Image compilation courtesy of Peter Lloyd Jones Lab, Institute for Medicine and Engineering, University of Pennsylvania.

Figures 3.2 and 3.3 Jonathan Asher, A.J. Chan, and Kenta Fukanishi; Jenny E. Sabin and Peter Lloyd Jones, Seminar "Nonlinear Biosynthesis"; Sabin+Jones LabStudio, Institute for Medicine and Engineering and School of Design, Department of Architecture, University of Pennsylvania, 2007.

Figure 3.4 Jonathan Asher, A.J. Chan, and Kenta Fukanishi. Photograph by Jenny E. Sabin for FIBER exhibition, organized by Jenny E. Sabin with Peter Lloyd Jones and Philip Beesley, 2008.

Figures 3.5–3.8 Christopher Lee and Andrew Lucia; Jenny E. Sabin and Peter Lloyd Jones, Seminar "Nonlinear Biosynthesis"; Sabin+Jones LabStudio, Institute for Medicine and Engineering and School of Design, Department of Architecture, University of Pennsylvania, 2007.

Figures 3.9–3.12 Jonathan Asher; Independent Study with Jenny E. Sabin; Sabin+Jones LabStudio, Institute for Medicine and Engineering and School of Design, Department of Architecture, University of Pennsylvania, 2008.

Figures 3.13 and 3.14 Image courtesy of Peter Lloyd Jones Lab, Institute for Medicine and Engineering, University of Pennsylvania.

Figures 3.15–3.20 Shuni Feng, Joshua Freese, and Jeffrey Nesbit; Jenny E. Sabin and Peter Lloyd Jones, Seminar "Nonlinear Systems Biology and Design"; Sabin+Jones LabStudio, Institute for Medicine and Engineering and School of Design, Department of Architecture, University of Pennsylvania, 2008.

Figures 3.21–3.27 Shuni Feng and Joshua Freese; Independent Study with Sabin and Jones; Sabin+Jones LabStudio, Institute for Medicine and Engineering and School of Design, Department of Architecture, University of Pennsylvania, 2009.

Figures 3.28–3.32 Kara Medow, Kirsten Shinnamon, and Young-Suk Choi; Jenny E. Sabin and Peter Lloyd Jones, Seminar "Nonlinear Systems Biology and Design"; Sabin+Jones LabStudio, Institute for Medicine and Engineering and School of Design, Department of Architecture, University of Pennsylvania, 2008.

Figure 3.33 "Vasculogenesis." Peter Carmeliet, "History and Formation of Blood and Lymph Cells," in "Angiogenesis in Life, Disease and Medicine." *Nature* 438 (2005): 933, Figure 1. Reprinted by permission from Macmillan Publishers Ltd, *Nature*, copyright 2005.

Figure 3.34 "Capillary network formation proceeds along three main stages and tracking trajectories of EC plated on Matrigel." Guido Serini, Davide Ambrosi, Enrico Giraudo, Andrea Gamba, Luigi Preziosi, and Federico Bussolino, "Modeling the Early Stages of Vascular Network Assembly," *The EMBO Journal* 20(6) (2003): 1772, Figures 1 and 2.

Figure 3.35 "Appropriate EC Density Is Required for Capillary Network Formation." Guido Serini, Davide Ambrosi, Enrico Giraudo, Andrea Gamba, Luigi Preziosi, and Federico Bussolino, "Modeling the Early Stages of Vascular Network Assembly." *The EMBO Journal* 22(8) (2003): 1775, Figure 5.

Figures 3.36–3.40 Chi Dang, Alexander Lee, and Annabelle Su; Jenny E. Sabin and Peter Lloyd Jones, Seminar "Nonlinear Systems Biology and Design"; Sabin+Jones LabStudio, Institute for Medicine and Engineering and School of Design, Department of Architecture, University of Pennsylvania, 2009.

Figure 3.41 Image compilation courtesy of Peter Lloyd Jones Lab, Institute for Medicine and Engineering, University of Pennsylvania.

Figures 3.42 and 3.43 Qiao Song, Dale Suttle, and Ziyue Wei; Jenny E. Sabin and Peter Lloyd Jones, Seminar "Nonlinear Systems Biology and Design"; Sabin+Jones LabStudio, Institute for Medicine and Engineering and School of Design, Department of Architecture, University of Pennsylvania, 2009.

Figure 3.44 "Extracellular Matrix (ECM)"; Jordan Berta, 2016, see Neil A. Campbell, *Biology*, 4th edn. San Francisco: John Benjamin, 1996, p. 145, Figure 8.5.

Figures 3.45–3.51 Qiao Song, Dale Suttle, and Ziyue Wei; Jenny E. Sabin and Peter Lloyd Jones, Seminar "Nonlinear Systems Biology and Design"; Sabin+Jones LabStudio, Institute for Medicine and Engineering and School of Design, Department of Architecture, University of Pennsylvania, 2009.

Figure 3.52 Image compilation courtesy of Peter Lloyd Jones Lab, Institute for Medicine and Engineering, University of Pennsylvania.

Figures 3.53–3.59 Winglam Kwan, Mark Nicol, and Yohei Yamada; Jenny E. Sabin and Peter Lloyd Jones, Seminar "Nonlinear Systems Biology and Design"; Sabin+Jones LabStudio, Institute for Medicine and Engineering and School of Design, Department of Architecture, University of Pennsylvania, 2010.

Figure 4.1 Image compilation courtesy of Peter Lloyd Jones Lab, Institute for Medicine and Engineering, University of Pennsylvania.

Figures 4.2 and 4.3 Jackie Wong; Cecil Balmond and Jenny E. Sabin, Seminar "Form and Algorithm"; School of Design, Department of Architecture, University of Pennsylvania, 2006.

Figure 4.4 and 4.6–4.8 Megan Born and Andrew Ruggles; Jenny E. Sabin and Peter Lloyd Jones, Seminar "Nonlinear Biosynthesis"; Sabin+Jones LabStudio, Institute for Medicine and Engineering and School of Design, Department of Architecture, University of Pennsylvania, 2007.

Figure 4.5 Image courtesy of Peter Lloyd Jones Lab, Institute for Medicine and Engineering, University of Pennsylvania.

Figure 4.9 Megan Born and Andrew Ruggles with updated rendering by Jordan Berta; Jenny E. Sabin and Peter Lloyd Jones, Seminar "Nonlinear Biosynthesis"; Sabin+Jones LabStudio, Institute for Medicine and Engineering and School of Design, Department of Architecture, University of Pennsylvania, 2007.

Figure 4.10 "Illu Pulmonary Circuit," 2006. Public Domain. Available at: https://commons.wikimedia.org/wiki/File:Illu_pulmonary_circuit.jpg; DRosenbach at English Wikipedia. "Endotelijalna ćelija," 28 February 2012. English Wikipedia. Licensed under Creative Commons Attribution-Share Alike 3.0 Unported; Microscopy Courtesy of Peter Lloyd Jones at the University of Pennsylvania Institute for Medicine and Engineering.

Figure 4.11 Jared Bledsoe, Vahit K. Muskara, and Gillian Stoneback; Microscopy courtesy of Peter Lloyd Jones Lab; Jenny E. Sabin and Peter Lloyd Jones, Seminar "Nonlinear Systems Biology and Design"; Sabin+Jones LabStudio, Institute for Medicine and Engineering and School of Design, Department of Architecture, University of Pennsylvania, 2008.

Figures 4.12–4.17 Jared Bledsoe, Vahit K. Muskara, and Gillian Stoneback; Jenny E. Sabin and Peter Lloyd Jones, Seminar "Nonlinear Systems Biology and Design"; Sabin+Jones LabStudio, Institute for Medicine and Engineering and School of Design, Department of Architecture, University of Pennsylvania, 2008.

Figure 4.18 Image compilation courtesy of Peter Lloyd Jones Lab, Institute for Medicine and Engineering, University of Pennsylvania.

Figure 4.19 Victor J. Small, and Klemens Rottner, "Protruding and Contractile Domains of a Moving Cell," in "Elementary Cellular Processes Driven by Actin Assembly: Lamellipodia and Filopodia," *Actin-based Motility* 5, Figure 1.1. Reproduced with kind permission from Springer Science and Business Media.

Figure 4.20–4.24 Christopher Allen, Benjamin Callam, and Katherine Mandel; Jenny E. Sabin and Peter Lloyd Jones, Seminar "Nonlinear Systems Biology and Design"; Sabin+Jones LabStudio, Institute for Medicine and Engineering and School of Design, Department of Architecture, University of Pennsylvania, 2009.

Figure 4.25 Alex Mogilner, "Graded Radial Extension Model and Actin Growth/Myosin Contraction Mechanism," in "Mathematics of Cell Motility: Have We Got Its Number?" *Journal of Mathematical Biology* (2008): 105–134, Figure 8. With kind permission from Springer Science and Business Media.

Image Credits 375

Figures 4.26 and 4.27 Image compilations courtesy of Peter Lloyd Jones Lab, Institute for Medicine and Engineering, University of Pennsylvania.

Figures 4.28–4.35 Yan Cong, Bo Rin Jung, Chia Liao; Jenny E. Sabin, Shawn Sweeney, and Peter Lloyd Jones, Seminar "Nonlinear Systems Biology and Design"; Sabin+Jones LabStudio, Institute for Medicine and Engineering and School of Design, Department of Architecture, University of Pennsylvania, 2010.

Figure 4.36 Image courtesy of Peter Lloyd Jones Lab, Institute for Medicine and Engineering, University of Pennsylvania.

Figures 4.37–4.41 Ikje Cheon, Josef Musil, and Dane Zeiler; Jenny E. Sabin, Shawn Sweeney, and Peter Lloyd Jones, Seminar "Nonlinear Systems Biology and Design"; Sabin+Jones LabStudio, Institute for Medicine and Engineering and School of Design, Department of Architecture, University of Pennsylvania, 2010.

Figure 4.42 © 2010 from *Essential Cell Biology*, 3rd edn by Bruce Alberts, Dennis Bray, Karen Hopkin, Alexander D. Johnson, Julian Lewis, Martin Raff, Keith Roberts, and Peter Walter, "Actin Filaments Allow Eucaryotic Cells to Adopt a Variety of Shapes and Perform a Variety of Functions," *Essential Cell Biology*, 4th edn. New York: Garland Science, 2013, p. 591, Figure 17–28. Reproduced by permission of Garland Science/Taylor & Francis Group LLC.

Figures 4.43–4.46 Bruce Alberts, Alexander Johnson, Julian Lewis, David Morgan, Martin Raff, Keith Roberts, and Peter Walter, "Panel 16–2: The Polymerization of Actin and Tubulin," in *Molecular Biology of the Cell*, 4th edn. New York: Garland Science, 2002, pp. 912–913.

Figures 4.47–4.51 John Wheeler, Brian Zilis, and Philip Tribe; Jenny E. Sabin, Shawn Sweeney, and Peter Lloyd Jones, Seminar "Nonlinear Systems Biology and Design"; Sabin+Jones LabStudio, Institute for Medicine and Engineering and School of Design, Department of Architecture, University of Pennsylvania, 2010.

Figure 4.52 © 2010 from *Essential Cell* Biology, 3rd edn by Bruce Alberts, Dennis Bray, Karen Hopkin, Alexander D. Johnson, Julian Lewis, Martin Raff, Keith Roberts, and Peter Walter, "Actin-Binding Proteins Control the Behavior of Actin Filaments in Vertebrate Cells," in *Essential Cell Biology,* 4th edn. New York: Garland Science, 2013, p. 593, Figure 17–31. Reproduced by permission of Garland Science/Taylor & Francis Group LLC.

Figures 4.53–4.60 John Wheeler, Brian Zilis; Independent Study with Jenny E. Sabin; Sabin+Jones LabStudio, Institute for Medicine and Engineering and School of Design, Department of Architecture, University of Pennsylvania, 2011.

Figure 4.61 Image courtesy of Peter Lloyd Jones Lab, Institute for Medicine and Engineering, University of Pennsylvania.

Figures 4.62–4.74 Ringo Tse, Yifan Wu, Wenda Xiao; Jenny E. Sabin, Shawn Sweeney, and Peter Lloyd Jones, Seminar "Nonlinear Systems Biology and Design"; Sabin+Jones LabStudio, Institute for Medicine and Engineering and School of Design, Department of Architecture, University of Pennsylvania, 2010.

Figure 5.1 Image courtesy of Peter Lloyd Jones Lab, Institute for Medicine and Engineering, University of Pennsylvania.

Figure 5.2 Adapted from: A. Taraseviciute, B.T. Vincent, P. Schedin, and P.L. Jones, "Quantitative Analysis of Three-Dimensional Human Mammary Epithelial Tissue Architecture Reveals a Role for Tenascin-C in Regulating c-Met Function," *American Journal of Pathology* 176(2) (2010): 827–838. Image courtesy of Peter Lloyd Jones Lab, Institute for Medicine and Engineering, University of Pennsylvania.

Figure 5.3 Wei Wang, Jenny E. Sabin, Peter Lloyd Jones; Image Courtesy of Sabin+Jones LabStudio, Institute for Medicine and Engineering and School of Design, University of Pennsylvania, 2007, 2008.

Figures 5.4–5.7 Image Sequence courtesy of Peter Lloyd Jones Lab, Institute for Medicine and Engineering, University of Pennsylvania.

Figures 5.8–5.24 Wei Wang, Jenny E. Sabin, and Peter Lloyd Jones; Image Courtesy of Sabin+Jones LabStudio, Institute for Medicine and Engineering and School of Design, University of Pennsylvania, 2007, 2008.

Figures 5.25 and 5.26 Photograph and model by Jenny E. Sabin; Wei Wang, Jenny E. Sabin, Peter Lloyd Jones; Image Courtesy of Sabin+Jones LabStudio, Institute for Medicine and Engineering and School of Design, University of Pennsylvania, 2007, 2008.

Figures 5.27–5.33 Misako Murata and Austin McIerny; Jenny E. Sabin and Peter Lloyd Jones, Seminar "Nonlinear Biosynthesis"; Sabin+Jones LabStudio, Institute for Medicine and Engineering and School of Design, Department of Architecture, University of Pennsylvania, 2007.

Figures 5.34–5.38 Misako Murata; Independent Study with Sabin and Jones; Sabin+Jones LabStudio, Institute for Medicine and Engineering and School of Design, Department of Architecture, University of Pennsylvania, 2008.

Figures 5.39–5.46 JaeYoung Lee, Jae-Won Shin, and Shou Zhang; Jenny E. Sabin and Peter Lloyd Jones, Seminar "Nonlinear Systems Biology and Design"; Sabin+Jones LabStudio, Institute for Medicine and Engineering and School of Design, Department of Architecture, University of Pennsylvania, 2008.

Figure 5.47 "Lumen Formation in MCF-10A acini." Christy Hebner, Valerie M. Weaver, and Jayanta Debnath, "Modeling Morphogenesis and Oncogenesis in Three-Dimensional Breast Epithelial Cultures" *Annual Review of Pathology: Mechanisms of Disease* (2007): 319, Figure 3.

Figures 5.48–5.56 David Ettinger, Pablo Kohan, and Huishi Li; Jenny E. Sabin and Peter Lloyd Jones, Seminar "Nonlinear Systems Biology and Design"; Sabin+Jones LabStudio, Institute for Medicine and Engineering and School of Design, Department of Architecture, University of Pennsylvania, 2008.

Figure 5.57 Mina J. Bissel and Celeste M. Nelson, "Of Extracellular Matrix, Scaffolds, and Signaling: Tissue Architecture Regulates Development,

Homeostasis, and Cancer," *Annual Review of Cell and Developmental Biology* 22(1) (2006): 287–309, Figure 3. Reproduced with permission of *Annual Review of Cell and Developmental Biology*, 2006 Volume 22, Issue 1 © by Annual Reviews, www.annualreviews.org

Figure 5.58 "Formation of Adhesion Zippers: A Prelude to AJs." Reprinted from Valeri, Vasioukhin, Christoph Bauer, Mei Yin, and Elaine Fuchs, "Directed Actin Polymerization Is the Driving Force for Epithelial Cell–Cell Adhesion," *Cell* 100(2) (2000): 210, Figure 1. Copyright 2000, with permission from Elsevier.

Figure 5.59 "Cross Section of a Polarized Gland Architecture." Christy Hebner, Valerie M. Weaver, and Jayanta Debnath, "Modeling Morphogenesis and Oncogenesis in Three-Dimensional Breast Epithelial Cultures," *Annual Review of Pathology: Mechanisms of Disease* (2007): 322, Figure 4. Reproduced with permission of *Annual Review of Cell and Developmental Biology*, 2008, Volume 3, Issue 1 © by Annual Reviews, www.annualreviews.org

Figures 5.60–5.63 Chun Fang, Emaan Farhoud, and Gregory Hurcomb; Jenny E. Sabin and Peter Lloyd Jones, Seminar "Nonlinear Systems Biology and Design"; Sabin+Jones LabStudio, Institute for Medicine and Engineering and School of Design, Department of Architecture, University of Pennsylvania, 2009

Figure 5.64 Photograph by Jenny E. Sabin.

Figure 5.65 "Reduction of Cell Motility to Cell Outlines." Erica S. Savig, Mathieu C. Tamby, Jenny E. Sabin, Peter Lloyd Jones; Image Courtesy of Sabin+Jones LabStudio, Institute for Medicine and Engineering and School of Design, University of Pennsylvania, 2007, 2010.

Figure 5.66 "A Spatio-Temporal Framework for Cell Signatures." Erica S. Savig, Mathieu C. Tamby, Jenny E. Sabin, and Peter Lloyd Jones; Image Courtesy of Sabin+Jones LabStudio, Institute for Medicine and Engineering and School of Design, University of Pennsylvania, 2007, 2010.

Figure 5.67 "Deconstructing Cell Morphologies with Geometric Analysis." Erica S. Savig, Mathieu C. Tamby, Jenny E. Sabin, and Peter Lloyd Jones; Image Courtesy of Sabin+Jones LabStudio, Institute for Medicine and Engineering and School of Design, University of Pennsylvania, 2007, 2010.

Figure 5.68 "Derivation of New Quantitative Parameters." Erica S. Savig, Mathieu C. Tamby, Jenny E. Sabin, and Peter Lloyd Jones; Image Courtesy of Sabin+Jones LabStudio, Institute for Medicine and Engineering and School of Design, University of Pennsylvania, 2007, 2010.

Figure 5.69 "Newly Derived Quantitative Parameters Show Statistically Significant Differences between Substrates." Erica S. Savig, Mathieu C. Tamby, Jenny E. Sabin, and Peter Lloyd Jones; Image Courtesy of Sabin+Jones LabStudio, Institute for Medicine and Engineering and School of Design, University of Pennsylvania, 2007, 2010.

Figure 5.70 "A Cell Choreography Revealed." Erica S. Savig, Mathieu C. Tamby, Jenny E. Sabin, and Peter Lloyd Jones; Image Courtesy of Sabin+Jones LabStudio,

Institute for Medicine and Engineering and School of Design, University of Pennsylvania, 2007, 2010.

Figures 5.71 and 5.72 "Radically Abstracted Cell Signatures." Erica S. Savig, Mathieu C. Tamby, Jenny E. Sabin, and Peter Lloyd Jones; Image Courtesy of Sabin+Jones LabStudio, Institute for Medicine and Engineering and School of Design, University of Pennsylvania, 2007, 2010.

Figure5.73 "Interpretation of Cell Signatures." Erica S. Savig, Mathieu C. Tamby, Jenny E. Sabin, and Peter Lloyd Jones; Image Courtesy of Sabin+Jones LabStudio, Institute for Medicine and Engineering and School of Design, University of Pennsylvania, 2007, 2010.

Figure 5.74 "Newly Defined Cell Behaviors." Erica S. Savig, Mathieu C. Tamby, Jenny E. Sabin, and Peter Lloyd Jones; Image Courtesy of Sabin+Jones LabStudio, Institute for Medicine and Engineering and School of Design, University of Pennsylvania, 2007, 2010.

Figure 5.75 "Data Visualization Developments for Mass Cytometry." Erica S. Savig, Mathieu C. Tamby, Jenny E. Sabin, and Peter Lloyd Jones; Image Courtesy of Sabin+Jones LabStudio, Institute for Medicine and Engineering and School of Design, University of Pennsylvania, 2007, 2010.

Figure 5.76 Albrecht Dürer, *Draughtsman Drawing a Recumbent Woman*, 1525. Woodcut. Public Domain. Available at: www.wikiart.org/en/albrecht-durer/draughtsman-drawing-a-recumbent-woman-1525 (accessed September 8, 2015).

Figure 5.77 Andrew P. Lucia, Jenny E. Sabin, and Peter Lloyd Jones, "Memory, Difference, and Information: Generative Architectures Latent to Material and Perceptual Plasticity," 15th International Conference on Information Visualisation (IV), 2011, pp. 379–88. © 2015 IEEE. Reprinted, with permission, from Proceedings of the 15th International Conference on Information Visualisation (IV), 2011.

Figures 5.78–5.80 Image courtesy of Andrew P. Lucia.

Figure 5.81 Edgar Degas, *Dancers Practising at the Barre*, 1876–1877. A study in oil paint on green paper. London, British Museum. Available at: www.britishmuseum.org/explore/highlights/highlight_objects/pd/e/edgar_degas,_dancers_practisin.aspx (accessed August 30, 2015).

Figure 5.82 Erica S. Savig, Mathieu C. Tamby, Jenny E. Sabin, and Peter Lloyd Jones; Image Courtesy of Sabin+Jones LabStudio, Institute for Medicine and Engineering and School of Design, University of Pennsylvania, 2007, 2010.

Figure 5.83 Image courtesy of Peter Lloyd Jones Lab, Institute for Medicine and Engineering, University of Pennsylvania.

Figure 5.84 Image courtesy of Andrew P. Lucia.

Figure 5.85 Andrew P. Lucia, Jenny E. Sabin, and Peter Lloyd Jones, "Memory, Difference, and Information: Generative Architectures Latent to Material and Perceptual Plasticity," 15th International Conference on Information Visualisation (IV), 2011, 379–88. © 2015 IEEE. Reprinted, with permission, from

Proceedings of the 15th International Conference on Information Visualisation (IV), 2011.

Figure 5.86 Andrew P. Lucia, Jenny E. Sabin, and Peter Lloyd Jones, "Memory, Difference, and Information: Generative Architectures Latent to Material and Perceptual Plasticity," 15th International Conference on Information Visualisation (IV), 2011, 379–88. © 2015 IEEE. Reprinted, with permission, from *Proceedings of the 15th International Conference on Information Visualisation* (IV), 2011. Cellular images courtesy of Peter Lloyd Jones Lab, University of Pennsylvania.

Figures 5.87 and 5.88 Andrew P. Lucia, Jenny E. Sabin, and Peter Lloyd Jones, "Memory, Difference, and Information: Generative Architectures Latent to Material and Perceptual Plasticity." 15th International Conference on Information Visualisation (IV), 2011, 379–88. © 2015 IEEE. Reprinted, with permission, from *Proceedings of the 15th International Conference on Information Visualisation* (IV), 2011.

Figures 5.89 and 5.90 Andrew Lucia, Jenny E. Sabin, Peter Lloyd Jones; Image Courtesy of Sabin+Jones LabStudio, Institute for Medicine and Engineering and School of Design, University of Pennsylvania, 2007, 2010.

Figures 5.91 and 5.92 Andrew P. Lucia, Jenny E. Sabin, and Peter Lloyd Jones, "Memory, Difference, and Information: Generative Architectures Latent to Material and Perceptual Plasticity." 15th International Conference on Information Visualisation (IV), 2011, 379–88. © 2015 IEEE. Reprinted, with permission, from *Proceedings of the 15th International Conference on Information Visualisation* (IV), 2011.

Figure 5.93 Andrew Lucia, after Leon Battista Alberti, *Visual Pyramid*, 1435–1436.

Figure 5.94 James J. Gibson, "The Ambient Optic Array from a Wrinkled Earth Outdoors Under the Sky," in *The Ecological Approach to Visual Perception*. New York: Psychology Press, 2015.

Figure 5.95 Image courtesy of Keith Neeves, Chemical and Biological Engineering, Colorado School of Mines.

Figure 5.96 K.B., Neeves, S.F. Maloney, K.P. Fong, A.A. Schmaier, M.L. Kahn, L.F. Brass, and S.L. Diamond, "Microfluidic Focal Thrombosis Model for Measuring Murine Platelet Deposition and Stability: PAR4 Signaling Enhances Shear-resistance of Platelet Aggregates," *Journal of Thrombosis and Haemostasis* 6(12) (2008): 2193–2201.

Figures 5.97–5.99 Image courtesy of Keith Neeves, Chemical and Biological Engineering, Colorado School of Mines.

Figure 5.100 Keith B Neeves and Scott L. Diamond, "A Membrane-based Microfluidic Device for Controlling the Flux of Platelet Agonists into Flowing Blood," *Lab on a Chip Lab Chip* 8(5) (2008): 701–709.

Figure 5.101 Image courtesy of Keith Neeves, Chemical and Biological Engineering, Colorado School of Mines.

Figure 6.1 Data provided by Shu Yang Group, University of Pennsylvania. Image originally published in J. Li, J. Shim, J. Overvelde, J. Deng, X. Zhu, K. Bertoldi, and S. Yang, "Switching Reflective Color to Transparency in Photonic Membranes via Pattern Transformation and Shape Memory Effect," *Soft Matter*, 8(40) (2012): 10322–10328.

Figures 6.2–6.9 John Amato Ross, Gabriel Wilson Salvatierra, and Elena Sophia Toumayan; Jenny E. Sabin, Seminar "Bioinspired Materials and Design"; Department of Architecture, Cornell University, 2012. Data provided by Shu Yang Group, University of Pennsylvania.

Figure 6.10 Image courtesy of Kaori Ihida-Stansbury, University of Pennsylvania, via personal communication.

Figures 6.11 and 6.12 Image courtesy of Kaori Ihida-Stansbury, University of Pennsylvania, via personal communication. Sean Kim, Ryan Peterson, Maria Stanciu; Jenny E. Sabin, Seminar "Bioinspired Materials and Design"; Department of Architecture, Cornell University, 2012.

Figures 6.13–6.17 Sean Kim, Ryan Peterson, and Maria Stanciu; Jenny E. Sabin, Seminar "Bioinspired Materials and Design"; Department of Architecture, Cornell University, 2012.

Figure 6.18 Image and data provided by Shu Yang Group and Kaori Ihida-Stansbury, University of Pennsylvania.

Figure 6.19 Image and data provided by Shu Yang Group and Kaori Ihida-Stansbury, University of Pennsylvania; Joon Hyuk Choe, Danlu Li, and Tzara Peterson; Jenny E. Sabin, Seminar "Bioinspired Materials and Design"; Department of Architecture, Cornell University, 2012.

Figures 6.20–6.26 Joon Hyuk Choe, Danlu Li, and Tzara Peterson; Jenny E. Sabin, Seminar "Bioinspired Materials and Design"; Department of Architecture, Cornell University, 2012.

Figures 8.1 and 8.2 Image courtesy of Peter Lloyd Jones Lab, Institute for Medicine and Engineering, University of Pennsylvania.

Figure 8.3 Sketches by Jenny E. Sabin; Image Courtesy of Sabin+Jones LabStudio, Institute for Medicine and Engineering and School of Design, University of Pennsylvania, 2008.

Figure 8.4 Jenny E. Sabin, Andrew Lucia, and Peter Lloyd Jones; Image Courtesy of Sabin+Jones LabStudio, Institute for Medicine and Engineering and School of Design, University of Pennsylvania, 2008.

Figures 8.5 and 8.6 Photograph by Jenny E. Sabin. Jenny E. Sabin, Andrew Lucia, and Peter Lloyd Jones; Image Courtesy of Sabin+Jones LabStudio, Institute for Medicine and Engineering and School of Design, University of Pennsylvania, 2008.

Figure 8.7 Photograph by Jenny E. Sabin, 2005; Image courtesy Jenny E. Sabin.

Figures 8.8 and 8.9 Jenny E. Sabin, Andrew Lucia, and Peter Lloyd Jones; Image

Courtesy of Sabin+Jones LabStudio, Institute for Medicine and Engineering and School of Design, University of Pennsylvania, 2008.

Figure 8.10 Photograph by Jenny E. Sabin; Jenny E. Sabin, Andrew Lucia, and Peter Lloyd Jones; Image Courtesy of Sabin+Jones LabStudio, Institute for Medicine and Engineering and School of Design, University of Pennsylvania, 2008.

Figure 8.11 Jenny E. Sabin, Andrew Lucia, and Peter Lloyd Jones; Image Courtesy of Sabin+Jones LabStudio, Institute for Medicine and Engineering and School of Design, University of Pennsylvania, 2008.

Figures 8.12 and 8.13 Photograph by Jenny E. Sabin; Jenny E. Sabin, Andrew Lucia, and Peter Lloyd Jones; Image Courtesy of Sabin+Jones LabStudio, Institute for Medicine and Engineering and School of Design, University of Pennsylvania, 2008.

Figures 8.14 and 8.15 Photograph courtesy of Lira Nikolovska, SIGGRAPH 2008; Jenny E. Sabin, Andrew Lucia, and Peter Lloyd Jones; Image Courtesy of Sabin+Jones LabStudio, Institute for Medicine and Engineering and School of Design, University of Pennsylvania, 2008.

Figures 8.16–8.20 Photograph by Jenny E. Sabin; Jenny E. Sabin, Andrew Lucia, and Peter Lloyd Jones; Image Courtesy of Sabin+Jones LabStudio, Institute for Medicine and Engineering and School of Design, University of Pennsylvania, 2008.

Figure 8.21 Jenny E. Sabin, Andrew Lucia, and Peter Lloyd Jones. Image courtesy of Sabin+Jones LabStudio, University of Pennsylvania, 2009; Figure originally featured as part of *ACADIA 10: LIFE in:formation, On Responsive Information and Variations in Architecture* (proceedings of the *30th Annual Conference of the Association for Computer Aided Design in Architecture (ACADIA),* New York, 2010.

Figure 8.22 Image courtesy of Peter Lloyd Jones Lab, Institute for Medicine and Engineering, University of Pennsylvania.

Figure 8.23 Adapted from: A. Taraseviciute, B.T. Vincent, P. Schedin, and P.L. Jones, "Quantitative Analysis of Three-Dimensional Human Mammary Epithelial Tissue Architecture Reveals a Role for Tenascin-C in Regulating c-Met Function," *American Journal of Pathology* (2010) February; 176(2)(2010): 827–838. Image courtesy of Peter Lloyd Jones Lab, Institute for Medicine and Engineering, University of Pennsylvania.

Figures 8.24 and 8.25 Photograph by Jenny E. Sabin; Jenny E. Sabin, Andrew Lucia, and Peter Lloyd Jones. Image courtesy of Sabin+Jones LabStudio, University of Pennsylvania, 2009.

Figures 8.26 and 8.27 Jenny E. Sabin, Andrew Lucia, and Peter Lloyd Jones. Image courtesy of Sabin+Jones LabStudio, University of Pennsylvania, 2009; Figure originally featured as part of *ACADIA 2010: LIFE in:formation, On Responsive Information and Variations in Architecture [Proceedings of the 30th Annual Conference of the Association for Computer Aided Design in Architecture (ACADIA)].*

Figures 8.28–8.30 Photograph by Jenny E. Sabin; Jenny E. Sabin, Andrew Lucia, and Peter Lloyd Jones. Image courtesy of Sabin+Jones LabStudio, University of Pennsylvania, 2009; Figure originally featured as part of *ACADIA 10: LIFE in:formation, On Responsive Information and Variations in Architecture* (proceedings of the *30th Annual Conference of the Association for Computer Aided Design in Architecture (ACADIA),* New York, 2010.

Figures 8.31–8.33 Photograph by Jenny E. Sabin; Jenny E. Sabin, Andrew Lucia, and Peter Lloyd Jones. Image courtesy of Sabin+Jones LabStudio, University of Pennsylvania, 2009.

Figure 8.34 Photo by Shane Burger; Smart Geometry Workshop 2010, Nonlinear Systems Biology and Design with SG cluster champions Jenny E. Sabin, Peter Lloyd Jones, Andrew Lucia, and Erica Savig.

Figures 8.35–8.40 Diagrams by workshop participant, Jacob Riiber; Smart Geometry Workshop 2010, Nonlinear Systems Biology and Design with SG cluster champions Jenny E. Sabin, Peter Lloyd Jones, Andrew Lucia, and Erica Savig.

Figures 8.41 and 8.42 Photo by Jenny E. Sabin; Smart Geometry Workshop 2010, Nonlinear Systems Biology and Design with SG cluster champions Jenny E. Sabin, Peter Lloyd Jones, Andrew Lucia, and Erica Savig.

Figure 8.43 Photo by Shane Burger; Smart Geometry Workshop 2010, Nonlinear Systems Biology and Design with SG cluster champions Jenny E. Sabin, Peter Lloyd Jones, Andrew Lucia, and Erica Savig.

Figures 8.44 and 8.45 Diagrams by workshop participant, Jacob Riiber; Smart Geometry Workshop 2010, Nonlinear Systems Biology and Design with SG cluster champions Jenny E. Sabin, Peter Lloyd Jones, Andrew Lucia, and Erica Savig.

Figure 8.46 Photo by Shane Burger; Smart Geometry Workshop 2010, Nonlinear Systems Biology and Design with SG cluster champions Jenny E. Sabin, Peter Lloyd Jones, Andrew Lucia, and Erica Savig

Figures 8.47 and 8.48 Photo by Jenny E. Sabin; Smart Geometry Workshop 2010, Nonlinear Systems Biology and Design with SG cluster champions Jenny E. Sabin, Peter Lloyd Jones, Andrew Lucia, and Erica Savig.

Figures 8.49–8.51 Photo by Jenny E. Sabin; DHUB, Barcelona, Spain; Working Prototypes exhibition featuring Nonlinear Systems Biology and Design cluster, 2010.

Figure 9.1 Images courtesy Shu Yang (PI) (materials science), Jan Van der Spiegel and Nader Engheta (Co-PIs) (electrical and systems engineering), Kaori Ihida-Stansbury, Peter Lloyd Jones (Co-PIs) (cell biology), University of Pennsylvania; Jenny E. Sabin (Co-PI) and Andrew Lucia (Senior Personnel) (architecture), Cornell University, 2010–2014.

Figure 9.2 Images courtesy Sabin+Jones LabStudio and Peter Lloyd Jones Lab, University of Pennsylvania; Jenny E. Sabin, Mark Nicol, and Peter Lloyd Jones, 2010–2011.

Figure 9.3 Images courtesy Shu Yang Group, Kaori Ihida-Stansbury, and Peter Lloyd Jones, University of Pennsylvania.

Figure 9.4 Images courtesy Sabin+Jones LabStudio, University of Pennsylvania; Jenny E. Sabin, Mark Nicol, Peter Lloyd Jones, 2010–2011.

Figure 9.5 Jenny E. Sabin (Co-PI), Andrew Lucia (Senior Personnel), Simin Wang, Cornell University, 2013; Image courtesy Sabin Design Lab.

Figure 9.6 Image adapted with permission from D. Chandra *et al.* "Biomimetic Ultrathin Whitening by Capillary Force Induced Random Clustering of Hydrogel Micro Pillar Arrays," *ACS Applied Material Interfaces* 1(8) (2009): 1698–1704. Copyright 2009 American Chemical Society.

Figure 9.7 a and b Images originally published in S. Wang, A. Lucia and J. Sabin, "Simulating Nonlinear Nano-to-Micro Scaled Material Properties and Effects at the Architectural Scale," in Simulation for Architecture and Urban Design (SimAUD), April 7–13, 2013.

Figure 9.7c Images courtesy Sabin+Jones LabStudio, University of Pennsylvania; Jenny E. Sabin, Mark Nicol, Peter Lloyd Jones, 2010–2011.

Figure 9.8 Image originally published in Jie, Li, Jongmin Shim, Justin Deng, Johannes T. B. Overvelde, Xuelian Zhu, Katia Bertoldi, and Shu Yang, "Switching Periodic Membranes via Pattern Transformation and Shape Memory Effect," *Soft Matter* 8(40) (2012): 10322.

Figures 9.9–9.13 Image originally published in Wang *et al.* (2013).

Figures 9.14–9.18 Originally published in J. Sabin, A. Lucia, Ott, and S. Wang, "Prototyping Interactive Nonlinear Nano-to-Micro Scaled Material Properties and Effects at the Human Scale" in *Symposium on Simulation for Architecture and Urban Design (SimAUD 2014) 2014 Spring Simulation Multi-Conference (SpringSim'14); Tampa, Florida, USA, 13–16 April 2014*, ed. David Jason Gerber, Red Hook, NY: Curran, 2014.

Figure 9.19 Photograph by Jenny E. Sabin

Figures 9.20–9.22 Jenny E. Sabin (Co-PI), Andrew Lucia (Senior Personnel), Cornell University, 2014.

Figure 9.23 Image courtesy Sabin Design Lab, Cornell University; Jenny E. Sabin, and Andrew Moorman, 2015.

Figure 9.24 Sabin Design Lab, Department of Architecture, Cornell University, 2014. Principal Investigator: Jenny E. Sabin; Design Research Team: Martin Miller, Daniel Cellucci and Andrew Moorman, Giffen Ott, Max Vanatta, David Rosenwasser, Jessica Jiang; As part of the CCA 2014 BIENNIAL: Intimate Cosmologies: The Aesthetics of Scale in an Age of Nanotechnology.

Figure 10.1 Image courtesy Nike Inc.; Jenny Sabin Studio; Architectural Designer: Jenny E. Sabin; Design and Production Team: James Blair, Simin Wang, Martin Miller, Meagan Whetstone, Brian Heller, and Nicola McElroy; Commissioned by Nike Inc. for FlyKnit Collective, 2012.

Figures 10.2, 10.4 and 10.5 Jenny Sabin Studio; Architectural Designer: Jenny E. Sabin; Design and Production Team: James Blair, Simin Wang, Martin Miller,

Meagan Whetstone, Brian Heller, and Nicola McElroy; Commissioned by Nike Inc. for FlyKnit Collective, 2012.

Figures 10.3, 10.6 and 10.7 Images courtesy Nike Inc.; Jenny Sabin Studio; Architectural Designer: Jenny E. Sabin; Design and Production Team: James Blair, Simin Wang, Martin Miller, Meagan Whetstone, Brian Heller, and Nicola McElroy; Commissioned by Nike Inc. for FlyKnit Collective, 2012.

Figures 10.8 and 10.9 Jenny Sabin Studio; Architectural Designer: Jenny E. Sabin; Design and Production Team: James Blair, Simin Wang, Martin Miller, Meagan Whetstone, Brian Heller, and Nicola McElroy; Commissioned by Nike Inc. for FlyKnit Collective, 2012.

Figures 10.10 and 10.11 Image courtesy Nike Inc.; Jenny Sabin Studio; Architectural Designer: Jenny E. Sabin; Design and Production Team: James Blair, Simin Wang, Martin Miller, Meagan Whetstone, Brian Heller, and Nicola McElroy; Commissioned by Nike Inc. for FlyKnit Collective, 2012.

Figures 10.12–10.14 Photo by Simin Wang; Jenny Sabin Studio; Architectural Designer: Jenny E. Sabin; Design and Production Team: James Blair, Simin Wang, Martin Miller, Meagan Whetstone, Brian Heller, and Nicola McElroy; Commissioned by Nike Inc. for FlyKnit Collective, 2012.

Figure 10.15 Image courtesy Nike Inc.; Jenny Sabin Studio; Architectural Designer: Jenny E. Sabin; Design and Production Team: James Blair, Simin Wang, Martin Miller, Meagan Whetstone, Brian Heller, and Nicola McElroy; Commissioned by Nike Inc. for FlyKnit Collective, 2012.

Figure 10.16 Diagrams courtesy AKT II Engineering, Daniel Bosia.

Figures 10.17–10.19 Jenny Sabin Studio; Architectural Designer: Jenny E. Sabin; Design and Production Team: James Blair, Simin Wang, Martin Miller, Meagan Whetstone, Brian Heller, and Nicola McElroy; Commissioned by Nike Inc. for FlyKnit Collective, 2012.

Figures 10.20 and 10.22 Photo by Simin Wang; Jenny Sabin Studio; Architectural Designer: Jenny E. Sabin; Design and Production Team: James Blair, Simin Wang, Martin Miller, Meagan Whetstone, Brian Heller, and Nicola McElroy; Commissioned by Nike Inc. for FlyKnit Collective, 2012.

Figure 10.21 Photo by Jenny E. Sabin; Jenny Sabin Studio; Architectural Designer: Jenny E. Sabin; Design and Production Team: James Blair, Simin Wang, Martin Miller, Meagan Whetstone, Brian Heller, and Nicola McElroy; Commissioned by Nike Inc. for FlyKnit Collective, 2012.

Figure 10.23 Sky and Water installation at University of Maryland, 2011, copyright Tobi Kahn. Sky and Water installation, first shown in the exhibition, *Aligned: Paintings* by Tobi Kahn at the University of Maryland Art Gallery, 2011.

Index

Page locators in italics refer to figures

2-D cell choreography, 203, *204*
2-D deployable structure, 172, *174*, 175
2-D tissue scaffold, 178–86, *180–85*
3-D cell and tissue architecture, 55–56, *66*, 129, *139*, 362; polymerization, *143*, 145–47
3-D designer microenvironments, 49, 53
3-D printed material, 25, 50, 239, 268–69; cell choreographies, 201, *209*; chemo-attraction, *105*, *106*; context-dependent cell deformation, *133*, *134*; Ground Substance project, 295–96, *295–98*, 298; knitting similar to, 343, 347; mammary gland, *169*, *170*, 292–98; mesh components, *74*; mutations and accumulation, *153*; nonstandard parts, 290, *290*, 30; Penrose tiling analysis, *121*; surface deformation, *81*, 83, *83*, *84*
3-D scissor mechanisms, 36, 172
3-D visualizations: cell choreographies, 203, *204*, 208, *209*, *211*, 213
3-D Voronoi diagram, *164*, 164–65, *166*
4-D context, 50, 53, 109, *115*
spatial signatures, 123–31, *124–25*, *127–30*

acini structures, 158–60, *162–64*, 164–66, *166*; deployable structures, 172; surface roughness, 158, 166, *168*, 179; surface tension, 166–67, *168*
actin, 96, 103, 125, *125*, 127, 137; polymerization, 140–47, *140–47*
AD, Architecture and Science, 269–70
adaptive and resilient networks: dynamic material systems, 103–6, *103–6*
adaptive architecture, 18, 20, 350–51, *351*; human-generated, 21; near living adaptive materials, 19; networking, 74, *77*, *78*

adaptive environments, 311, 336. *See also* eSkin project; skins
adaptive technologies, 320
Addington, Michelle, 18, 229n10
adherens junctions, 170–78, *171*, *173–78*, *193*
adhesion, *64*, 123, *125*, 126, 131, *193*, 194–95, 202, 210, *232*, 233–38; cadherins, 171–72, 175; cellular adhesion molecules (CAM), 96
Advanced Geometry Unit (ARUP London), 36
aesthetics, 216
affordances, 17–18, 30n2; in counter-distinction to the solutionism, 17, 28; thermodynamic environments, 21
agent-based algorithm, 129
Aiolova, Maria, 18
airway network, 56–58, *57*
algorithms: context-dependent cell deformation, *134*; geometric and site-based information, *161*; human-space interaction, 128–29; ice dancing, 110, *110*, *111*; motility, 98–99, 110–11, 114, 126, *127–31*; motion capture, 126; object-environment interaction, 127–28. *See also* visual exercises (generative and algorithmic studies)
Allen, Christopher, 123–31
The Alphabet and the Algorithm (Carpo), 268–69
ambient information array, 226–27, *228*
Ambrosi, Davide, 87–88
amplified effects, 20
analogical models, 13–14
analogy, 7, 13; crystal, 10–11
Anderson, Chris, 263n3
angiogenesis, 57, 60, 67, *81*, *88*

apoptosis (programmed cell death), 159–60, 188–92
Archilab, 273, 276–77
architectural prototyping, 257
architecture: code in context, 31, 35–37, 305, 307–9; dynamic reciprocity, 40–42; horizontal and collaborative authorship, 268–69; matrix, 270–71; soft textile-based, 350; topological, 269–70
Architecture Bits conference, 48–49
Aristotelian science, 260–61
Aristotle, 4, 39
Armstrong, Rachel, 19
array: ambient information, 226–27, *228*; light-borne phenomena, 241–42, *242, 318, 318*; nano-/micro-scale pillar arrays, 317–18, *318, 319*
Ars Electronica, 288
art, in medicine, 356
Art Nouveau, xiii
arteriogenesis, *88*
Asher, Jonathan, 59, 61
Aulonia, 10, *10*
autonomous approaches, 4, 36, 217
avant-garde, 3, 5

Balmond, Cecil, 31, 36, 37, 43n12, 49
Baranski, Jan, 49
basal lateral polarity, *171,* 172
basement membrane, 57–58, 67, 81–82, 94, *293*; acini structures, 158–60, *162–64,* 172, 175
Bateson, Gregory, 215, 216, 229nn3, 8
Beesley, Philip, 18, 20–22, 28
behavior: abstraction of, 208–10, 300–301; scale-free networks, 67–69; tensile connections, 103
behavioral modeling, 67–70, 72, 79–80, 199
behavioral "signatures," 107
Behrens, Peter, 7
Benjamin, David, 18
Berlage, H.P., 6–7, 11
Bernstein, Emily, 290
Betz, Oliver, 26
biaxial symmetry, 10–11
Big Data environment, 259–60, 263n3, 281–82, 358–59

bimorphism, 160
binary data, 83, 85, *119,* 271
binary oppositions, 5
bioconstructivisms, 3–16
bioconstructivist theory, 11
bio-crystallisation, 8–10
biodata, dynamic, 278, *279*
BioInspired materials and design, 239–56; Branching Morphogenesis project, 271, 273–89, *274–75, 277–88*; dynamic degeneration and reconstruction of ECM, *246–50,* 246–51; Ground Substance project, 290–98, *290–98*; skin systems and dynamic boundary conditions, 251–56, *251–56*; structural color at the architectural scale, *240–45,* 240–46
"BioInspired Materials and Design" course, 239–40
biology, 31, 364; matrix, 22, 26, 36; nonlinear processes, 56, 103; systems, 52–53
biomedicine, 48, 56, 86. *See also* breast cancer; mammary gland; pulmonary hypertension
biomimetic principles, 314–15
biomimicry, *24,* 106, 136, 266–67, 273; bionic research pavilion, 25–28, *26, 27, 28, 29*; limits of, 49; raindrop, 32
BIOMS group, 18, 19, 28; breathing membrane, 23–24, *25*
bionic research pavilion, 25–28, *26, 27, 28*
"Bionics of Animal Constructions" (University of Stuttgart/Tubingen University), 26
biosynthesis, 49, 266–67, 270–71; modeling a complex multi-variable system, 59–65
biotechnics, 11–12
Bissell, Mina, 40
Blake, William, 216
Bledsoe, Jared, 117–23
blood clotting models, *232–34,* 232–36, *236*; interstitial space of clot, 236–38, *237*; membrane microfluidic model, 235–36, *236*
Blumenbach, Johann Friedrich, 4
BodyBlanket, 278, 335

Born, Megan, 112–16
Bosia, Daniel, 345
boundary conditions, 251–56, *251–56*, 294. *See also* edge behaviors
Brainard, George, 357
Branching Morphogenesis, 271, 273–89, *282*, 335; images from, *274–75, 277–88*; "time-sheets" or tapestries, *280*, 281, *285*
branching structures, 36–37, 50, 58–59, 61–62, 87, 114, 305; simulation, 96–98, *97. See also* fractal branching
Brayer, Marie-Ange, 273, 276, 277
breast cancer, 157–59, 165–69, *167, 168*, 171–72, 179, 189, *194*, 194–95, 292
breathing membrane, 23–24, *25*
British Pavilion or Seed Cathedral (2010 World Expo), 32–33, *34*
Brock, Amy, 125–26
Brown, Scott, *217*
Buffon, Count, 4
building skins. *See* eSkin project; skins
Bussolino, Federico, 87–88

Cache, Bernard, 268, 271
cadherins, 171–72, 175
calculus, 263
Calinescu, Matei, 5
Callam, Benjamin, 123–31
Callimachus, xii
cancer, 214, 360. *See also* breast cancer
capillary formation, 56–58, *57*, 87–94
Carnoy, 10, *10*
Carpo, Mario, 268–69, 270–71, 281
Cartesian logics/grid, 207, *207*, 216, 227, 242, *244*, 281
Cassman, 47
categories, 4
causality, 260, 263
cell choreographies, 201–14, *203–7, 209, 211–13*, 293; abstraction of behaviors, 208–10; biological interpretation, 210–11, *211–12*; bridging of geometry from image to numbers, 205–7, *205–10, 209*; cell elongation, *207*, 208, *212*; (de)constructing cell morphology, 205; revealed, 206–8, *207*; video-microscopy, 202–3, *203–6*, 207, 210

cell density, 88–89, 195–97
cell membrane, *96,* 131, 182, 306. *See also* basement membrane
cell-cell interactions, 59, 171–72, 175, 178, 185, *193*, 193–97, 287
cellness, 267
cells, 51; 3-D architecture, 55–56; attraction, *63, 64*; centroid, 129, 148, 160, 165–66, *166,* 182, 205, *206–12,* 208–12, 324, *325*; cohesion, 83–84, 210, *211*; edge behaviors, 111–16, 202, *205, 206,* 208–12, *209, 211, 212,* 221, *221*; elongated phenotype, 82; epithelial, 157–58, *158*; filaments, 96; internal stiffness, 125; lung mesenchymal, 67; receptors, 231–32; segmentation/detection, 221. *See also* extracellular matrix (ECM); motility
cellular adhesion molecules (CAM), 96
cellular automata, 36
"Central Dogma," 39
chain responses, 20
Chan, A.J, 59–65, *61*
Chang, Eric, 125–26
change, observation of, 215–30; usefulness of information, 216–17, *217*
Charter-Sphere Dome, *33*
chemical-physical loop, 40
chemo-attraction, 87–88, *90,* 103–6, *103–6, 104*
Cheon, Ikje, 137–40
Choe, Joon Hyuk, 251–56
Choi, Young-Suk, 81–86
Chu, Karl, 36
CITA (Royal Danish Academy of Fine Arts), 18
clay, 3-D, *295*, 295–96
"A Clearing through the Forest" (Winter), 355
closed loop systems, 23
clustering effect, 69–72, *70, 71, 75*
code, 56; context, 305; in context, 31, 35–37, 305, 307–9, *307–10*; digital and physical, 56; digitally woven curves, 83; DNA, 58; protein, 58; Prx-1 gene, 67, *68,* 72, 79, 81–86; self-renewal and, 179
cohesion, 83–84, 210, *211*

collaborative endeavors, 48, 148, 275, 361–64; multi-year and multi-phase research plan, 361–62. *See also* trans-disciplinary research practice
collagen, 124; fibrillar and nonfibrillar, 202, *203, 204,* 206, *207,* 208, *209,* 210, *211, 212*; native or denatured, 119, 122–24; type I, 56, 109, 202; vascular injury model, 232–33, *233–34*
College within the College, 357
color, use of, 53, 87, 126, 129, 160, *162*; structural color, *24,* 31
color channels, 160, *162*
ColorFolds, *23,* 331–34, *333*
complex multi-variable systems, 59–65, *60–66*
complexity, 55, 276, 355
component-based construction system, *27,* 92–93
computer models, 72–74; complex multi-variable systems, 59–65, *60–66*; potential parameters, 59–60
Cong, Yan, 131–36
connectivity, 61, 67–72, *71, 71,* 75, 88; responsive surface specificity, 192–97, *193–97*; tensile connections, 103–6, *103–6*
constellar associations, 10–11
constructivism, 4, 6; international, 11
context: 4-D, 50, 53, 109, *115*; cell structure, 89; code in context, 31, 35–37, 305, 307–9, *307–10*; of computational matter, 25, *26,* 28; ECM, 72; field of change, 223; observation of change and, 216–17, *217*; of sustainable architecture, 17–18; tensile connections, 103; transformations, 270. *See also* motility
context-dependent cell deformation, 131–36, *131–36*
Conway, John, 36
Corinthian capital, xii–xiii
Cornell Architecture, 18, 22
Cornell University, 19, 50, 52
corrugated sheet metal, 34
counter-revolutions, 6
crafts-based media, 22
Crick, 34, 38
crisis applications, 25

critical concentration (CC), 141
crystalline form, 8–9, 10–11, 12; acini structure, modeling, 165–67, *166*
Cupkova, Dana, 18
curved surfaces, 3, 10, 26, 28, 74; 3-D print models, 83, *83, 84*; spline curves, *81,* 82–83, *85*
Cuypers, P.J.H., 6
cytoskeleton, 31, 40–41, *40–42, 41,* 94, 187, 306–7; changes over time, 123–24; internal cell stiffness, 125; tensile connections, 103; types, 307; vascular smooth muscle cells (SMCs) and, 109

Dance and Space (Wong et al.), *110,* 110–11, *111*
dancing. *See* ice dancing
Dang, Chi, 87–94
Darwin, Charles, 267
data acquisition, *217,* 263n4; costs, 259; new science of making, 259–64; unlimited availability, 259
data visualization, 215, 218, *218,* 223; datascapes, 36, 199, 226, 239, 257, 265, 271, 273, 282, 285, 288; Branching Morphogenesis, 273, *278,* 282; myThread Pavilion, 335–51, *336–42, 344–46, 348–51*
Davies, Peter F., 47, 49, 155, 186–88
De Architectura (Vitruvius), xii
de Bazel, K.P.C., 6
de Groot, J.H., 6
Decker, Martina, 20
deformation, *63, 73, 74, 75,* 270, *320*; 2-D tissue scaffold, 183–86, *184, 185*; chemo-attraction and, *104, 105*; context-dependent, 131–36, *131–36*; Idealized, 241, *241*; lattice, *189–91*; in pulmonary hypertension, 124
Degas, Edgar, 220, 221
Delaunay, Boris Nikolaevich, 160
Delaunay Tessellation, 51, 159–60, *164,* 164–65, *293*; 3-D, 166, *167*
Deleuze, Gilles, 14–15
democratic conceptions, 6
deployable structures, 36, *152–55,* 153–55, 159; adherens junctions, 170–78, *171, 173–78*; key elements, 170–72

design: as complex multi-variable system, 59
Design and Computation gallery (SIGGRAPH), 273
design computation tools: criteria and goals, 51–53, *53*; design goal metaphor, 47–48; introduction, 47–51; matrix architecture, 53–54. *See also* networking
design research in practice: new model, 31–44
design research models, 17–30; high-risk experimental models, 18
design science revolution, xiv
design studio, 56
design tools, 6, 35–36; 3-D CNC tools, 266; conventional tooling languages, 215–16; mammary gland acini structures, *158*, 158–60, *162–64*; translation, 87. *See also* BioInspired materials and design
deterministic modeling systems, 36
developmental biology, 39
developmental geometries, 187
Di Cristina, Giuseppa, 269–70
difference, 215–20, *218, 219*; mechanisms of, 215–17; probabilities through time, 218, *219*, 222, *223*, 223–25; weak signal, *219*
digital complexity, 27, 55
Digital Culture in Architecture (Picon), 276
digital design tooling, 35–36
digital fabrication, 266, 362. *See also* microfabrication; physical exercises (3-D print and digital fabrication outputs); robotic fabrication methods
digital handcraft, 265, 271
digitized Jacquard loom, *279*, 280
disciplinary traditions, 216
disease processes, 17, 50, 56, 310. *See also* breast cancer; pulmonary hypertension (PH)
dissipative structures and diffusion, 21
di-tetrahedron structure, *181*, 181–83, *182*
DNA, xiv, 34, 38–41, *41*, 179; overview of function, 306–7; complementary interactions, 179; lung vascular development, 58
double-skin façade systems, 20

drawing, *215, 220*, 220–21
Dürer, Albrecht, *215*
Dymaxion Automobile, 32
Dynamic Degeneration and Reconstruction of the ECM, 240
dynamic reciprocity, 40–42, 51–52

Easterling, Keller, 15
"ECM environment," *252*, 252–53
The Ecological Approach to Visual Perception (Gibson), 18, 226–27, *228*, 230n13
ecological building materials and design, 314
ecoLogicStudio, 18
economies of scale, 262–63
edge behaviors, 111–12, *149*; cell choreographies, 202, *205, 206*, 208–12, *209, 211, 212*; dynamic boundary conditions, 251–56, *251–56*; "empty circle" property, 160; object detection, *221*, 221–22; segmentation, 221, *221*; in vitro studies, 112–16
El Lissitzky, 11–12
elastic connection, 98, *99*
elastin, 109
Ellingsen, Eric, 49
Elytra beetle, 26
Elytron, 26
emergency housing, 25
Emlein, Anne, 343, *344*
emotivity, 356
Empire (Hardt and Negri), 5–6
emplaced architectures, 21
"empty circle" property, 160
endothelial cell (ECs), 59, 81–82, 86, 87, 94, *103*; pulmonary artery, 108–9; RFL-6 or MFLM-4, 67; subendothelial matrix, *232*, 232–33
energy consumption, 17, *300*
energy efficiency, 17
Energy Minimization via Multi-Scalar Architectures: From Cell Contractility to Sensing Materials to Adaptive Building Skins, 22–23, 313–14. *See also* eSkin
Engel, Dwight, 273
Engheta, Nader, 314, 320, *321*, 328
engineering, 262, 263

entrepreneurship, 356–57
envelope, 20
envelope, of building, 20
epigenesis, 4, 38–40, *39*, 48, 55
epigenetic landscape, *39*, 269, 271
Epiphyte Chamber, *20, 21,* 21–22
epistemology, 4
epithelial cells, 157–59, *158,* 171–72, 179, 188, *193*
eSkin project, xiv, *22,* 22–23, 28, 106, 239, 313–34; background on materials research and scaled prototypes, 317–20; ColorFolds, 331–34, *333*; incident angles, *321, 322,* 323–24; interface design for scaled eSkin components, 326–34, *327–28, 330–33*; interface design for simulation, 323–26, *325, 326*; interpreting simulation data, 322–23, *323*; local adaptation, 320, *330, 331*; nano-/micro-scale pillar arrays, 317–18, *318, 319*; phases, 316; prototyping optical properties at the nano/micro scale, 320–22, *321, 322*; second skin, 315; simulated, digital, real-time and interactive prototype, 319–20. *See also* light-borne phenomena; structural color
essentialism, 13
Ettinger, David, 188–92
Euclidian mathematics, 4–5
Euler, Leonhard, 10
Eulerian circuit, *181,* 181–83, *184*
evolution: forms and types of, 7
evolutionary considerations, 185, 186–88
existing built fabric, 19, 316
experimental material systems, 53
extracellular matrix (ECM), 37, 40–42, *41,* 55–56, 94, 306, 362; in Branching Morphogenesis project, *274*; communication and connectivity, 61–65; components, *96*; context, 72; dynamic degeneration and reconstruction, 246–51; "ECM environment," *252,* 252–53; elastic connection, 98, *99*; external signals, 231; influence, 55–56, *97, 98, 99,* 100; lung endothelial cell formation, 56–58, *57,* 81–82, 86; mammary gland structure, 157, *158,* 165, 167, 175–76, 179–80, *182,* 182–85, *183,* 187–89, 192–95, *193, 194*; modeling a complex multi-variable system, 59–65, *60–66*; non-networked (simulated), 61–62; non-visible variables, 59, *60*; obstruction, 98, *100*; phenotypic determinant, 56; pillared and non-pillared, *251,* 251–52, *252, 253*; pulmonary artery, 108–9; virtual, *316. See also* basement membrane; Dynamic Degeneration and Reconstruction of the ECM; matrix architecture; motility; networking; surfaces

façade systems, 20
Fang, Chun, 192–97
Farhoud, Emaan, 192–97
feedback loops, 20–21, 72, *81,* 175, 178
Feng, Shuni, 66–78
Fiber exhibition, physical prototype, *62*
fiber reinforced polymer structures, 25
Fibonacci Sequence, 36–37
"Fibroblast Adaptation and Stiffness Matching to Soft Elastic Substrates" (Solon, Levental, and Sengupta, et al.), 125–26
fibronectins, 96
Fierro, Annette, 273
filopodia, 123, *124, 125,* 126, *171,* 172, 204, *207,* 208; "tail" and, 137; tripodal formation, *207, 208,* 212
filopodial polarization, *207, 208,* 210, *212*
filopodial propulsion, 208, *212*
filters, materials as, 21
"fingering" effect, *254,* 254–56, *255*
flexible surfaces, 3
Flexner Report (1910), 355–56, 358
fluid dynamics, 254, *254*
flux, 235–36, *236*
FlyKnit shoe, 350
form, 220–21, 266–67; dynamics of living organisms, 8–9; mathematicals and, 4; seven technical, 11–12
form and measure, 202–14
formalization, 259, 260
formative drive, 4
form-finding techniques, 345

form-fitting material systems, 336
fractal branching, 36–37, 58, 70, 116n1; cell edge, 113–14. *See also* branching structures
Francé, Raoul H., 11
Franklin, Rosalind, 34
Freese, Joshua, 66–78
Freshwater Pavilion (1994–1997), 3
Fuchs, Rebecca, 290
Fukunishi, Kenta, 59–65
Fuller, R. Buckminster, 9–11, 13–14, 31, 32, *33,* 40, 42; nature, view of, 267–68

Galileo Galilei, 5
Gamba, Andrea, 87–88
Game of Life (Conway), 36
generative, 14–15
generative approaches, 36
generative architectural design, 38
Generative Components (GC), 73, *81,* 82, 84–86, 114, *181,* 182, *277*
generative fabrication, 53
GenerativeComponents model, *81*
genes: Prx-1, 67, *68,* 72, 79, 81–86
genome/code, 37–40; ECM influence on, 55–56
geodesic domes, 9, 10–11, 32, *33*
geography: objectile, 268, *269,* 271
geometric abstraction, 51
"Geometric Determinants of Directional Cell Motility Revealed Using Microcontact Printing" (Brock, Chang, Ho, et al.), 125–26
geometric filtering, 113–16, *113–16,* 120
geometry, 199; as *a priori* architectonic system, 7; architectural authorship, 268–69; cell choreographies, *205–7,* 205–10, *209;* edges, 222; Idealized Deformation, *241,* 241–42; in microfabrication, 237; structural information content, 215, 227
Gestaltung, 11, 13, 15
Gibson, James J., 18, 30n2, 218, *218,* 226–27, 230n13
Giraudo, Enrico, 87–88
global systems, 67, 72
goal-based research, 266–67

gradient extraction, *218*
granularity, 261, 263
Grew, Nehemiah, 11
Ground Substance project, 271, 290–98, *290–98,* 310; 3-D printed materials, 295–96, *295–98,* 298
Guattari, Félix, 14–15
Gutierez, Maria-Paz, 19, 23, 24–25

Haeckel, Ernst, xiii, 7–10, *8*
Hardt, Michael, 5–6
heat pressing, 240–41
Heatherwick, Thomas, 32–33, *34*
Heatherwick Studio, 32–33, *34*
hemophilia, 236
Hensel, Michael, 19
hexagons, 9–10, 175
Ho, Chia-Chi, 125–26
Hoberman, Chuck, 36
Hoberman Associates, 36
Homeostatic Façade System, 20
Howard, Thomas C., 33
hub communities, 69, *69*
hubs, 68
human beings: as biodynamic model, 336–38, *338, 339;* design goal metaphor, 47–48
human bio-data, 336–39, *338,* 343
human interactions, 20–21
human-space interaction, 128–30
Hunt, Jamer, 87
Hurcomb, Gregory, 192–97
Huxley, Thomas, 39
hybrid ecosystems, 19
hybrid materials, 267, 282
Hylozoic Architecture group, *20, 21*
Hylozoic Ground, 20
hyper communicative landscape, 21–22
hyper-communication, 21–22, 74

ice dancing, *110,* 110–11, *111,* 201
Ihida-Stansbury, Kaori, 240, 246, 314
image differencing, 223, *223,* 229n10
immanence, 3–4, 5–6, 7
immersive environment, 21
immersive environments, *20, 21,* 28
immunoreceptor function, 353
in silico test systems, 59, *237,* 271

in vitro test systems, 59, 231; breast morphogenesis, 158–60, *159,* 179; pulmonary artery smooth muscle cell, 112–16
in vivo systems, 60, 81
incident angles, *321, 322,* 323–24
Indium Tin Oxide (ITO) glass, 326, 328
individual, 261
indoor environmental quality, 17
inductive method, 259–60
influence, 55–56, *97,* 98, *99,* 100
information: change, *220,* 220–22, *221, 222*; defined, 215, 217–18; distillation of patterns within field of change, 222–29, *223–28*; patterns, 215, 217, 222–29, *223–28*; probabilities, 218, *219,* 222, *223,* 223–25; relativistic issue, 217, *219,* 220; as spatial and temporal structure, 217–20, *218, 219*; subjectivity, 216, 220, 226; sympathetic model, 218, *219*; usefulness of, 216–17
information theoretical framework, 215, 217–18, 229–30n12
Ingber, Donald, 41–42, 44n23, 150, 153
Institute for Computational Design (University of Stuttgart), 18, 25–28, *26, 27, 28, 29*
Institute for Lightweight Structures, 13
Institute for Medicine and Engineering (IME), 37, 43n13, 47, 50
Institute of Building Structures and Structural Design (ITKE) (University of Stuttgart), 25–28, *26, 27, 28, 29*
Institute of Medicine and Engineering, University of Pennsylvania, 231
integration, 52
integrins, 96, 103, *193*
intercellular connectivity, 192–97, *193–97*
intermediate filaments, 307
intermediates, doctrine of, 4
international constructivism, 11
internet, 67, 72
irregularities, 12
isomorphism, 9–10, 11, 13

Jenny Sabin Studio, 352; myThread Pavilion, 335–51, *336–42, 344–46, 348–51*

Joachim, Mitchell, 18
Jones, Peter Lloyd, 37, 48–49, 55, 186–88, *200, 299, 301,* 314; Branching Morphogenesis project, 273–89; Ground Substance, 290–98; on medical education, research and practice, 352–60
Jones Lab, 246, 281. *See also* Sabin+Jones LabStudio
Jung, Bo Rin, 131–36

Kahn, Tobi, 353–54, *354*
Kant, Immanuel, 4
Karamanian, Vanesa, 112–16
KATS: Cutting and Pasting – Kirigami in Architecture, Technology, and Science, 332
Kayne, Benji, 349
Kettner, Raymond, 83
Kim, Sean, 246–51
kirigami-inspired folding patterns, 23, 331–34, *333*
knitting, 343; WHOLEGARMENT knitting, 339–40, *341,* 347, 350
knowledge, 80
Kohan, Pablo, 188–92
Kolatan, Ferda, *200*
Kretzer, Manuel, 20
Kunstformen der Natur (Haeckel), xiii, 7–9, *8*
Kwan, Winglam, 103–6
Kwinter, Sanford, 21–22, 79–80, 186–88, 199–200, *200,* 269
"Landscapes of Change," 269

LabStudio. *See* Sabin+Jones LabStudio
Lachterman, David, 4–5
Laguerre tessellation, 346
Lake, Matthew, 273
lamellipodia, 119, 123, *125*
laminin, 158, 165
"Landscapes of Change" (Kwinter), 269
Lauweriks, J.L.M., 6
Le Ricolais, Robert, 12–13, 34, *35,* 40, 42, 160, 192–93, 197, 198
"Le Toles composees et leurs applications aux constructions metalliques legeres" (Le Ricolais), 40

Lee, Alexander, 87–94, 281
Lee, Christopher, 59–65, 61, 273
Lee, JaeYoung, 178–86
LEED, 17
Levental, Ilya, 125–26
Levy, Aaron, 49
Li, Danlu, 251–56
Li, Huishi, 188–92
Liao, Chia, 131–36
Libeskind, Daniel, 36
light-borne phenomena, 216–17, 220, 223; arrays, 241–42, *242*, 318, *318*; image differencing, 223, *223*; nano-/micro-scale optical properties, 320–22, *321, 322*; optical interference, 240, 317–20, *318–20*. See also eSkin project; structural color
The Living, 18
Lonsdale, Kathleen, 12
Lucia, Andrew, 59–65, 273–89, *299, 301,* 314
luminal space, 157, 158, 160, 188–92, *189–92,* 194
lung endothelial cells, 56–58, *57,* 81–82, 86, 282, 285; in Branching Morphogenesis project, 271, *274*
lung morphology, parameters, 59–61
Lynn, Greg, 74

machines, 11; abstract, 14–15
Magnetic Resonance Imaging (MRI), 288
making, science of, 259–64, 268–69; inductive method, 259–60
mammary gland, 157–200, *291*; acini structures, 158–60, *162–64,* 164–67, *166*; adherens junctions, 170–78, *171, 173–78, 193*; apoptosis (programmed cell death), 159–60, 188–92; breast cancer, 157–59, 165–69, *167, 168,* 171–72, 179, 189, *194,* 194–95, 292; epithelial cells, 157–59, *158,* 171–72, 179, 188, *193*; lumen, 108–9, 117, 158, *171,* 175, 179, 188, *193, 194,* 194–95, 291–92; luminal space, 157, 158, 160, 188–92, *189–92*; tenascin-C, 158–59, *159,* 165, *167–68,* 171

mammary gland structure: extracellular matrix (ECM), 157, *158,* 165, 167, 175–76, 179–80, *182,* 182–85, *183,* 187–89, 192–95, *193, 194*
Mandel, Katherine, 123–31
mapping: image differencing, *223–25,* 223–28; perspectival space, *215,* 227, *227*; through dynamic Penrose tiling, *117–22,* 117–23
mapping techniques, 107, 113–14, 199
material agency, 240, 267
Material Dynamics Lab (NJIT), 20
material principles, 276
material systems, 53
materialism, 3
materiality, 276, 362
materials science, 22
materials selection, 17
mathematics: Euclidian, 4–5; as immanent, 5–6
mathesis universalis, 6, 13
matrix, 79–80; architecture, 53–54; comments on role of, 79–80. See also extracellular matrix (ECM)
matrix architecture, 270–71
matrix biology, 22, 26, 36
McIerny, Austin, 170–78, 273
McQuaid, Matilda, 48
mechanical forces, 231
"Mechanical Signaling and the Cellular Response to Extracellular Matrix in Angiogenesis and Cardiovascular Physiology" (Ingber), 150, 153
mechanistic theories, 4
Mediated Matter Group (MIT), 18
medical education, 352–60
medical education, research and practice, 352–60
Medicine Plus, 356–57
Medicine+Design, 357
Medow, Kara, 81–86, 290
MEDstudio@JEFF, 52–53, 353, 356, 357–58
membrane microfluidic model, 235–36, *236*
memory, 276
Menges, Achim, 25, 28, 262
Mennan, Zeynep, 15

Mertins, Detlef, 32, 47, 270
Merz, 11–12
meshwork, 9, 20–21, *70, 71,* 113, 263. *See also* Delaunay Tessellation; Voronoi Diagram
metalloproteinase, 67
metaphysics, 4
MFLM-4 cells, 67
microenvironment, 40, 50, 180, *193,* 194, *292, 293,* 305. *See also* extracellular matrix (ECM)
microfabrication, 231–38, *232–34, 236–37*; measuring platelet adhesion in a microfluidic vascular injury model, *232–34,* 232–35; models of the interstitial space of a blood clot, 236–38, *237*; scaling analysis, 237–38; simulating the release of platelet agonists, 235–36, *236*
microfilament network, 307
microfluidics, 233–35, *234–35*; membrane microfluidic model, 235–36, *236*
microlithography, 317–18, *318*
micropatterning, 233
microtubules, 307
Migayrou, Frédéric, 276
"Modeling the early stages of vascular network assembly"(Serini, Ambrosi, Giraudo, Gamba, Preziosi, and Bussolino), 87–88
modernity, 3, 5–6
molecular biology, 39
Moran, Marta, 273
morphogenesis, 48, 148, 205, *205*; breast, 158–60, *159, 175*; lung, 59–61
morphology, 267
Mössel, Ernst, 9
motility, 50, 56, 107–55, 201; algorithms, 98–99; boundary conditions, 253, *253*; cell choreographies, 201–14, *203–7, 209, 211–13*; "cell dances," 111; change, observation of, 215–30; context-dependent cell deformation, 131–36, *131–36*; dynamic transformation of structure, 137–40; edge behaviors, 111–16; experimental strategy, 202–3; filtering and extracting, 203; fleeting and fractile nature, 112–16, *113–16*; geometric tooling, 203–10; ice dancing, *110,* 110–11, *111,* 201; internal logic, *150*; mapping through dynamic Penrose tiling, *117–22,* 117–23; mutations and accumulation, 148–55, *148–55*; pathophysiology of pulmonary hypertension, 108–9; space and structure of movement, 110–12; spatial signatures, 123–31, *124–25, 127–30*
Motility Project (LabStudio), 216–17, *217*
Muir, John, 355
Muller-Sievers, Helmut, 4
multifunctional building membranes, 23–24, *24, 25,* 74
multiplicity, 4–5, 6, 8, 12–13
multitude, 6
Murata, Misako, 170–78, 273
Museum of Contemporary and Modern Art, Seoul, *20,* 21, *21*
Musil, Josef, 137–40
Muskara, Vahit K., 117–23
myosin, *125*
myThread Pavilion, 271, 335–51, *336–42, 344–46, 348–51*; cone family topology, *346*; seam pattern, *348*; site plan, *349*

N of One Medicine, 359–60
nanometer scale, 41, 271
nano-/micro-scale, 364; interface design for scaled eSkin components, 326–34, *327–28, 330–33*; optical properties, 320–22, *321, 322*; pillar arrays, 317–18, *318, 319*
nanoparticle solution, 326–28
nanostructures, 240
nasci, as term, 11
National Science Foundation (NSF) EFRI SEED grants, 22–23, 313–14, 331–32
native or denatured environment, 119, 122–24, 126, *127, 128,* 131, *132–36,* 137, 148, *149,* 202
natural materials, 261–62
natural sciences, 37, 260, 261
Naturalizing Architecture, 276–77
nature: design and human survival, 35–36; as muse, 31–35
nearest neighbor relationship, 36
Neeves, Keith, 112–16, 231

Neeves Lab in Chemical and Biological Engineering, Colorado School of Mines, 231
negative space, 95, *110*
Negri, Tony, 5–6
neighborhood and community conditions, 69
Nesbit, Jeffrey, 66–78
net structures, 14
networked mesh systems, 20–21
networking, 50, 55–106; adaptive and resilient networks: dynamic material systems, 103–6, *103–6*; adaptive component system, 74, *77, 78*; advanced dynamic models, 73–74; algorithm principles, *61*; Branching Morphogenesis project, *274*, 280, *280, 281*; dynamic surface modeling, *81*, 81–86; elongated cell phenotype, 82; flexible scaffolds and emergent space, *94–97*, 94–102, *99–102*; global systems, 67, 72; matrix, role of, 79–80; modeling a complex multi-variable system, 59–65, *60–66*; Nonlinear Systems Biology and Design course models, 300, *303, 304*; part-to-whole relationships, 56–57; percolative transition and network formation, 87–94, *89–93*; pseudo code, *75*; real-world environmental inputs, 61, *61*; rules, *75*; scale-free networks, 66–78, *67–71, 73–78*; translation, *60*, 69–70, 72–74, *76*, 87. See also extracellular matrix (ECM)
Net-Zero Energy Commercial Building Initiative (CBI), 19
Neutra, Richard, 34–35
new geometries, *116*, 205, *206*
new science of making, 259–64, 281
New Territories, 36
"New Views of Humankind" exhibit, 288
Nicol, Mark, 103–6
Nike, 336
Nike FlyKnit Collective, 335–37, *336*
Nike FlyKnit shoe
Nike+ FuelBand technology, *339*, 338–339
Nike Inc., 341
Nike Innovation Kitchen, 336, 341

Nike Stadium NYC, 347, *349*
Nolan, Gary, 214
non-dimensional numbers, 237
non-geometric issues, 216
nonlinear biological processes, 56, 103; modeling a complex multi-variable system, 59–65, *60–66*
nonlinear biological systems, 22, 27, 47–48
"Nonlinear Biosynthesis," 49
"Nonlinear Systems Biology and Design" course (2010), 50, 239, 281, 299–310, *300–309*; fabrication and production, 300; scripting and simulation, 300; working prototypes, 299, 302, *302*, 305–8
Nonlinear Systems Organization (NSO), 37, 43n12, 48
nonlinear thinking, 36
non-visible variables, 59, 60
nucleotide hydrolysis, 140–41, *141*

object detection, *221*, 221–22, *222*, 229n9
object-environment interaction, 127–28
objectile, 268, 269, 271
objectivity, 215–17, 220
observation, 216–17, *217*
observing mechanism, 223, 226
obstruction, 98, *100*
On Growth and Form (Thompson), xiii, 8–10, 267
openings, 34
optical interference, 240, 317–20, *318–20*
ordering principles, 48
oscillating register, 21
Otto, Frei, 3, 13–14, 270
Oxman, Neri, 18

parallel design thinking, 55, 271
parameters, 79–80; formal expression, *92*, 92–93; lung morphology, 59–61
parametric and associative design methods, 188–92, *189–92*, 198
Parsons, Ronnie, 155
part-to-whole relationships, 37, 42, 50, 282–83; cell formation, 56–58, *57*, 86, 89; mammary gland, 159; networking projects, 56–57

Pasquero, Claudia, 18
passive architecture, 23
patterns, 215; intuitive recognition, 216, 225, 275, 276, 279; distillation of within field of change, 222–29, *223–28*; micropatterning, 233–35, *234–35*; scale-free context, 217
pavilions: bionic research pavilion, 25–28, *26, 27, 28*; British Pavilion or Seed Cathedral (2010 World Expo), 32–33, *34*; Freshwater Pavilion (1994–1997), 3; myThread Pavilion, 335–51, *336–42, 344–46, 348–51*. See also Branching Morphogenesis
PDMS (polydimethylsiloxane), *315*, 318, 321–22
PennDesign, 47, 49, 110, 201, 295
Penrose tiling, *117–22*, 117–23
perceptual psychology, 217–18
percolative transition and network formation, 87–94, *89–93*; phase changes, 91
personal architecture, 128–31
personal architecture and medicine, 311
personalized medicine, 107–8, 112
personification, 210
perspectival space, *215, 227*, 227
Peterson, Ryan, 246–51
Peterson, Tzara, 251–56
Die Pflanze als Erfinder (Francé), 11
phase space, 199
Philip Beesley Architect Inc., 18
photonic crystals, 240–41
physical exercises (3-D print and digital fabrication outputs), 50, 55
physical stretching, 240–41
Picon, Antoine, 276
pilot scale, 237
Piper, John, *39*, 269
pixel-based data, 160, *162–63*, 217, 221, 223–24, *224, 226*, 324, *325*
platelet agonists, 235–36, *236*
platonic oneness, 9, 32
pluralism, 6
pneumatic structures, 14
Poletto, Marco, 18
polymer structures, 233; bionic research pavilion, 25–28, *26, 27, 28, 29*

polystyrene; endothelial cell (ECs) cultured on, 59
positioning mechanism, 155
postmodern philosophy, 260, 263n5, 264n9, 271
post-scientific revolution, 259–60
potential parameters, 59–60
pragmatics, 14–15
predetermination, 3, 6, 55
pre-formationism, 39–40
Preziosi, Luigi, 87–88
Prigogine, Ilia, 264n9
probabilities, 218, *219*, 222, *223*, 223–25
process-driven research, 37, 52, 266–67
Processing 1.0, 181
Processing inputs, *81*, 83
programmable matter, 18, 20, 28, 198
programmed cell death. *See* apoptosis (programmed cell death)
proportional systems, 7
protocell fields, 21
proximity, 59–60, 62, 64, *65*, 69, 160
Prx-1 gene, 67, *68*, 72, 79, 81–86
pulmonary hypertension (PH), 107–8; design tools, scientific implications, 126; dynamic degeneration and reconstruction of ECM, *246–50*, 246–51; mapping through dynamic Penrose tiling, *117–22*, 117–23

quantification, viii, 50–51, 112–16, 158–59, 205, 213, 225

radiolarians, 7–13, *8, 10, 34*, 198; difference between organic forms, 12
raindrop, 32
random networks, 68
rapid prototyping (RP) technologies, 51
rating systems, 17
real-world environmental inputs: networking, 61, *61*
receptor clustering, 306
receptors, 180, 231–32
reciprocity, 21–22, 267; biological, 100; dynamic, 40–42, 51–52
recombination, 14–15
reductionism, 217
regularity, 9, 10–11, 12

repetition, 6–7, 9, 11, 12
Report of the Scientific Results of the Voyage of HMS Challenger (Haeckel), 7
Resch, Ron, 175
research and design units, 18
resource consumption, 19
responsive architecture, 19–20, 314–15
Reynolds number, 237
RFL-6 cells, 67
Rhino software, *81, 83–85, 85,* 203, 205, 250
rhythm, 12
rigidification, 3, 13–14, 155
Riiber, Jacob, *301, 303*
Ritterbush, Phillip, 10–11
robotic fabrication methods, 25–28, *26, 26, 27, 28*
Roche, François, 36
Rothemund, Paul W. K., 179
Ruggles, Andrew, 112–16
rule-based systems, 36, 37–38, 241
rules: chemo-attraction, *104*; of proximity, 59–60, 62, 64, *65,* 69; of scale-free networks, 67–69, *75*; translational, *95,* 96
Ruy, David, 49

Sabin, Jenny E., 37, 42, 48–49, *299, 301,* 314; BodyBlanket, 278; Branching Morphogenesis project, 273–89; comments on role of matrix, 79–80; Ground Substance, 290–98; interview with Tykocinski and Jones, 352–60; myThread Pavilion, 335–51, *336–42, 344–46, 348–51; StringWeave2,* 83
Sabin Design Lab (Cornell University), 18, 22, 50, 52, 290, 295, 331–34, *333,* 352, 361. *See also* Sabin+Jones LabStudio
Sabin+Jones LabStudio, 17–30, 22, *38,* 49, 56, 159, 199, 352, 361; Branching Morphogenesis, 271, 273–89, *274–75, 277–88*; formation of, 37–38; Ground Substance project, 271, 290–98, *290–98,* 310; methodological approach, 53; Motility Project, 216–17, *217*; "Nonlinear Systems Biology and Design" course (2010), 50, 239, 281, 299–310, *300–309*; primary goal, 55–56. *See also* networking; surface design

Saffman-Taylor instability, *254,* 254–55
Salvatierra, Gabriel Wilson, 240–46
Savig, Erica Swesey, 49, 201, *299, 301, 303*
scaffolds: 2-D folding, 179–81, *180, 182–86, 184, 185*; for artery repair in PH, 246, *247*; Eulerian circuit, *181,* 181–83, *184*; flexible scaffolds and emergent space, *94,* 94–102; tissue, self-organization and formation of dynamic surfaces, 178–86, *180–85*
scale-free networks, 66–78, *67–71, 73, 73–78,* 130; change, observation of, 217; clustering effect, 69–72, *70, 71, 75*; conditional phases, *71*; connectivity, 67–72, *71*; neighborhood and community conditions, 69; transitional method, 70, *71*
scale-less physical models, 244
scaling analysis, 237–38
Schue, Allison, 49
Schwitters, Kurt, 11–12
Science in Energy and Environmental Design (SEED), 22–23
scientific method, 34, 201, 259–60, 281–82
scientific representation, 87
scripting logics, 52, *151,* 153, 207, 239, 271; interface design for simulation, 323–26, *325, 326*
Seed Cathedral (2010 World Expo), 32–33, *34*
segmentation, 221, *221*
self-active matter, 23, *24,* 28
Self-Assembly Lab (MIT), 18, 20
self-generation (*autopoiesis*), 3, 5, 6, 13, 36, 144–45, 270; in cells and tissues, 56; *nasci,* 11
Self-Generation: Biology, Philosophy and Literature around 1800 (Muller-Sievers), 4
self-organization, 31, 37–38, 106, 131, 178–79, 181–83, 273, 354; scaffolds and open space, 100; self-renewal, 178–79; tissue scaffolds, 178–86, *180–85*
semiotics, 14–15
Semper, Gottfried, 7
Sengupta, Kheya, 125–26
sensing and control technologies, 320
Serini, Guido, 87–88

seven technical forms, 11–12
Shaffer, Ben, 341
Shannon, Claude, 218
shape memory materials, 20
shear forces, 231–35, *232, 233,* 236, *236*
Shima Seiki, 341, 347
Shin, Jae-Won, 178–86, 240, 246
Shinnamon, Kirsten, 81–86
Shu Yang Group (University of Pennsylvania), 240–41, *241*
Sidney Kimmel Medical College, 52–53, 353, 356, 357–58
Sierpinski triangle, 183
SIGGRAPH, 273, *287*
simulation, 61–62, 316–17, 362
skeletons, 8–9
skins, 22–23, 172–74, 239, 362–63; dynamic boundary conditions, 251–56, *251-56. See also* eSkin project
Smart Geometry, Nonlinear Systems Biology and Design, 271
Smart Geometry workshop, 290
smooth muscle cells (SMCs), *221, 225,* 315. *See also* vascular smooth muscle cells (SMCs)
Snelson, Kenneth, 40, 42
Snooks, Roland, 106, 136
soap bubbles, 9
soft lithography, 233
software packages, 198
solar radiation, 244, *244*
Solon, Jérôme, 125–26
soluble molecules, 231–32, *232*
solutionism, 17, 28
Song, Qiao, 94–102
space frames, 12–13, 34, *35*
spatial considerations, 80; microfabrication, 231–38, *232-34, 236-37*
spatial signatures, 123–31, *124-25, 127-30, 149*; cell choreographies, 201, *204,* 208, *209,* 214; human-space interaction, 128–30; ice dancing, *110,* 110–12, *111,* 201
spatial structure, 217–18
"Special Topics in Construction: Bio-Inspired Materials and Design" seminar, 50

spheres: composite, 14
spicules, 9
Spinoza, Baruch, 6
spline curves, *81,* 82–83, *85*
sport, 335
Spuybroek, Lars, 3, 13–15
Stanciu, Maria, 246–51
Stock Exchange in Amsterdam (1901), 6
Stoneback, Gillian, 117–23
StringWeave2, 83, 85
structural color, *24,* 31; at the architectural scale, study, *240-45,* 240–46; Idealized Deformation, *241,* 241–42; "Long Duration Extremes," 244; "Momentary Duration Extremes," 244; in nature, 318, 326; Physical-Exertion dataset, 241, *241*; Subject to Heat dataset, *241,* 241–42, *242. See also* eSkin project
structural information content, 215, 227
structuralism, 267
structures: branching, 37, 50, 58–59, 61–62, 87, *97,* 114, *274,* 305; deployable, 36, *152-54,* 153–55; fiber reinforced polymer, 25; polymer bionic research pavilion, 25–28, *26, 27, 28, 29*
style, 6
Su, Annabelle, 87–94
subendothelial matrix, *232,* 232–33
subjectivity, 216, 220, 226
surface design, 51, 53; adherens junctions and structural deployability, 170–78, *171, 173-78*; dynamic surface modeling, *81,* 81–86, *83-86*; geometries of change, 159–69, *161-69*; mammary gland, 157–200, 291, *291,* 292; tessellation, *81,* 83–86
surface-bound molecules, 231–32, *232*
surfaces, 3, 14, *23,* 37; deformation, *81, 82, 83, 83, 84*; nurbs-based, 165, *166, 346*; responsive specificity, 192–97, *193-97*; self-organization of tissue scaffolds, 178–86, *180-85*; tectonic, 14, 28; tessellated, 323. *See also* extracellular matrix (ECM)
sustainability, 17–19, 22, 28–29, 198, 266, 313–16. *See also* eSkin project
sustainability fatigue, 19
sustainable site development, 17

"Sustaining Sustainability" symposium, 19
Suttle, Dale, 94–102
Sweeney, Shawn, 103, 136, 145
sympathetic model of information, 218, *219*
synergistic optimization, 23
systemic ecology, 17
systems biology, 52–53

tail behavior, 137, *137–39*
Tamby, Mathieu, 49, 112–16, 201
Taraseviciuete, Agne, 49, 170–78
tectonic, crisis of, 276
tectonic surfaces, 14, 28, 276
temporal considerations, 186, 203, *204–7,* 205–12, *209, 211–12*; Branching Morphogenesis, 275, *279, 280*; change, observation of, 217–20, *218, 219*; probabilities through time, 218, *219,* 222, *223,* 223–25
temporal structure, 217–18
tenascin-C (TN-C), 158–59, *159,* 165–66, *167–68, 291*; adherens junctions and structural deployability, *171,* 171–72, 178; responsive surface specificity, 192–97, *194–97*; self-organization of tissue scaffolds, 179
tensegrity, 40–42, 44n23, 140, 153, 307–8
tensile connections, 103–6, *103–6,* 210
tension networks, 34, *35, 76,* 306–7; cell attraction, *63, 64*
Terreform ONE, 18
tessellation, *81, 83–86,* 331; Laguerre, 346. *See also* Delaunay Tessellation; Voronoi Diagram
tetrahedron structure, 181, *181*
textiles, 23, *23,* 335–37
"The Theory of Affordances" (Gibson), 18
thermodynamic environments, 21, 28
Thompson, D'Arcy Wentworth, xiii, 8–10, 12, 267
Thomsen, Mette Ramsgard, 18
A Thousand Plateaus (Deleuze and Guattari), 14–15
thrombin flux, 235–36
Tibbits, Skylar, 18, 20

time, as variable, 126. *See also* 4-D context; temporal considerations
"time-sheets" or tapestries, *280,* 281, *285*
tissue cultures, 107–8
tissue scaffolds, 178–86, *180–85*
tissueness, 267
tools, generative, 6
topological architecture, 269–70
topologically free cells, 136
Toumayan, Elena Sophia, 240–46
trabecula, 26
trabecula of Elytra beetle, 26
transcendence, 5–6
trans-disciplinary research practice, 17–18, 31–32, 37–38, 315, 332, 340–41; responsive architecture, 19–20. *See also* collaborative endeavors
transformational, 14–15, 269–70
transformative research, 17–30
translation, 28, 34, 53, 241–42; networking, *60,* 69–70, 72–74, *76,* 87
treadmilling, 141, *141*
triangular niche unit, 180, *180*
triangulation, 180, *180–83, 182,* 187
Tribe, Philip, 140–47
tripodal formation, *207,* 208, *212*
Troxler Fading, *219,* 222, *222*
Tse, Ringo, 148–55
Tubingen University, 26
tumor formation, 56
Turner, J. Scott, 268
Tykocinski, Mark L., 352–60

uncertainty, 218, *219,* 224
universalization, 11, 14, 15, 32
University of Pennsylvania Graduate School of Design, 37, 43n12
University of Stuttgart, 18, 25–28, *26, 27, 28*
ur-animals (protozoa), 8
ur-plants (protophytes), 8
U.S. Department of Energy (DOE), 19
U.S. Green Building Council, 17

Van der Spiegel, Jan, 314, 320, 321, 328
vascular cells, 50; percolative transition and network formation, 87–94, *89–93*
vascular injury model, *232–34,* 232–35

vascular smooth muscle cells (SMCs), 108–9, 112–16, 117, *124*, 137; cell choreographies, 201–14; tail behavior, 137, *137–39*
vascular system, 87
vasculogenesis, *88*
vector gradient, *218*
vectors, 98–99, *99*
Victoria and Albert Museum, 36
visual exercises (generative and algorithmic studies), 50, 55, 107, 148, 316; biomedicine, 107, 109, 111–12; branching structures simulation, 96–98, *97*; cell choreographies, *204–7*, 208–9, *209*, *211–13*; endothelial cellular networking, *60, 61*; Penrose tiling, *117–22,* 117–23. See also individual case studies
visualization, 362; data visualization, 215
visualization and simulation, 50, 53
Vitra Museum, 33
Vitruvius, xii
volumes, 89–92, 158, 179; apoptosis, 188–90, *189, 190, 192*; platelet aggregates, *236*
volumetric models, 89–93, *92*; connectivity, performance, and time, 52; structural color, *245,* 246
Vom Geheimnis der Form und der Urform des Seines (Mössel), 9
von Willebrand disease, 235
von Willebrand factor (VWF), 232–33
Voronoi Diagram, 51, 159–60, *164,* 164–65; 3-D, *164,* 164–65, *166*

Waddington, Conrad H., 12, 39, *39,* 269, 271
Wang, Wei, 49, 159, 170–78
water, 253–54
water savings, 17
Waterloo Architecture, 18
Watson, 34, 38
weak signal, *219*
Wei, Ziyue, 94–102
Weidlinger, Paul, 9
Weiss, Paul, 40
Wheeler, John, 140–47
Whitehead, Alfred North, 12
Whole Garment knitting, 339–40, *341,* 347, 350
wholeness (One), 32
Winter, Steve, 355
Wolff, Caspar Friedrich, 4
Wolfram, Stephen, 36
Wong, Jackie, *110,* 110–11, *111*
World Expo (Shanghai, 2010), 32–33, *34*
World Trade Center (New York), 14
Wu, Yifan, 148–55

xenoarchitecture, 36
Xiao, Wenda, 148–55
X-ray log scanning, 262

Yamada, Yohei, 103–6
Yang, Shu, xiv, *24,* 148, 240, 246–47, 314, 362
Yang Lab, 317, 320, 324, 328, 332

ZCorp 510 color 3-D printer, 129, 290, 295, *295, 297,* 298, 301
ZCorp rapid prototyping technology, 129
Zederbauer, Emmerich, 9
Zeiler, Dane, 137–49
zero energy input, 23
zero-energy commercial buildings, 19
Zhang, Shou, 178–86
Zilis, Brian, 140–47